THE RFF READER IN ENVIRONMENTAL AND RESOURCE POLICY

SECOND EDITION

WALLACE E. OATES, EDITOR

RESOURCES FOR THE FUTURE • WASHINGTON, DC, USA

Printed in the United States of America

No part of this publication may be reproduced by any means, whether electronic or mechanical, without written permission. Requests to photocopy items for classroom or other educational use should be sent to the Copyright Clearance Center, Inc., Suite 910, 222 Rosewood Drive, Danvers, MA 01923, USA (fax +1 978 646 8600; www.copy-right.com). All other permissions requests should be sent directly to the publisher at the address below.

An RFF Press book
Published by Resources for the Future
1616 P Street NW
Washington, DC 20036-1400
USA
www.rffpress.org

Library of Congress Cataloging-in-Publication Data

The RFF reader in environmental and resource policy / edited by Wallace E. Oates.—2nd ed.
    p. cm.
  Rev. ed. of: The RFF reader in environmental and resource management. c1999.
  "An RFF Press book"—T.p. verso.
  Includes bibliographical references.
  ISBN 1-933115-16-5 (alk. paper) — ISBN 1-933115-17-3 (pbk. : alk. paper)
  1. Environmental policy. I. Oates, Wallace E. II. Resources for the Future. III. RFF reader in environmental and resource management. IV. Title.

  GE170.R447 2005
  333.7--dc22   2005029361

The paper in this book meets the guidelines for permanence and durability of the Committee on Production Guidelines for Book Longevity of the Council on Library Resources.

This book was originally designed by Diane Kelly. The second edition was typeset by Peter Lindeman. Cover design by Marc Alain Meadows, Meadows Design Office Inc., www.mdomedia.com.

About Resources for the Future
and RFF Press

Resources for the Future (RFF) improves environmental and natural resource policymaking worldwide through independent social science research of the highest caliber. Founded in 1952, RFF pioneered the application of economics as a tool for developing more effective policy about the use and conservation of natural resources. Its scholars continue to employ social science methods to analyze critical issues concerning pollution control, energy policy, land and water use, hazardous waste, climate change, biodiversity, and the environmental challenges of developing countries.

RFF Press supports the mission of RFF by publishing book-length works that present a broad range of approaches to the study of natural resources and the environment. Its authors and editors include RFF staff, researchers from the larger academic and policy communities, and journalists. Audiences for publications by RFF Press include all of the participants in the policymaking process—scholars, the media, advocacy groups, NGOs, professionals in business and government, and the public.

Contents

Part 3. Environmental Regulation

Part 4. Environmental Accounting and Statistics

Part 7. Energy Policy for the Twenty-First Century

Part 8. Global Climate Change

Part 9. Thinking About Sustainable Development

Part 10. Environmental Policy in Developing and Transitional Countries

Part 11. New Horizons in Environmental Management

Part 12. An Historical Perspective

Preface

The original motivation for *The RFF Reader* had its source in the many articles from *Resources* that I had been using as supplementary readings in my undergraduate course in environmental economics at the University of Maryland. *Resources* is a quarterly publication of Resources for the Future (RFF) that contains articles providing brief, yet quite incisive, nontechnical treatments of a range of central issues: analytical techniques for the evaluation of environmental programs, useful findings from specific pieces of empirical research, and balanced assessments of important policy programs. Through the years, I had found that these shorter pieces could usefully supplement the more lengthy (and sometimes ponderous) discussion in many of the standard texts. And it occurred to me that a collection of these articles in book form could provide a supplementary reader that many instructors might find valuable for their courses—in environmental economics, environmental policy, and in interdisciplinary programs

But, as it worked out, the first edition of the *Reader* reached an even broader audience. While these articles can be helpful to students, they can also provide insightful treatments of environmental issues to the larger community concerned with environmental and resource management. In particular, the papers make clear the important contribution that an economic perspective can make to the understanding of these issues and to the design and evaluation of environmental policy. Thus, the first edition of the *Reader* served the purpose of providing concise economic analyses of environmental problems both for students and for a somewhat more general audience.

But things change rapidly on the environmental front. And, since the publication of the first edition of the *Reader* in 1999, much has happened to our understanding of the science, economics, and politics of environmental issues. The *Reader* had thus become a bit dated, and it seemed the time had come to revise and update its contents. This volume is the result of that effort. A couple of specific comments concerning the new edition may be helpful. In the first edition, nearly all the articles (in fact, all but two) came directly from various issues of *Resources*. In the revised edition, I have taken somewhat greater liberty in the selection of chapters to draw on some other outlets for RFF writings. But the basic principles remain the same. The articles are concise, typically four or five pages, and explain in a nontechnical way either a methodological approach to the analysis of environmental problems or the analysis of an actual issue (such as some form of air or water pollution or global climate change). The great bulk of the articles continue to come from *Resources*. However, in their effort to communicate their findings more broadly, RFF researchers have reached out to a number of other outlets, including other RFF publications, and books and periodicals published elsewhere. I have taken advantage of this wider range of sources in this second edition of the *Reader*.

As I did in the first edition, I have grouped the papers into sections that roughly parallel the structure of a typical course in environmental economics. Following my introduction and overview of the economic perspective on environmental issues are papers on science and its role in environmental policymaking. The next two sections take up basic methodological issues in the valuation of environmental amenities (what economists call "benefit-cost analysis") and in the design and implementation of policy measures for environmental regulation. Many of the later sections address specific environmental problems such as global climate change, biodiversity, the management of the world's fisheries, and the pressing challenges of environmental deterioration in developing and transitional countries.

I have expanded the revised edition in several ways that I hope will prove helpful. There is a new section on environmental accounting and statistics that sets forth the need for a broader and more systematic way to assemble data on the environment and that explores the proposal to incorporate measures of environmental use and services into our construction of Gross Domestic Product (GDP). I have also introduced a new section devoted exclu-

sively to energy issues, since energy problems figure so prominently in our environmental prospects for the twenty-first century. To highlight some of the new and more innovative applications of economic analysis, there is a section on "New Horizons in Environmental Management," including papers that examine what economists have to say, in one case, about the troubling issue of growing antibiotic resistance and, in the other, about the problem of invasive species. The *Reader* concludes with a piece of my own from *Resources* that looks back over the past forty years and explores the changing role of economics in environmental policymaking. This is itself a fascinating story that begins with a virtually total disregard of economic considerations in the early environmental legislation in the 1970s, but evolves into a setting in which economic analyses of environmental programs and the use of economic incentives in actual policy measures have become a routine part of policymaking.

To make the new edition current, I have, in some instances, simply discarded articles from the first edition and, where possible, replaced them with more recent pieces. In all cases, authors have been offered the opportunity to revise their papers or, in an option more frequently taken, to include an "Update Box" along with the original paper, each of which discusses, in a brief paragraph or two, any significant new developments (e.g., new findings or legislation) since the article was originally published. Authors also provide a short list of "Suggested Readings" for those who would like to explore the issue in greater depth. Finally, I want to call attention to the rich source of current information and analyses of environmental issues that is available on the RFF web site (http://www.rff.org).

For those unfamiliar with the organization, RFF is an independent, nonprofit institution, located in Washington, DC, that undertakes research and public education on environmental and natural resource issues. Through the years, RFF has been the source of much fundamental and important research, both basic and applied, into environmental policymaking. Indeed, the very first recipients of the prestigious Volvo Prize in

Environmental Science were Allen Kneese and John Krutilla of RFF for their path-breaking work in the analysis of environmental problems. I have been most grateful over many years to have a continuing association with RFF as one of their University Fellows, and I am delighted to have this opportu-

nity to assemble a collection of RFF readings that I hope many will find as valuable as I have.

Wallace E. Oates
University of Maryland and
Resources for the Future

An Economic Perspective on Environmental and Resource Management
An Introduction

Wallace E. Oates

The central concern of economics is the allocation of scarce resources. The basic problem is one of using our limited means to provide an array of goods and services that satisfies peoples' preferences in an efficient and equitable manner. It doesn't require much reflection to realize that our environmental resources are scarce. Clean air and water, the diversity of species, and perhaps even a stable global climate are clearly not available in unlimited supply, irrespective of human activities. Perhaps economics has something useful to say about the management of our environment.

This is indeed the case. I shall suggest here that economics has three basic and important messages for environmental protection. First, economic analysis makes a compelling case for the proposition that an unfettered market system will generate excessive pollution. A market system, in a sense, "overuses" many of the services provided by the environment, resulting in excessive environmental degradation. Thus, economics makes a basic and persuasive case for the need for public intervention in the form of environmental regulation.

Second, economics provides some guidance for the setting of standards for environmental quality. It provides one approach to answering the question: How clean should the environment be? In fact, this approach is simply a straightforward application of the general economic principle that any activity should be extended to the point where the marginal benefits equal the marginal costs.

And third, once we have determined the standards or targets for environmental quality (and even if, incidentally, this determination is made irrespective of marginal analysis), economics has some important things to say about the design of the policy instruments to achieve these standards. In particular, economic analysis suggests how we can structure policy measures so as to realize our environmental goals in the most effective and least-cost ways.

In this introduction, I want to explain and explore these three ideas, for nearly all the papers in this book draw on this conceptual framework in one way or another. In fact, it is the purpose of this volume to show how basic economic analysis can help us to understand the causes of environmental degradation and to design policies to protect and improve the environment.

Free Markets and the Environment

Economists have a deep appreciation of the market system. Guiding the individualized choices of both consumers and producers, a system of markets has the capacity to channel our limited resources into their most highly valued uses. In pursuing their own gain, individuals (as Adam Smith put it) "are led by an invisible hand" to promote the social good.

Markets generate and make use of a set of prices that serve as signals to indicate the value (or cost) of resources to potential users. Any activity that imposes a cost on society by using up some of its scarce resources must come with a price, where that price equals the social cost. For most goods and services ("private goods" as economists call them), the market forces of supply and demand generate a market price that directs the use of resources into their most highly valued employment.

There are, however, circumstances where a market price may not emerge to guide individual decisions. This is often the case for various forms of environmentally damaging activities. In the first half of this century at Cambridge University, A.C. Pigou set forth the basic economic perspective on unpriced goods (encompassing pollution) in his famous book, *The Economics of Welfare*. Since Pigou, many later economists have developed Pigou's insights with greater care and rigor. But the basic idea is straightforward and compelling: the absence of an appropriate price for certain scarce resources (such as clean air and water) leads to their excessive use and results in "market failure."

The source of this failure is what economists call an "externality." A good example is the classic case of the producer whose factory spreads smoke over an adjacent neighborhood. The producer imposes a real cost on the community in the form of dirty air, but this cost is "external" to the firm. The producer does not bear the cost of the pollution in the same manner that it does for the labor, capital, and raw materials that it employs. The price of labor and such materials induces the firm to economize on their use, but, because the health and aesthetic costs of foul air are borne by the larger community, there is no such incentive for the firm to control its smoke emissions and thereby conserve clean air. The point is simply that whenever a scarce resource comes free of charge (as is typically the case with our limited stocks of clean air and water), it is virtually certain to be used to excess.

Many of our environmental resources are unprotected by the appropriate prices that would constrain their use. From this perspective, it is hardly surprising to find that the environment is overused and abused. A market system simply doesn't allocate the use of these resources properly. In sum, economics makes a clear and powerful argument for public intervention to correct market failure with respect to many kinds of environmental resources. Markets may work well in guiding the production of private goods, but they cannot be relied upon to provide the proper levels of "social goods" (like environmental services).[1]

But if we can't rely on markets to "manage" our environmental resources, what principles should we employ to regulate their use? To this I turn next.

The Setting of Standards for Environmental Quality

There is a basic economic principle that indicates the efficient level of any economic activity: extend that activity to the level at which the benefits from an additional "unit" of the activity equal the costs.

Economists sometimes refer to these extra units as "incremental" or "marginal." Thus, the condition for the economically correct level of any activity can be stated simply as the equality of marginal benefits with marginal cost.

The intuition here is straightforward. So long as higher levels of a particular service yield additional (marginal) benefits that exceed the additional (marginal) costs, we are obviously better off providing the additional units of the service than not providing them. But it clearly would not be a good idea to go past the point where marginal benefits equal marginal cost, for any units past this point would cost more than they are worth (i.e., marginal cost would exceed marginal benefits for such units).

The moral of this exercise for environmental policy, from the standpoint of economic efficiency at least, is that we should set standards for environmental quality such that the benefits at the margin from further tightening the standards exactly equal the marginal cost of pollution control (often called "marginal abatement cost"). Note that this implies that, in general, the economically efficient level of pollution is not zero. The cost of a perfectly pure environment would simply be too much to make it worthwhile. Economics is, in a sense, rather pragmatic when it comes to setting standards for things. It recognizes that tradeoffs and compromises are needed in order to make the best use of our limited resources.

While this guidance for the setting of environmental standards seems straightforward and sensible in principle, it is not so easy to implement. Consider, for example, the case of improved air quality. In considering the benefits from a proposal to introduce a more stringent standard for clean air, we must somehow quantify the improvement in well-being that comes with the associated reduced levels of illness and increased longevity. And this, along with any other benefits (such as reduced damages to materials and wildlife), must be compared to the additional abatement costs that the measure would entail. Such quantitative analyses are not easy, but neither are they impossible. Part II of this volume presents a series of short essays that take up some of the difficult problems that arise in "benefit-cost analysis."

It is interesting in this regard that the early major pieces of environmental legislation in the United States almost completely ignored the economic approach to the setting of environmental standards. The Clean Air Act Amendments of 1970, which still embody the basic principles for air-quality management in the U.S., literally directed the Environmental Protection Agency to set standards for air quality so stringent that *no one anywhere in the United States* would suffer any adverse health effects from air pollution. The courts have consistently held that, since this law was silent on the role of costs in setting air quality standards, these costs may *not* be taken into account. Two years later, the U.S. Congress declared in the Clean Water Act Amendments of 1972 that the goal was the complete elimination of "all discharges into the navigable waters by 1985."

Some of these extreme strictures have been relaxed in later legislation and/or their implementation has been modified by presidential executive orders. For example, in Executive Order 12291, President Reagan required benefit-cost studies for all major new regulatory measures (as President Carter had done under an earlier executive order). Such systematic studies of the benefits and costs of proposed programs continue in the executive branch. Moreover, there have been overtures in Congress to override provisions in laws that prohibit costs from being considered, but they have not come to a vote. In fact, we find ourselves presently subject to a somewhat puzzling and conflicting set of requirements. First, some legislation (under, for example, certain sections of the Clean Air and Clean Water Acts) prohibits taking costs into consideration in the setting of environmental standards. In sharp contrast, other legislation (including parts of the Toxic Substances Control Act and the basic pesticide law, the Federal Insecticide, Fungicide, and Rodenticide Act) *requires* a balancing of benefits against costs. Second, and more confusing, even where a law prohibits regulators from considering costs, they must still conduct benefit-cost analyses for major rules!

This is not to say that the findings from a benefit-cost study should constitute the sole criterion for deciding whether or not to undertake a new environmental program. The complexities in estimating benefits and costs and the importance of other objectives suggest that it would probably be unwise to institute a rigid rule requiring that any proposed program pass a benefit-cost test. At the same time, such analyses surely provide important information that should be an integral part of the decisionmaking process.

The Choice of Policy Instruments

Once we have set specific targets for environmental programs, there remains the critical and challenging task of designing a set of regulatory measures to attain the targets. Here again, economics provides some valuable insights. In particular, it is important that a regulatory regime achieves its targets effectively and in the least costly way. A good system of regulatory instruments will both minimize abatement costs in the short run and provide incentives over the longer term for firms to discover and introduce yet better techniques for controlling pollution.

It is here that a set of incentive-based policy instruments has real appeal. Our earlier discussion suggested that excessive pollution results from the absence of an appropriate price to induce controls on waste emissions. The implication is that we can correct the resulting market failure through the introduction of the missing price. Economic analysis thus points directly to a concrete policy proposal: the introduction of a surrogate price in the form of a unit tax on polluting waste emissions. Such a tax can play the role of the missing price by providing the needed incentive to polluters to economize on their use of the environment; cost-minimizing polluters will respond to the tax by cutting back on their polluting waste emissions.

For example, suppose we have set a standard for air quality requiring that sulfur emissions in a particular region be cut by 50 percent. One way to achieve this goal would be to introduce a tax per pound of sulfur emissions and simply raise the tax to a sufficiently high level to induce a 50 percent reduction in sulfur discharges. Such a regulatory strategy has some appealing properties. It is

straightforward to show that a system of effluent taxes can attain the target at the minimum total cost to society. In addition, the system provides firms with an incentive over the longer term to seek new and cheaper ways to control waste emissions, for such R&D efforts by polluting firms can reduce their tax bills and increase their profits. Systems of environmental taxes (or "green taxes" as some call them) effectively redirect the powerful profit motive of the market to the protection of the environment.

The environmental-tax approach is not the only way to mobilize economic incentives on behalf of the environment. An interesting alternative is a system of tradeable emissions permits (sometimes called a "cap-and-trade system"). Under this approach, the environmental authority issues a limited number of permits, each of which allows a certain number of pounds of pollutants per year to be emitted into the environment. The total number of such permits is limited to ensure that the predetermined standard for environmental quality is attained. But these permits have the important property that they can be traded: polluting firms buy and sell them in a market. Firms with relatively high costs of pollution control can purchase rights to emit a particular pollutant from those who can control their emissions more inexpensively. In this way, standard cost-minimizing behavior leads to a least-cost pattern of pollution control efforts among firms. Likewise, such cap-and-trade systems (like environmental taxes) provide a longer-run incentive for the discovery and introduction of more effective and less costly control technologies. Such systems are currently in use in the United States for the control of sulfur and nitrogen-oxide emissions into the atmosphere (and are discussed in chapters nine and ten). Moreover, some form of an emissions-trading system is under serious consideration for use on a global scale to address the problem of global warming. In fact, in 2005, the European Union introduced the largest and most ambitious emissions-trading system in the world to meet the European commitment under the Kyoto Protocol for the reduction of carbon-dioxide emissions to curtail global climate change.

As with the setting of environmental standards,

the economic approach to the choice of policy instruments was essentially ignored in the early pieces of environmental legislation. Instead, environmental authorities employed so-called command-and-control (CAC) techniques for pollution control. Such CAC regimes often consisted of directives to individual polluters that specified, at times in considerable detail, the precise forms of control measures that were to be adopted. Many studies have documented the unnecessarily high costs that these programs have imposed on polluters and the economy by restricting flexibility in polluter responses. Not only this, but such measures typically provided little in the way of incentives for efforts to develop more effective control technologies, since firms are basically concerned with complying with the relevant directives.

Over time, we have come to appreciate the need for attaining our environmental objectives efficiently. For one thing, if we can keep control costs down, we will be in a position to do more in the way of environmental cleanup. There has, in consequence, been a growing interest in the use of incentive-based policy instruments, including not only taxes and systems of transferable permits, but such things as deposit-refund systems and various forms of legal liability that can, in certain instances, give polluters appropriate inducements for adopting control measures. Even where the CAC approach is still used, it is recognized that it is important *not* to specify control technologies, but to allow firms some flexibility in determining the most effective and least-cost way to comply with the limit that the regulatory authority imposes on its emissions.

Some Further Issues

In this introduction, I have focused attention on three basic ideas or lessons that economic analysis provides for environmental and resource management. Economics does, of course, have interesting and important things to say on other matters. In Part 5 of this volume, for example, there are four essays addressing the issue of "environmental federalism," the question of how to assign regulatory responsibility for environmental management among the different levels of government. These essays suggest some principles for making this assignment.

More generally, the reader will find that the three ideas discussed in this introduction manifest themselves in a wide variety of forms in the essays that make up this book. Even for an issue like biodiversity (addressed in Part 6), one with important ethical content, policy choices involve the use of our scarce resources—and thus inescapably have economic dimensions. Perhaps the most challenging of all—because of its enormous potential consequences, scientific uncertainties, and distant time horizon—is the issue of global climate change. Part 8 offers five essays on this critical issue. In view of its importance and inherent complexity, researchers at RFF have, and are, devoting a major effort to the study of climate change and the range of available policy responses on both a national and global scale. This revised edition of *The RFF Reader* introduces a new section (Part 7) on "Energy Policy for the Twenty-First Century." In view of the crucial role of energy for the economy, national security, and the environment (including global climate change), energy policy has a high priority on the national and global agenda. The essays in Part 7 explore systematically various sources of energy and their potential for addressing future energy needs.

Part 10 of *The RFF Reader* turns to the analysis of pressing environmental issues in developing countries and nations that are making the transition from formerly socialist regimes to more democratic and market-oriented systems. The course of environmental management in the developing world is clearly going to have a profound impact on the future of the global environment. And here the issue of scarce resources presses especially hard on the capacity to introduce ameliorative measures for the environment. Sensible goals and efficient policy measures may be even more important in this setting than in the industrialized world.

Part 11, a new section to *The RFF Reader* contains essays that apply economic analysis to two newly emerging issues: antibiotic resistance and the problem of invasive species. These essays call attention to the fact that what are seemingly solely scientific issues have important economic dimensions that need to be explored. They provide examples of the relevance of basic economic analysis to the challenging variety of new environmental issues.

Over the past 30 years, a growing appreciation of the importance of the economic perspective on environmental and resource management has emerged. Our experience with environmental legislation and policy has made it clear that ignoring the lessons of economics takes a heavy toll on our efforts to clean up the environment and to do so in a relatively efficient way. This last point is becoming increasingly important as we try to improve environmental quality yet further. We have come a long way in cleaning up the air in our cities and our polluted lakes and rivers. This has been relatively easy in the sense that there exist straightforward measures for improving the environment when *initiating* cleanup programs. However, in the current lingo, we have picked the "low-hanging fruit," and we now must invest in more complex and expensive methods for further environmental improvements. This means that there will be an even larger premium both on the selection of sensible environmental targets and the design of cost-effective regulatory measures to attain these targets. Economic analysis has much to contribute to this effort. In the final essay in this volume (Part 12), I offer reflections on the evolution of the role of economics in environmental policymaking. It is encouraging to find that, in contrast to the early major legislation on the environment, current debate over environmental measures routinely employs economic analysis both in the discussion of appropriate standards for environmental quality and in the selection of policy instruments for attaining these standards.

Note

1. Two qualifications are worthy of note here. First, there are some cases where voluntary negotiations among a small group of affected parties can effectively resolve an "externality." Such cases are the subject of the famous paper by Ronald Coase, "The Problem of Social Cost," *Journal of Law and Economics* (October 1960): 1-44. Although the Coasian treatment has gotten considerable attention in the literature, its applicability remains limited. The major environmental problems, including, for example, urban air pollution and water pollution, cannot be addressed through voluntary market mechanisms; they require public regulatory intervention. Second (and closely related), one might envision a system where markets are supplemented by a perfect tort system such that polluters are fully liable for the costs of any damages that they impose on society. Such liability could, in principle, provide the needed incentives for efficient levels of pollution abatement. Liability rules, in fact, have an important role to play in environmental protection, but the various imperfections inherent in any practicable legal system for environmental protection leave a large role for regulatory measures.

Part 1

Science and Environmental Policy

1

What the Science Says
How We Use It and Abuse It To Make Health and Environmental Policy

James Wilson and J.W. Anderson

In an ideal setting, the best science would flow, pure and undefiled, into the policymaking process. But this isn't what happens. The best assurance of good public policy lies not only in scientific knowledge per se, but in open debate, caution, and a regulatory system capable of self-correction.

Environmental policy is always based on science—up to a point. But defining that point is often a matter of fierce dispute and political combat. Then the quality of the science involved becomes an issue.

Decisions are easiest when threats and benefits are immediately visible to the naked eye. No one questioned, for instance, the proposition that burning soft coal in fireplaces and furnaces meant smoky skies over St. Louis. When people got sufficiently tired of the smoke, as they finally did in 1937, this source of home heating was outlawed with no argument over causation. But much of the modern environmental protection movement has been a response to menaces that are invisible, indirect, and detectable only through advanced technology. The effect has been to draw subtle and complex scientific issues into the arenas of politics.

The debates burn hottest where scientific uncertainty is the greatest and economic stakes are the highest. Scientific uncertainty comes in many forms.

About-Face on Thresholds

When science changes, environmental regulation has great difficulty adapting. One dramatic example is the issue of carcinogens' thresholds—whether there are doses below which carcinogens have no adverse effect on health. On that one, the consensus among scientists has reversed twice in less than fifty years.

Originally published in *Resources,* No. 128, Summer 1997.

Until the 1950s, it was a settled principle of toxicology that every poison had a threshold below which the dose was too slight to do harm. But with rising anxiety about the environmental causes of cancer, especially in the context of the debates about nuclear radiation and weapons testing, it began to seem more prudent to assume that carcinogens generally had no thresholds. One result was the famous Delaney Clause that Congress wrote into the 1958 Food, Drug, and Cosmetics Act.

The Delaney Clause banned all carcinogens from any processed food. At the time Congress, like the experts advising it, was under the impression that carcinogens were few and readily identifiable. But over time research found more and more substances that, if fed to rats in sufficiently massive amounts, could cause cancer. Some were naturally present in common foods—including orange juice. At the same time the increasing sophistication of measuring techniques identified traces of widely used pesticides and fungicides in many foods.

The regulatory system generally responded to these unwelcome findings by ignoring them. But at the same time the science was changing. Improved understanding of the processes by which cancers originate and develop made it seem increasingly likely that thresholds exist after all. The regulators themselves became convinced of that, although the Delaney Clause remained the law. The Food and Drug Administration quietly whittled away at the clause until the courts told them to go no further.

There's a high cost to society when government must enforce laws that make no sense to the people charged with enforcing them. It engenders cynicism among the regulators, and among the public it erodes confidence in both the law and its enforcement. But while Congress increasingly understood that the law was unenforceable, it refused to consider any reform that might be attacked as lowering the standard of health protection.

Lawsuits Force the Issue

A lesson for science policy lies in the way this paralysis was ended. It wasn't the advance of science that did it, although the science was certainly advancing. Instead, as often happens in environmental affairs, the issue was forced by litigation—in this case, litigation brought by people who wanted the Delaney Clause enforced more literally. In 1992 a federal appellate court decision raised the prospect that the Environmental Protection Agency would be required to ban many widely used pesticides, with drastic implications for farmers' crops and retail food prices. That got the attention of Congress, and in 1996 it replaced Delaney's flat ban with a more realistic standard of "reasonable certainty" of no harm. According to its authors, the phrase was intended to mean a lifetime risk of cancer of no more than one in a million. With this change, the law is now back in conformity with scientific opinion and the regulators' actual practice.

Opinion Masked as Science

If it is possible to draw up a list of the circumstances that generate strife over the application of science to policy, along with changing science, disputes among scientists must also be near the top. To many laymen, certainty and precision is the essence of science: as they understand it, a scientific question can have only one right answer. But especially in matters of public health, it is often essential to make policy decisions long before the science is entirely clear. When people's lives and welfare are at stake, it is not possible to wait until every technical doubt has been resolved.

The situation is frequently aggravated by scientists who underestimate the uncertainties in their own work, leading them to blur the line between science and policy. Endless examples have turned up in the congressional hearings in 1997 on the EPA's proposals to revise the air quality standards for ozone and particulate matter. The EPA's Clean Air Science Advisory Committee (CASAC) set up a special panel of experts on ozone, and the panel came to general agreement that, within the range of standards under discussion, there was no "bright line" to distinguish any of them as being "significantly more protective of public health" than the others. Setting the standard, they said, was purely a policy choice. But the law specifically authorizes CASAC panels to offer policy advice, and more than half of the panel went on to offer EPA their various and conflicting personal opinions as to where the standard should be set. CASAC is deliberately organized to represent a wide range of views and interests.

The policymakers, most of them trained as lawyers, seized whichever of these personal opinions agreed with their own and cited them as the voice of science itself. In congressional hearing after hearing, EPA's Administrator, Carol Browner, defended her proposed standards as merely reflecting "the science." Her adversaries then quoted back to her the opinions of scientists who disagreed, some of them members of CASAC and others officials of the Clinton administration.

A more productive way to approach policy choices is to acknowledge uncertainty and take it explicitly into account. Do you go on a picnic if the weather report forecasts a 60 percent chance of rain? Do you commit society to a complex new air quality regulation if there's a 40 percent chance that it will not provide health benefits as intended? Attempting to quantify risk is an important step in making policy decisions. Unfortunately, it violates the current style of politics, in which it is safer to minimize responsibility and discretion by suggesting that decisions are determined solely by the science.

But which science? Toxicology looks for the mechanisms of damage to health at the molecular level, in terms that can be demonstrated in the laboratory, and tends to dismiss anything less specific as mere speculation. Epidemiology, on the other hand, sees reality in the statistical associations between the presence of a pollutant and the evidence of damage. As Mark Powell has pointed out in his RFF discussion paper on EPA's use of science in setting ozone policy, the tension within the agency between the toxicologists and the epidemiologists is as old as EPA itself. On clean air, CASAC is similarly divided.

In the current round of debate over clean air rules, the policymakers who support tighter standards cite the epidemiologists. Those who resist tighter standards cite the toxicologists. At present the differences between the two specialties' positions on particulate matter is substantial, and there is no one view that represents settled and accepted scientific truth.

Science as Proxy for Other Issues

In the vehement debates over science, scientific uncertainty often becomes the proxy for other issues—in the case of the Clean Air Act, for the forbidden subject of economic costs. The act prohibits EPA from taking costs into account in setting standards. Opponents of proposed regulations, unable to pursue their argument that the costs will outweigh any prospective benefits to health, go after the scientific basis of the regulations instead.

Confusion also arises when science asks the wrong question—sometimes because the law requires it. Here again the Clean Air Act provides examples. To take a prominent one, the act wants science to tell the regulators what effects each of six common pollutants has on human health. Since the pollutants are regulated separately, the health effects have to be studied separately. Scientists have been trying to tell the regulators for some years that it would be far more useful to investigate these pollutants mixed together, in the "soup" that people actually breathe, because the presence of one compound can affect the impact of another. But Congress has never responded to that advice because the concept of mixtures doesn't fit easily into the existing statutory framework for regulation. When environmental reality collides with statutory tradition, it's not always the statute that gives way.

Sometimes the Wrong Battle

Science, or what seems to be science, can sometimes be flatly wrong. The process of scientific inquiry is self-correcting over time. That is its greatest strength. But policy doesn't always wait for the corrections.

The Superfund program originated, notoriously, in response to mistaken and exaggerated scientific judgment. The Love Canal, in Niagara Falls, NY, had been well-known locally as a toxic chemical dump that was leaking insecticide into Lake Erie. But it suddenly became a national news story and a symbol of a new range of hidden environmental dangers, when in the summer of 1978 the state's health commissioner declared it a threat to the health of people living there. It was an election year in New York, and suddenly politicians at all levels, including President Carter, were competing to show concern and protect the residents. The following year a scientific consultant to the local homeowners association reported findings that indicated a wide range of threats to health. Then another consultant engaged by EPA reported evidence of high rates of chromosome damage among residents. Those claims established the atmosphere in which Congress began to draft the Superfund legislation.

Subsequently, review panels within EPA severely criticized the contractor's chromosome report, and a special committee of scientists set up by the governor of New York dismissed all of the health findings as inconclusive. But by the time that happened, the Superfund bill was approaching final passage.

It would be pleasant to think that some mechanism might be invented to allow the best science to flow, pure and undefiled, directly into policy. But that's hardly realistic, amidst the turbulence of rapidly developing science and especially in a field that, like environmental and health protection, has emerged as one of the leading battlegrounds of national politics. The best assurance of good public policy seems to lie in open debate, caution, and a regulatory system capable of self-correction.

Research Needs Funding

One point on which improvement is both possible and badly needed is the funding of scientific research relevant to regulatory decisions. Private and public spending in this country to meet the federal requirements for pollution control and abatement during the mid-1990s was in the range of $140 billion a year. Congress gives EPA less than half of one percent as much to spend on all its scientific and technological work for all purposes, a sadly disproportionate effort to ensure that environmental rules have the best possible scientific base.

It's not only the general pressure to cut the budget that inhibits adequate spending on science to support environmental regulation. Concerns about global warming have led to substantial outlays of federal science money on other purposes, and on other agencies than EPA. Currently, the EPA science budget is only about 10 percent of total federal spending on environmental scientific research and development.

The purpose of balancing the budget is to enhance the economy's efficiency and promote future growth. But budget cuts won't help the economy if they lead to the waste of resources on misguided policy.

2

Using Science Soundly
The Yucca Mountain Standard

Robert W. Fri

Nuclear waste is piling up at electric power plants and weapons production facilities. When the federal government decided to build a national repository for radioactive waste, it was hoped that "sound science" could resolve the concerns about the safety of such a facility. In the author's view, science can surely help, but some of the most difficult problems are bigger than science alone can solve.

Using "sound science" to shape government regulation is one of the most hotly argued topics in the ongoing debate about regulatory reform. Of course, no one is arguing that the government should rely on *unsound* science for its decisions. But supposing, as some reform advocates apparently do, that even the best science will sweep away regulatory controversy is equally foolish.

My experience as the chair of a National Research Council (NRC) committee that studied the scientific basis for regulating high-level nuclear waste disposal drove home this conclusion for me. I learned that science alone could resolve few of the key regulatory questions. More often, science could only offer a useful framework and starting point for policy debates. And sometimes, science's most helpful contribution was to admit that it had nothing to say.

A Short History of Nuclear Waste Regulation

Both commercial generation of electric power and government production of nuclear weapons result in high-level (long-lasting and highly radioactive) nuclear waste. At present, these wastes are stored at nearly a hundred sites around the United States, but federal policy mandates that the wastes ultimately be placed in a mined underground geologic repository. In 1987, Congress decreed that the first such repository be located at Yucca Mountain, which is near Las Vegas, Nevada.

The basic idea of geologic disposal is to use permanent natural barriers as a principal means of isolating nuclear waste

Originally published in *Resources*, No. 120, Summer 1995.

from the environment. Over time, however, some of the radioactive material will escape from even the best repository. At Yucca Mountain, for example, the casks in which nuclear waste will be initially stored will eventually break down, allowing the waste to migrate to the water table, which is located several hundred feet below the repository, and contaminate the flow of groundwater away from the repository site.

This process may take many thousands of years, but the nuclear waste will retain some of its radioactivity for more than a million years. Once the groundwater is contaminated, then the people who use it for drinking and irrigation will be exposed to radionuclides. Given this inevitability, the goal at Yucca Mountain is to design a repository that will limit to an acceptable level over very long periods of time the human health effects associated with nuclear waste releases.

Developing a standard that defines this acceptable level is one of Washington's longest running regulatory dramas. After ten years of work, the U.S. Environmental Protection Agency (EPA) first promulgated a standard in 1985. But following a successful court challenge in 1987, the standard was remanded to the agency for revision. Before EPA could issue the new standard, however, Congress enacted the Energy Policy Act of 1992, which mandated a new and different process for setting the standard for the proposed repository at Yucca Mountain.

Congress clearly wanted to curtail the debate over the standard. To do this, it reposed considerable faith in sound science. It required the National Academy of Sciences (through the National Research Council) to evaluate the scientific basis for a Yucca Mountain standard and directed EPA to promulgate a new standard "based on and consistent with" the findings of the academy. At the time, the idea of constraining regulators with the findings of a scientific panel was unfamiliar to the agency and the academy. Since a similar idea is afoot in regulatory reform, the Yucca Mountain experience may be instructive for that debate.

The Yucca Mountain Standard

Developing a standard that specifies a socially acceptable limit on the human health effects of nuclear waste releases involves many decisions. As the NRC committee learned in evaluating the scientific basis for the Yucca Mountain standard, a scientifically best decision rarely exists. The trick is to make the best use of the science that is available.

The first decision that EPA faces is how to measure safety. This decision entails setting a socially acceptable limit on some aspect of the repository's performance. As a technical matter, for example, the limit could be stated in terms of how much radioactivity the repository releases per year, how much radiation people will be exposed to as a result of releases, or people's risk of dying from this exposure. The committee recommended to EPA a standard stated in terms of risk of death.

The evolving scientific understanding of the relationship between radiation doses and the health effects that they cause certainly influenced this recommendation. Over the years, successive scientific reviews typically have concluded that a given dose of radiation may cause more deaths than scientists had previously believed. As a result of this trend in science, it makes sense to state the standard as a limit on the number of additional deaths attributable to releases from the repository. Doing so would mean that the standard would not have to change as the science continues to evolve. This observation also weighed heavily in the committee's preference for a risk-based standard.

Although a scientific fact lies behind it, this recommendation is clearly not dictated by science. Changing a standard to incorporate new information is technically not a problem. The preference for a stable, risk-based standard rests on the belief that changing so controversial a standard as one that specifies the acceptable level of human health effects associated with nuclear releases is socially, politically, and administratively undesirable.

This intersection of science and policy permeates the other decisions that have to be made in setting the standard for determining whether the Yucca Mountain repository would adequately protect human health. In particular, EPA has to specify what level of protection is to be afforded, to whom, and over what time period. For only one of these decisions does science provide reasonably conclusive guidance.

Establishing the level of risk that the standard will allow is a question of policy, not science. In other contexts, however, EPA and other organizations have set limits on a variety of nuclear risks that range from one additional death per hundred thousand persons to one in a million. At best, this information provides a scientifically defensible starting point for debating the acceptable level of risk at Yucca Mountain. It certainly does not predestine the outcome. Acknowledging this reality, the NRC committee could only recommend a reasonable range of risks for EPA to consider in crafting its regulatory proposal.

To determine whether a repository provides the acceptable level of protection, the risk that repository releases could impose on a specific individual or group must be calculated. How this person or group is defined can determine whether the standard is met. It has a particularly significant effect on whether the standard is met at Yucca Mountain, because the geology of the site lends itself to the creation of spots—for example, places in a groundwater plume—at which radiation tends to concentrate. A clever opponent of the repository could define the person to be protected as someone drawing water for drinking and irrigation only from one of these hot spots. An advocate for the repository would naturally assume that the affected parties were located at a safe distance from these areas.

As a matter of policy, the NRC committee preferred to avoid these extreme assumptions. Given this policy, it looked to science (or at least to careful scientific thinking) to contribute a methodology for calculating compliance with the standard that resists extreme cases. The methodology that the committee chose was the "critical group method," which calculates the average risk to a member of the group at greatest risk.

Guidance for the time period over which the standard should provide protection is provided by

the fact that radioactivity associated with high-level nuclear waste will not dissipate for more than a million years. Ideally, then, compliance with the standard would be tested over the full duration of this period in order to determine the time at which the greatest effect on human health occurs. Whether this determination is possible depends on the ability of scientists to evaluate the behavior of the repository over very long periods of time.

Here, for a change, is a question of science rather than policy. The committee answered it by saying that compliance assessment is feasible for most physical and geological aspects of repository performance on the order of a million years at Yucca Mountain. Still, this answer is based on the expert scientific judgment that the fundamental geologic structure will be relatively stable for this long, not on the testable hypotheses of scientific method. Thus, other experts might reach a different conclusion.

Running Out of Science

The NRC committee was able to recommend the foregoing elements of the standard with at least one foot in the realm of science. Unfortunately, however, science can contribute little to answering three of the most controversial questions that bothered Congress about the standard in the first place. For three of these questions, the scientific basis for decisionmaking essentially does not exist.

What Is a Negligible Risk?

The main concern of a standard for a nuclear waste repository is to protect populations living near the repository. In principle, however, a very large and dispersed population could be affected by releases of nuclear waste. In the case of Yucca Mountain, radioactive carbon dioxide gas could escape from nuclear waste canisters and be inhaled by people living far away from the repository. The carbon-14 problem, named after the radioactive isotope present in the waste, is one of the most vexing problems with which EPA must deal. Because carbon-14 releases from Yucca Mountain would be mixed with the global atmosphere, the health risk to any one individual is exceedingly small. On the other hand, the number of people exposed worldwide over the life of the repository is astronomical. If we multiply the very small risk by this very large number of people, we can calculate that many additional deaths could occur over a very long time period.

But how do we interpret a number computed in this way? No adverse health effects may occur at the very low doses of carbon-14 to which people would be exposed; but lacking data to show that this would be the case, experts in the field say that the prudent course is to assume that health effects will occur. Making this assumption could produce a scenario that leads either to abandoning the Yucca Mountain site or to spending a great deal of money to contain carbon dioxide gas.

To the dismay of policymakers, science cannot make this problem go away. Faced with this dilemma, the committee could only observe that the risk to any one individual in the global population would be very small—perhaps ten thousand times lower that the one-in-a-million level at which the basic standard might be set. A responsible decisionmaker could conclude that such risks are so negligible that they should not affect the design of the repository, but he or she would have to do so without much definitive guidance from the scientific community.

Can We Guard against Future Human Intrusion at a Repository?

One way to project significant human exposure to radiation releases from repositories is to assume that someone intrudes after they close. For example, a future oil explorer could drill into a waste canister and bring radioactive material directly to the surface. In crafting its charge to the NRC, Congress specifically asked whether any scientific basis exists for evaluating this risk or for assuming that it can be prevented.

The answer to both questions is no. The committee found no scientific basis for predicting the behavior of humans thousands of years into the future. Since neither the probability of human intrusion nor the effectiveness of preventive measures is predictable, the committee concluded that these

issues should not be considered in the assessment of compliance with a risk-based standard. (We did, however, offer an alternative analysis to test the resilience of the repository to an assumed intrusion.)

In this case, the absence of a scientific basis is probably a help to decisionmaking. Admitting the limits of science should greatly reduce the considerable analysis and controversy lavished on speculation about the likelihood of human intrusion. I should note, however, that if regulators were deciding whether to dispose of waste at scattered surface sites instead of in a geologic repository, as at Yucca Mountain, analyzing the risks of human intrusion might be crucial.

What Assumptions Do We Make about Exposure Scenarios?

In all of the above issues, the committee walked the line between science and policy without dissent. But consensus failed when it came to specifying the exposure scenario to use in calculating compliance with the standard.

The exposure scenario describes how radiation that is released from the repository passes through the biosphere to expose humans. The scenario thus must specify whether and how water wells are drilled into the groundwater underlying Yucca Mountain, whether the water is used for drinking or irrigation, how much of a person's food intake is contaminated by this irrigation, and so on. Science can put bounds on many of these assumptions; for example, people can drink only so much water, and plants retain radionuclides at predictable rates. Developing exposure scenarios, even for the distant future, is therefore not entirely a blue-sky exercise.

Still, science cannot predict human behavior. This consideration is important in the Yucca Mountain case, because the area is sparsely settled—one good reason for locating a repository there. Given this, what should an exposure scenario assume about whether someone is present to be exposed to any release that might occur?

Remember that the committee recommended a standard that would protect the people at greatest risk, while avoiding the trap of extreme assump-

Update

Ten years after our National Research Council report was issued, the Yucca Mountain Standard is still in flux. The Environmental Protection Agency proposed a standard, as required by Congress, "based on and consistent with" our report. In 2004 the D.C. Circuit Court of Appeals remanded the standard to EPA on the grounds that its proposal was not consistent with our recommendations. Specifically, the court objected to EPA's decision to evaluate compliance with the standard over a 10,000-year period rather than out to the time of greatest risk as we recommended. At this writing, EPA has not yet indicated how it plans to deal with the court's decision. It is interesting to note, however, that the compliance assessment period is the recommendation for which our committee had the strongest scientific basis.

In the first edition of this volume, I wrote a sidebar that is still relevant to the radioactive waste disposal issue and is worth repeating here. It said that I had "failed to mention what is perhaps the most important role of 'sound science'–to guide policymakers toward the most useful questions. Asking whether long-term radioactive waste disposal creates a risk that exceeds some threshold turns out to be not a very good question . . . Perhaps a better question would be comparative: given that the nuclear waste has to go somewhere, where is the least risky place to put it?" In our post-9/11 world, this question is perhaps even better than it was six years ago.

bution of population in the distant future is no small accomplishment. Indeed, doing so would considerably circumscribe the current debate about Yucca Mountain.

Even within this narrowed range of options, however, members of the committee disagreed on the exact population-distribution assumption that should be used. One member felt strongly that the exposure scenario should assume that a subsistence farmer will always be living at the place where exposure to radiation will be highest over the life of the repository. The other members believed that the physical features of the site naturally lead to a dispersed population and that the exposure scenario should take account of this fact.

These alternative views can excite considerable passion on the part of their proponents. In my view, however, such controversy obscures two crucial points. One is that the population-distribution assumption cannot be resolved on the basis of science. No one can predict where people will live in the future; therefore, regulators must make a judgment call in choosing an assumption about population distribution in the exposure scenario. The other point, noted above, is that the debate is over a fairly narrow range of assumptions. Despite the passion attendant on it, this debate is far more manageable than the open-ended debate to which EPA might be exposed if the committee had not narrowed the range of assumptions.

The Role of Science in Regulatory Decisions

The lessons that the NRC committee learned in studying the scientific basis for the Yucca Mountain standard may be important to those involved in the regulatory reform debate. The chief lesson is that the soundest science rarely provides black-and-white answers for regulatory decisionmaking; it only brightens a bit the familiar gray space in which decisions are made.

To be sure, science can sometimes have a conclusive effect on a regulatory decision. In the Yucca Mountain case, the conclusion that the standard should be applied without time limit rests almost entirely on expert scientific judgment. By contrast,

tions. It would be inconsistent with this principle to base the exposure scenario on, say, the expectation that millions of people will move into the Yucca Mountain neighborhood. A more reasonable assumption is that farmers scattered about the area will comprise the population at greatest risk. Insisting on such a cautious but reasonable approach to narrow the range of assumptions about the distri-

the current EPA standard applies only over a 10,000-year duration. Accepting the scientific judgment of the Yucca Mountain study would thus have a profound effect on the design of the standard.

Admitting that science has nothing to say also can powerfully affect decisionmaking. For example, the committee found no scientific basis for evaluating the probability of human intrusion. Therefore, it concluded that the issue should not be considered in assessing compliance with a risk-based standard. If EPA accepts this conclusion, a significant line of argument that could distract the regulatory debate will be closed off.

Mostly, however, the Yucca Mountain study shows that science is helpful, but not conclusive, in arriving at reasonable decisions—such as setting the acceptable level of protection, defining the people to be protected, and specifying the exposure scenarios to be used for compliance analysis. In these instances, the committee avoided asserting that sound science provided a complete answer, but did try to use scientific judgment to define a reasonable starting point and a bounded range of options for EPA to consider. In this way, science can be quite helpful in fostering constructive debate.

Finally, the Yucca Mountain study indicates that science cannot protect public officials from hard decisions. Advocates of the Yucca Mountain repository would like nothing better than for science to make the carbon-14 problem go away. But science cannot do that; it can only note that the risk from carbon-14 emissions to an average individual in the global population is exceedingly small. Whether these risks are so small as to be negligible is a tough political call that science cannot—and should not—make.

In short, the Yucca Mountain study clearly illustrates that excessive faith in the power of sound science is more likely to produce messy frustration than crisp decisions. A better goal for regulatory reform is the sound use of science to clarify and contain the inevitable policy controversy.

Suggested Reading

National Research Council. 1990. *Rethinking High-Level Radioactive Waste Disposal*. Washington, DC: National Academy Press.

———. 1994. *Science and Judgment in Risk Assessment*. Washington, DC: National Academy Press.

———. 1995. *Technical Bases for Yucca Mountain Standards*. Washington, DC: National Academy Press.

Part 2

Valuation of the Environment and Benefit-Cost Analysis

3 Economics Clarifies Choices about Managing Risk

A. Myrick Freeman III and Paul R. Portney

Environmental decisionmaking inherently involves risk. And the choices we make also embody basic tradeoffs. Economics can help in making these decisions by providing a systematic framework for weighing the benefits against the costs of the available alternatives.

Government officials making decisions about such issues as allowable pesticide residues in foods, nuclear reactor safety standards, and air quality standards face a difficult problem. On the one hand, there is the evident desire of the public to reduce the risks inherent in modern life. On the other hand, reducing these risks is costly. So choices about risk policy involve tradeoffs. Risk management refers to the process through which a variety of considerations—scientific, legal, political, economic (benefits and costs), and even philosophical—are taken into account and a decision is reached concerning an environmental regulatory problem.

Economics can contribute in a number of ways to managing risks to health and the environment. At a basic level, economics can help to inform decisionmakers about how much various regulatory approaches or pollution control options will cost society. Upon first blush, this might seem pedestrian and straightforward. In fact, it might even appear that engineers rather than economists are better able to make such determinations, especially when the options under consideration involve primarily structures and equipment.

But appearances are deceiving. One of the real, albeit subtle, virtues of economics is its focus on what are called *opportunity costs*—that is, what society must give up in the form of other desirable things in order to pursue a desired goal such as reduced environmental risk. Under some circumstances, expenditures for pollution control equipment, cleaner fuels, or the like will closely approximate true opportunity costs. However,

Originally published in *Resources,* No. 95, Spring 1989.

often this correspondence between money expenditures and opportunity cost is lacking. For example, rules on private behavior such as mandatory recycling of household wastes or limits on eating fish caught by sports fishermen involve no direct money outlays, but they impose costs in the form of time or reduced satisfaction. An economic perspective on costs provides valuable insights about the nature and magnitude of these forgone opportunities.

An important criterion for the rational management of risk is that any reduction in risk be accomplished at the lowest possible economic cost. Economic analysis can help to identify the least costly way to accomplish a particular reduction in environmental risk. Used in this way—how we can accomplish X for as little as possible—the application of economics goes by the name of cost-effectiveness analysis.

Besides helping to identify and properly measure costs, economics can help us to understand how these costs (as well as benefits) are distributed among the population. For instance, we might be interested in knowing whether residents of rural areas would bear a disproportionate share of the costs of an acid rain control program. Or, we might wish to know whether financing Superfund cleanups via direct budgetary outlays is more or less regressive in its impacts than financing those same cleanups through taxes on manufacturing firms. Again, we might want to determine whether the favorable effects of a policy are distributed equally among current and future generations. Economics can help us answer these questions.

Normative Guidance

Using economics merely to supply information about the costs of different options is one of its less controversial applications in risk management. The challenge comes when economics is used to answer questions like: What should we do about the problem of pesticide residues in foodstuffs? Which cleanup strategy is best at the XYZ site? Here economics is being asked to go beyond the purely informational—beyond describing what would happen here or there—and instead is being asked

to provide normative guidance to decisionmaking—that is, to help us answer the question, What *ought* we do?

To answer normative questions like these, economists generally rely on a branch of economics known as benefit-cost analysis. (See the box, "Benefit-Cost Analysis and Risk.") Economists view benefit-cost analysis as akin to common sense. This is because after peeling away the analytical veneer, formal benefit-cost analysis essentially asks: If we pursue a particular policy option, what good will come of it and what will we have to sacrifice to get it? It is a simple extension to ask whether the former is worth the latter.

Although benefit-cost analysis can clarify the pros and cons of taking particular actions, its application to the problems of environmental risk management has not gone smoothly. It is not embraced in any major environmental statutes except the Toxic Substances Control Act of 1976 and the Federal Insecticide, Fungicide, and Rodenticide Act of 1972. Nor is it the rule in other regulatory statutes protective of public health (for example, those having to do with occupational safety and health or with consumer products). In fact, the balancing of benefits and costs appears to be *prohibited* when the Environmental Protection Agency sets most standards for air and water pollution and the regulation of active or abandoned hazardous waste disposal sites. Similarly, the well-known Delaney clause in the Federal Food, Drug, and Cosmetic Act of 1938 explicitly prohibits the head of the Food and Drug Administration from considering the health benefits associated with certain food additives if these additives are known or suspected of causing cancers in humans. Moreover, although the last three presidents have issued executive orders mandating that benefit-cost analyses accompany any new proposed or final regulations, federal regulatory agencies have often resisted, and Congress has battled to have these presidential orders weakened.

In addition to the political unease over benefit-cost analysis, there is also more than a little public concern about its use in environmental decisionmaking. This concern is harder to document, but it shows up often in public meetings, opinion polls, and everyday discussions.

Political Unease

Political reservations about using benefit-cost analysis to help make risk management decisions are based on several concerns.

Distributional Issues

Benefit-cost analysis is in one sense distributionally neutral. That is, a dollar's worth of benefits (or costs) count the same regardless of the economic position, geographic location, or other characteristics of the individuals to whom they accrue. This can spell trouble in political circles.

Consider, for instance, the case of acid rain. Emissions of sulfur and nitrogen oxides from coal-fired utility and industrial boilers, as well as from mobile sources, are believed to be responsible for damages to aquatic ecosystems, forests, agricultural products, materials, and even human health. A variety of control measures are available and reasonably well understood. If risk managers decide to use the "polluter pays" principle, there would be very uneven geographic distributional effects. Because states in the Ohio River valley are emitters of large amounts of sulfur dioxide, they would bear a heavy share of the total costs of controlling emissions. Application of the polluter pays principle could cause electricity bills in those states to increase by as much as 15 to 20 percent. Such geographic concentration of costs has been one of the stumbling blocks to amending the Clean Air Act of 1970 to deal with acid rain.

Imprecise Information

Politicians view benefit-cost analyses of risk management options with suspicion for another reason. Estimates of benefits and costs must rest on a foundation of knowledge of the physical, biological, and engineering systems involved as well as the economic factors determining monetary values. For example, it must be possible to answer such questions as: How much will indoor radon concentrations be reduced by air filtration equipment? How

Update

Through the years, benefit-cost analysis has come to play an increasingly larger role in the regulation of environmental and other risks. First of all, every president since Richard Nixon has issued an Executive Order, or reaffirmed one issued by his predecessor, requiring that regulatory agencies like the Environmental Protection Agency or the Food and Drug Administration conduct a benefit-cost analysis on any proposed new major regulation (defined as one that would impose new costs on the economy of $100 million or more annually). Moreover, since 1981 the Office of Information and Regulatory Affairs (OIRA), within the White House Office of Management and Budget, has had the responsibility for monitoring agencies' compliance with this requirement and also for issuing guidance to the agencies on how these analyses are to be conducted. Occasionally, the Executive Office of the President has told regulatory agencies that they may not issue new regulations until and unless they improve the quality of the economic analysis accompanying the rule.

Having said all this, one should not conclude that benefit-cost analysis rules the day. Many of the most important federal regulatory statutes direct agencies to set standards "to protect human health," or words to that effect. The Supreme Court ruled in 2001 (*Whitman vs. American Trucking Association*) that where a statute is silent as to whether costs can be taken into account when setting health standards, the statute must be interpreted as prohibiting that. This effectively has put the use of benefit-cost analysis off limits for many important regulatory decisions.

many fewer cases of lung cancer will there be if radon levels are reduced? How much will emissions be reduced by vapor recovery devices on gasoline pumps? What effect will this have on atmospheric ozone levels? What will be the impact on agricultural productivity of reduced ozone concentrations?

None of these questions is easy to answer. We sometimes have no more than well-educated guesses about the answers to these technological, physical, and biological questions. Some critics therefore believe that benefit-cost analysis can be rigged; that it will more often than not be used to justify a risk management decision that is taken not for analytical but rather for political or other reasons.

Myth of Abundance

One of the uses of benefit-cost analysis is to help ration scarce resources among competing ends. This use is a reflection of the fact that there are more things worth doing than there are resources with which to do them. While this sounds innocuous enough, politicians generally prefer to avoid making explicit such declarations. No politician is likely to gain much support for telling a group that, although they are bearing some environmental risk from problem X, the risk is relatively small and the money necessary to reduce it could be better spent elsewhere. *Even when it knows better,* the public likes to be told that its government is working to eliminate all environmentally transmitted risks. Sensing this, politicians shy away from analytical approaches based on the premise that resources are finite and priorities have to be set.

Public Unease

Politicians aside, the public has additional concerns about applying benefit-cost techniques to risk management problems.

Uncompensated Risk

Benefit-cost analysis is silent on the question of whether the losers from any risk management policy should be compensated. In practice, therefore, even policies that result in aggregate benefits in excess of costs could still leave some people worse off. It would be natural for the losers to oppose the policies. And this opposition could be quite vocal if the losses were concentrated among a relatively small group of people.

Nowhere is the issue of uncompensated risk more clearly visible than in the problem of proposed siting of LULUs (locally undesirable land uses) such as hazardous waste incinerators and low-level nuclear waste disposal facilities. Those around any proposed site will reject the argument that it is in the best interest of society for them to accept the increased risks that these facilities pose. And very often their opposition will prove successful. In an effort to deal with this impasse, analysts have begun to propose mechanisms for compensating the losers. (See the box, "Compensating the Losers.")

The "Right" To Be Risk-Free

Many citizens feel that they have a basic and inalienable "right" to be free from contaminants in the water they drink, the air they breathe, and the food they eat. They resent these rights being weighed against economic dislocations, balance-of-trade concerns, and other seemingly impersonal factors.

There is a ready response to such objections. First, even those rights guaranteed in the Bill of Rights are not absolute. For instance, one's freedom of speech is restricted when it comes to standing up in a crowded theater and shouting "Fire!" While no formal benefit-cost analysis supported these relatively mild restrictions on our basic rights, they are premised implicitly on the notion that completely unfettered speech or assembly may sometimes do more harm than good. In other words, the benefits of some restraints may be worth the costs.

To those who would argue that we have a right to be free from all environmental risks, the counterargument would run as follows. First, in a fundamental physical sense, we can never be free of such risks. Primitive woodburning puts harmful particulate matter in the air, and the human digestive system ensures that some wastes will always be with us. Thus, a no-risk world is simply impossible. Even if it were not, some risks would surely be judged to be so small in comparison to the costs of alleviating them that it would be best to accept them.

Expert versus Lay Opinion

Public opinion polls show a steady erosion of public

faith in experts. For a variety of reasons—some having to do with erroneous predictions in the past (for example, that nuclear power would become too cheap to meter), some having to do with generally increasing skepticism—the public seems less willing to be reassured that a particular risk, while real, is nonetheless quite small. This means that benefit-cost analyses, which depend critically on expert opinion or findings, will also have detractors among the populace. This becomes all the more likely when the experts themselves represent business concerns or, if university-based, derive part of their funding from corporations or trade associations. In cases where such suspicions are rampant, it becomes difficult to quell fears that experts feel are unwarranted. This divergence between expert and lay opinion cuts the other way, too. The public is often very slow to warm to concerns that experts may place near the top of their list of environmental risks.

Qualitative Dimensions of Risk

Risk analysts are sometimes puzzled when people react strongly to what may seem to be relatively small risks, yet appear to accept, or even seek out, risks such as skydiving and motorcycle racing. Such behavior is understandable. The risks associated with such sports are voluntarily borne, while one has little choice about the air one breathes while outside. Research during the last twenty years or so has demonstrated over and over again that such characteristics as voluntariness, familiarity, and dread influence the way individuals perceive and react to risks.

Facing Facts

Economics is the science of scarcity, and society is surely limited in the resources it can allocate to the control of environmental risks. Thus, it is important to think analytically about which risks we want to address first and how much control we wish to pursue. Like it or not, tradeoffs will be made when these risks are addressed. This follows directly from the observation above that society's resources are limited. Because this is so, we simply cannot eradicate any and all risks.

At some point decisionmakers will have to say to themselves that additional risk reductions will be so expensive that they are probably not worth additional effort. The virtue of economics is that it makes these decisions explicit. In other words, it forces decisionmakers to say openly, for example, that society cannot afford to spend $1 billion to save an additional life through more stringent regulation of substance X. While such acknowledgements are often painful, they do enable the public to see the tradeoffs that their elected officials are making and object if they disagree with them. Pretending that such tradeoffs do not have to be made only means that they will be made implicitly and out of the public eye.

One conclusion, then, is that the public and its political leaders would be well served if the public better understood economic methods and their application to problems of environmental risk management. The fact that this is a familiar refrain does not detract from its importance.

Sauce for the goose, however, is sauce for the gander. Just as it would behoove the public and its political leaders to better understand the economic approach to risk management, so too must economists understand why their message is so often ignored. While benefit-cost and cost-effectiveness analyses have their strengths, they also have weaknesses, some of which are nearly fatal in the political realm. Until economists do more than pay lip service to the importance of distributional concerns in real policymaking, for instance, they will remain peripherally involved in policy formulation at best.

Economists must also understand that the public cares about more than simply the statistical magnitude of risks. It is also concerned about the mechanisms through which these risks are transmitted, the degree to which the risks are voluntary, the benefits that accompany the risks, and other dimensions that are often disregarded in standard economic analyses. Until these concerns are acknowledged and incorporated in our economic models, economists may dismiss as irrational responses that make very real sense.

Suggested Reading

Arrow, Kenneth et al. Is There a Role for Benefit-Cost Analysis in Environmental, Health, and Safety Regulation? *Science* 272 (April 12, 1996): 221–2. Reprinted in *Economics of the Environment: Selected Readings*, 5th ed. (New York: Norton, 2005): 249–254.

Morgenstern, Richard D. (ed.). 1997. *Economic Analyses at EPA: Assessing Regulatory Impact.* Washington, DC: Resources for the Future.

Health-Based Environmental Standards
Balancing Costs with Benefits

Paul R. Portney and Winston Harrington

Several major pieces of environmental legislation in the United States call for health-based environmental standards on the grounds that everyone has a right to a safe environment. But, for many pollutants, there is no *safe* level. Moreover, environmental measures can involve high costs that need to be weighed against the benefits. An explicit balancing of benefits against costs makes decisionmakers face up to the tradeoffs that are inherent in environmental policymaking.

Balancing the pros and cons of a proposed action seems like a commonsense approach to decisionmaking. But often that is not the approach embodied in environmental legislation. In establishing health-based environmental standards under the Clean Air Act, the Safe Drinking Water Act, and several other major environmental laws, for instance, Congress all but explicitly prohibits the U.S. Environmental Protection Agency (EPA) from balancing the benefits of tighter standards against the attendant costs. Given the 104th Congress's strong interest in using benefit-cost analysis for federal regulation, why have previous legislatures excluded such balancing from the most important standard-setting decisions made by EPA?

Below, we identify two basic arguments that have been put forward for disregarding costs in environmental decisionmaking and raise counterarguments to both. While these arguments and counterarguments require a more thorough analysis than we can devote to them here, our hope is that we will stimulate a more open and enlightened debate about them than we have seen to date.

The Right-to-a-Safe-Environment Argument

The right-to-a-safe-environment argument is perhaps the most common response to those (like us) who would seek to balance benefits and costs in standard setting. This argument is, of course, based on the presumption that safe levels of environmental contaminants *can* be found, a presumption that is appar-

Originally published in *Resources*, No. 120, Summer 1995.

ent in our environmental laws. For instance, the Clean Air Act requires EPA to provide "an adequate margin of safety... requisite to protect public health" in setting National Ambient Air Quality Standards. From our perspective, the right-to-a-safe-environment argument has two flaws—the first scientific, the second philosophical.

From a scientific standpoint, the problem is that *no* safe level is likely to exist for most, if not all, pollutants. Rather, lower ambient concentrations of a particular pollutant almost always will imply lower risks of an adverse health effect (see Figure 1). In the case of air pollution, even very low levels of pollutants pose some risk of adverse reactions in children and the elderly with chronic respiratory disease.

If air quality standards are required by law to provide an adequate margin of safety, and if even weak concentrations of pollutants pose some risk to some individuals, it appears that only zero concentrations could be permitted under the law, for only zero concentrations would provide a "margin of safety" against adverse health effects. But totally eliminating ubiquitous air and water pollutants is impossible in a modern industrial society like ours (and would be impossible even in a primitive society, at least as long as fires were allowed!).

The philosophical problem with the right-to-a-safe-environment argument is whether it makes sense to treat risk-free levels of air and water quality—even if they could be identified—as inalienable rights, such as freedom of speech. Those who oppose a balancing approach to environmental standard setting often argue that we did (or do) no such balancing in establishing and protecting the basic freedoms that are guaranteed in the Constitution.

Figure 1. Two possible relationships between exposure to pollution and adverse health effects.

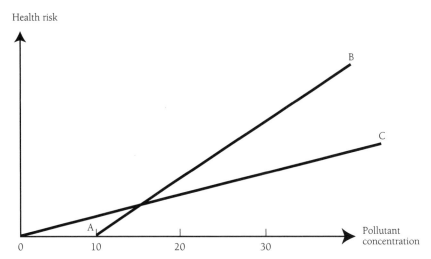

Through clinical, epidemiological, or animal toxicological studies, researchers try to identify a dose-response relationship between exposure to a harmful substance and adverse health effects—for example, incidences of cancer and asthma attacks. The line AB illustrates a case in which pollutant concentrations below ten units cause no such known effects. In this case, setting a pollution standard at some level below ten units would provide a "margin of safety" against adverse health effects. If the true dose-response curve is more like the line 0C, however, then any pollutant concentration above zero units would give rise to some risk. In this case, no safe level is likely to exist for the pollutant in question, making standard setting much more complicated.

But elevating environmental quality to the status of a constitutional right, as some have proposed, would remove neither the necessity for nor the desirability of balancing. Even the basic freedoms that are guaranteed in the Bill of Rights have been subjected to a very crude kind of balancing test. For example, we cannot stand up and scream "Fire!" in a crowded theater; libel laws constrain our ability to write whatever we want to write about a person; and other basic rights are constrained in varying degrees. Such restrictions on the basic rights of Americans reflect a clear balancing mentality—that is, a carefully considered view that some extensions of our fundamental rights could create greater problems (read "costs") than the additional freedoms ("benefits") that the extensions would provide. If the authors and guardians of our Constitution made and continue to make qualitative trade-offs concerning our basic rights, then we see no reason why the freedom to enjoy a clean environment would not be similarly qualified, even after the freedom's elevation to "right"-hood.

Let's suppose that we are to regard environmental quality as a constitutional right. In that case, should we create a constitutional "right" to affordable housing? This amenity is arguably of greater importance to the average citizen than a risk-free environment. What distinguishes the rights guaranteed in the Constitution from those that are not guaranteed? And into which group does the right to a clean environment belong?

We respond to the question about distinguishing rights by noting that the freedoms of speech, religion, and so on are freedoms that people can enjoy extensively without reducing the rights of others. They impose no costs except in those extreme cases where the law already makes restrictions. In contrast, a right to shelter would impose costs on others. In this light, the answer to the second question is clear: in its costliness, a right to a clean environment is more like the right to shelter than the right of free speech. If costly rights were guaranteed in the Constitution, the need for constitutional balancing would be the rule rather than the exception.

To put this argument another way, the need to balance environmental quality against other social objectives will not disappear just because we designate environmental quality a "right," but doing so may make balancing more difficult to achieve. For example, the right to a clean environment would conflict with constitutionally guaranteed rights to use and enjoy private property, as recent congressional debates about "takings" of property attest.

But suppose that environmental quality became a right and that we could identify safe levels of environmental contaminants. The question we would then have to ask is whether society could afford the expenditures that would be required to assure safe air (or water) quality for all citizens. To be sure, we should aspire to this goal; but just as we acknowledge that we have too few resources to accomplish other worthy goals, so too we might collectively decide that we cannot afford to reduce all air pollutants to safe levels everywhere. In view of the costs that might be involved, we might do better to expend at least some of our resources on other important social problems.

We illustrate this assertion using some numbers. By our accounting, we would guess that the nation will be spending at least $25 billion annually to control ground-level ozone by the year 2000. If our rough estimates are correct, we will soon spend about as much each year to comply with the ozone standard as we currently spend on all federal food stamp programs for the poor. Now spending the same amount on ozone control as on food stamps may be perfectly appropriate; after all, people in all walks of life are affected by poor air quality. But we believe that the allocation of resources is a subject about which there should be open and informed debate. In our opinion, we ought not to spend more on ozone control than we do on food stamps (or vaccinations, for that matter) simply because we can find a "safe" level at which to set an ozone standard.

This argument applies to other environmental standards. Even if it were possible to identify a safe level for, say, a drinking water contaminant, it doesn't follow that all communities should be required to meet that level under the Safe Drinking Water Act. Some communities might quite rationally decide to aim at a somewhat less ambitious

Update

The use of benefit-cost analysis (BCA) in regulatory matters originated in an Executive Order, which mandated a "Regulatory Impact Analysis" (RIA) to accompany every major rule-making proposal and which was signed by President Ronald Reagan within days of taking office in 1981.

A major part of the RIA was a requirement to estimate the expected benefits and costs of the proposal. At the time this article was written, those who had opinions regarding the use of BCA in regulatory matters (and most Americans probably did not know or care) could be divided into three groups: the supporters, the opponents, and the suspicious. This last group, which included many decisionmakers in Congress and executive branch agencies, had been moved by the opponents' arguments—the complexities of taking distributional matters into account, the difficulties of estimating benefits, the tendency to ignore certain classes of benefits where estimation was difficult or impossible, and the tendency to use BCA for some types of decisions and not others. These arguments, together with the fact that many regulators initially saw BCA as a bid by the White House and OMB to exert greater control over their decisions, made it easier for skeptics to believe that benefit-cost estimation was inherently biased against regulation. Its use was not only controversial, but at times a partisan political issue.

Today, it is probably safe to say that the attitudes of the hard-core supporters and opponents have not changed, and in some circles BCA is as controversial as it ever was. But BCA is now firmly established as an important part of the regulatory process, largely beyond politics. An important part of the change was the acceptance by the Clinton Administration of Reagan's executive order with only minor changes. Equally important, we believe, is that many of the career public servants who were so suspicious of BCA fifteen years ago are much less so today. Part of this change is undoubtedly turnover, but there has also been a change in attitudes among regulators, who have learned how to live with BCA and may even value the additional information that it provides.

The growing importance of BCA is evident in the better integration of the RIA process with the process of regulatory design. Fifteen years ago, it was common for BCA to be prepared late in the regulatory process, often after the major regulatory decisions had been made. Today, preparation of BCA begins simultaneously with other parts of the regulatory package and in the best of circumstances informs and is informed by them.

standard and use the cost savings from doing so to finance another public program. In fact, the current flap in Washington over so-called unfunded mandates—federal regulatory requirements that fall on lower levels of government rather than on corporations—hinges on this point. State and local governments resent being told that they must spend their scarce resources on priorities established in Washington when they face other problems that they sometimes feel are far more pressing.

The Costs-Are-Considered-Anyway Argument

It could be argued that federal regulators inevitably consider costs in real-world environmental decisionmaking, despite the apparent statutory prohibition against doing so. Nonetheless, some people would assert that we should maintain the principle of excluding costs. This argument has two variants.

According to the first variant, we ignore costs in selecting ambient environmental standards, such as standards for the quality of our air and water, but take them into account in writing discharge standards for individual sources of pollution, such as electricity-generation facilities that often put sulfur dioxide into the air or farms that use pesticides that run off into lakes, rivers, and streams. These discharge standards place limits on the amounts of various pollutants that pollution sources can emit into the environment; the limits are intended to bring air and water quality, for example, into attainment with ambient standards. Typically, the discharge standards direct pollution sources to install the "best available technology," when these technological requirements are "affordable" or "economically achievable." In this sense, costs do come into play, ensuring that unaffordable discharge standards will not be imposed.

But what if the affected pollution sources cannot "afford" the technological requirements that would be necessary to meet ambient environmental goals? Short of extending the deadlines for complying with these requirements, EPA has little choice but to close down the affected sources. In short, costs can be taken into account, so long as the ultimate goals of environmental policy will be met; but those costs mean nothing if health-based standards are not met.

Insisting on effective discharge standards is appropriate if truly important health values, ecological values, or both would be compromised. But suppose that all the firms in a particular industry could afford to install the most sophisticated—and, therefore, the most expensive—pollution-control equipment made. Not everyone would agree that they should be required to do so simply because they can afford it—particularly if the health benefits of installing the equipment were deemed to be of marginal significance (that is, would reduce risk very little). While several of our current environmental statutes imply that any affordable environmental goal should be required, we suspect that many people would disagree. And they might ask whether these same statutes are creating a disincentive to succeed by requiring profitable, well-managed firms to meet stringent technological discharge standards, while treating leniently firms or industries that are on the brink of bankruptcy.

The second variant of the costs-are-considered-anyway argument is both frustrating and harder to rebut. According to this variant, we do not have to change environmental laws in order to balance health considerations against economic and other considerations, because such balancing occurs sub rosa each time that EPA sets health-based standards.

So why, the argument goes, make balancing a requirement by law?

EPA *does* appear to take economic effects into account in setting some supposedly health-based standards. For instance, in 1978, when EPA promulgated the National Ambient Air Quality Standard for ozone, it stated that finding a literally "safe" ozone level was impossible and that setting a very tight ozone standard would significantly and negatively affect economic and social activities. For this reason, EPA rejected a zero-level standard. According to the documentation supporting the 1978 revision of the ozone standard, public health was the most compelling factor in the revision, but economic impact also was weighed.

If EPA acknowledges that economic impacts play at least some role in its setting of ambient standards under the Clean Air Act, and if this role is recognized and condoned, then it seems to us that Congress should amend the act, and other environmental laws as well, to explicitly allow balancing of health and economic considerations in standard setting. If that is current practice, and there exists general agreement that such practice is appropriate, then balancing should be explicitly encouraged in the law. Not to do so engenders cynicism about the seriousness of our national intentions as well as contempt for our laws. Moreover, if no "safe" levels of many environmental contaminants can be found (as we suggest above), we cannot understand how Congress can avoid making our environmental laws explicitly require that health effects be balanced against economic and other possible adverse consequences.

Balancing Benefit-Cost Information with Other Information

We do not intend to suggest that establishing ambient environmental standards should be set on the basis of a formal quantitative benefit-cost analysis alone. Several considerations hinder an attempt to do so.

First, despite great progress in understanding how individuals value better health, reduced risks of premature mortality, aesthetic amenities, and other environmental benefits, economists are still a long way from pinning down precisely the marginal benefits associated with proposed changes in ambient environmental standards. In particular, great uncertainty surrounds estimates of how many lives such changes will save, how many illnesses they will prevent, and how much ecosystem protection they will provide.

Second, the costs associated with tighter standards are much harder to estimate than the public—and even some economists—realize. One reason is that regulations can impose costs even when no one must make out-of-pocket compliance expenditures. This would be the case if a regulation led, for example, to the withdrawal from the market of a useful product. Another reason is that regulated parties often cannot foresee technological advances that will reduce their compliance costs.

Third, even if we knew the marginal benefits and costs associated with alternative environmental quality standards, we still would not know whether equating the two would result in the "right" standard. Among other things, we might wish to know just who the winners and losers would be under new standards. For instance, suppose that only millionaires benefited from a tighter air quality standard, while the poor paid all the costs. Even if the added benefits from the tighter standard greatly exceeded the costs, we might resist adopting the new standard unless we could find a way to redistribute some of the net gain. In short, distributional considerations and other nonquantifiable factors having nothing to do with economic efficiency also matter a lot in standard setting.

Objecting to formal benefit-cost analysis as the sole basis for public decisionmaking is easy enough. Determining how such analysis *should* be used is far more difficult. We believe, however, that an analogy drawn from decisionmaking in the private sector can be useful in making this determination.

Before making an important investment decision, a good corporate manager will gather reports on the financial soundness of the venture and the expected future profits. Rather than slavishly basing a final decision on these reports alone, the good manager will temper the analytical information with his or her own judgment and experience. The manager may decide, for example, to overrule an apparently unfavorable financial projection out of a conviction that the long-run health of the company requires entry into new markets that will not pay off for some time. Or he or she may decide that the profit potential does not outweigh the risks of the project. In short, the manager understands that analytical information will rarely be complete or accurate enough to base decisions entirely on it. Giving due weight to and acting on information from all sources is the essence of good decisionmaking, and one of the private sector's strengths is its ability to recognize and reward good decisionmaking.

In the public sector, decisionmaking differs in ways that may make the use of formal benefit-cost methods both more difficult and, arguably, even more important. First, benefit-cost analysis in the public sector will probably be neither as complete nor as precise as its private-sector counterpart. Second, the public manager may have to weigh additional objectives, such as the distribution of benefits, that do not easily fit into a formal benefit-cost analysis. Finally, success and failure in the public sector are much harder to identify, making any need to take corrective action that much more difficult to discern.

Since feedback from public-sector decisions is often weaker or more ambiguous than that from private-sector decisions, the methods and data used in making decisions become more important. While much of that information will be incomplete or imprecise, it will not be useless as long as its limitations are understood. If public decisionmakers are good at what they do, they will be able to weigh both the content and the quality of information about benefits and costs in the context of available information. Those who believe that decisions would be improved if benefit-cost information were denied to decisionmakers must harbor a pessimistic view of decisionmakers' abilities, a view that sits oddly with a generally expansionist view of the role of regulation.

Taking Economic Issues Seriously

Refusing to admit the need to consider costs may result from our collective desire to believe that difficult trade-offs need not be made. Well, we can't have it all. After more than twenty years of concerted efforts to meet our nation's environmental quality goals, we are still short of the mark in many areas. Moreover, since we now have acted upon the least expensive opportunities to reduce pollution, the remaining options are generally quite costly. Thus, providing all the protection we would like to provide is even less likely than it was two decades ago.

Nothing is wrong with wanting to provide maximum environmental protection to all citizens, just as we would like to provide all the other comforts of a happy and prosperous life. But something is wrong with denying that resources are scarce relative to our prodigious wants and that we must, accordingly, accept unpleasant trade-offs. Since in public rulemakings we openly acknowledge that we cannot find safe levels of environmental contaminants and since we admit the importance of economic considerations, shouldn't we revisit those portions of our environmental statutes that prohibit even the consideration of costs in standard setting? While economic considerations should never take primacy over public health or ecological concerns in policymaking, we believe that the answer to this question is an unambiguous yes.

5

Discounting the Future
Economics and Ethics

Timothy J. Brennan

How should we evaluate the current benefits from environmental polices relative to benefits that will occur far into the future—benefits that may accrue largely to future generations? There are both economic and ethical issues inherent in assessing policies with long-term effects. And it is important to understand the role of "discounting" (both its value and its limitations) in addressing this issue.

How much do we care about people whose lives won't begin until long after our own have ended? How much *should* we care about them? These questions come up when we contemplate environmental projects that benefit people who are separated by many years or even by generations from those who pay the costs. Whether the interests of future generations will be at all significant in determining how much we should limit carbon emissions, preserve the ozone layer, or protect endangered species depends on whether a dollar's worth of future benefits is worth less than a dollar's worth of present costs—what economists mean by *discounting*.

Much controversy surrounds the practice of discounting. Divisive caricatures of the discounting wars pit economists, who allegedly view the environment as just another capital asset, against ethicists, who look out for the interests of people born in the future, and environmentalists, who advocate the inherent, noneconomic values in sustaining nature. In reality, discounting battles rage even among economists. Two leading experts on the economics of public projects, William Nordhaus of Yale University and Joseph Stiglitz of the president's Council of Economic Advisers, disagree over the appropriate way to discount the future costs and benefits of climate change.

When an issue has defied resolution for so long, perhaps the difficulty is a misunderstanding of the fundamental questions. Indeed, the difficulty may be that all the seemingly contrary positions on discounting have some validity. One cannot hope to resolve discounting debates among economists or to

Originally published in *Resources*, No. 120, Summer 1995.

allay the intensifying criticisms of discounting from those outside economics, but reflecting on the central arguments and illuminating the relationships between their economic and ethical sides may add a little light to the heat.

What Is Discounting?

One way to understand how discounting works is to compare it with the compounding of interest on savings (see Figure 1). Most people are familiar with the way compound interest increases the value of one's savings over time, in an accelerating way. For example, $100 invested today at 6 percent interest will be worth $106 in a year. Because the 6 percent interest will be earned on not just the initial $100 but the added $6 as well, the gains in the second year will be $6.36. Over time, these compounding gains become substantial. At 6 percent interest, the $100 investment will be worth about $200 in twelve years, $400 in twenty-four years, and $800 in thirty-six years. It will be worth around $3,300 in sixty years and almost $34,000 in a hundred years. A penny saved is more than a penny earned; after a century, the penny becomes $3.40. In 1626, Dutch explorers bought Manhattan for a mere $24; if that sum had been invested at just over 6 percent per year, it would have yielded more than $40 billion in 1990—about the total income generated in Manhattan that year.

Discounting operates in the opposite way. While compounding measures how much present-day investments will be worth in the future, discounting measures how much future benefits are worth today. To figure out this discounted present value, we must first choose a discount rate to transform benefits a year from now into benefits today. If we choose the same discounting rate as the interest rate in the above example of compounding, $106 a year from now would be equal in value to $100 today. Discounting the benefits of a project that generates $200 in twelve years by a discount rate of 6 percent per year would tell us that those benefits are worth $100 today.

To economists, this is the same as saying that $100 invested at an interest rate of 6 percent will generate $200 in twelve years. For this reason, they often use the terms discount rate and interest rate interchangeably, although *discount rate* properly refers to how much we value future benefits today, while *interest rate* properly refers to how much present investments will produce over time.

The paramount consideration in assessing future environmental benefits is the size of the discount rate: The larger the discount rate, the less future benefits will count when compared with current costs. If the discount rate were 10 percent, $200 in twelve years would be worth only about $64 today; if the rate were 3 percent, the current value would be $140. At a zero discount rate, $1 of benefits in the future would be worth $1 in cost today. Differences in discount rates become crucial for benefits spanning very long periods.

The Obvious Cases for and against Discounting

The close relationship between interest rates and discount rates is the basis for the obvious case in favor of discounting. Suppose that an environmental program costing $100 today would bring $150 in benefits twelve years from now. If other public or business projects yield 6 percent per year, however, those future benefits of $150 would be "worth" only about $75 today after discounting. By investing the $100 today in one of these alternative projects, we could produce $200 in benefits in twelve years, leaving $50 more for the future.

Whether we view the environmental investment in terms of the present value of benefits ($75 as compared with $100) or in terms of an alternative investment that produces benefits of greater value ($200 as compared with $150), it fails the test of the market. Using a bit of economic jargon, we can call this market test the *opportunity-cost rationale* for discounting. Here, opportunity cost refers to the most value we can get by investing $100 in something other than the environment. According to the opportunity-cost rationale, we should discount future benefits from a current project to see if these benefits are worth *at least* as much to people in the future as the benefits they would have if we invested current dollars in medical research,

Figure 1. How much we value future dollars today: The effect of time and the discount rate.

Discounting operates in the reverse direction of compounding. While compounding measures how much present-day investments will be worth in the future, discounting measures how much future benefits are worth today. The illustration below shows how the discounted present value of future benefits can shrink to very small amounts as time goes on. Specifically, it shows how much $100 earned now and in 20, 40, 60, 80, and 100 years is worth today when a 3 percent discount rate is applied.

Along with the passage of time, increases in the discount rate also can dramatically shrink the discounted present value of future benefits. The illustration below shows how much $100 in benefits 100 years from now would be worth today at discount rates ranging from 0 to 6 percent.

When we see how small variations in the timing or discounting of future benefits can make large differences in deciding how much the benefits are worth today, it's easy to understand why discounting can lead to such heated policy debates.

education, more productive technology, and so on.

In effect, the opportunity-cost rationale tells us that our discount rate should be the market interest rate. Consequently, looking at the four factors that produce the interest rates that we see in financial markets will help explain what lies behind discount rates. The first factor is the level of economic activity. If investors want a lot of money for a lot of pro-

jects, they will have to pay a higher interest rate for loans; during slow economic times, investors will require fewer loans, leading to a lower interest rate. The second factor is inflation. Future dollars will be discounted if one cannot buy as much with them in the future as one can today. The third factor is risk; a guaranteed bird in the present hand may be worth a chancy two in the future bush. The fourth factor, and the most controversial one in environmental assessments, is what economists call *pure time preference*. This preference refers to the apparent fact that people require more than $1 in promised future benefits in order to be willing to give up $1 in goods today.

Critics of the opportunity-cost rationale often find that discounting leads to a present-day valuation of future environmental benefits that they believe is too low. Threats to life and nature from environmental degradation are notoriously hard to measure and, in the views of many, impossible to compare with the "mere" economic benefits that accrue from investing in a business project. Moreover, the benefits from a business investment might accrue to the wealthy or be frittered away today, while the benefits from an environmental project are likely to be distributed more widely across society and into the future.

Environmental benefits may or may not be overestimated in policy evaluations, and they may or may not be distributed more equitably than the returns from other investments. Those well-known criticisms, however, apply to cost-benefit tests in *any* context. The specific case against discounting fundamentally concerns pure time preference. A principle in most prominent ethical philosophies is that no individual's interests should count more than another's in deciding how social benefits should be distributed. If all men are created equal, as Thomas Jefferson wrote, there can be no justification for regarding the well-being of present generations as more important than that of future generations simply because of the difference in time. Given that principle, are we really justified in refusing to sacrifice $24 in 1995 if that $24 would bring "only" $4 billion—and not $40 billion—to people living in the year 2359? Substituting lives, or the capacity of wealth to save lives, for dollars makes this question even more vivid and pressing. How could a future life, no matter how distant, be worth less than a present one? Using the language of philosophers and lawyers, we might call the insistence that future lives be valued equally to present ones the *equal standing* argument against discounting future benefits.

Might Cases for and against Discounting Both Be Valid?

Suppose we ask whether present generations should sacrifice short-run economic growth to undertake a particular program to improve the environment and leave more resources for future generations. Proponents of opportunity cost, who would discount future benefits, might say no, but proponents of equal standing, who would not discount future benefits, might say yes.

When a question has two compelling yet contradictory answers, it may really combine two questions in one. A close look at the question "should we undertake this environmental policy now to benefit future generations?" reveals that it asks a question about obligation (what duty do we have to sacrifice today to benefit future generations?) *and* a question about description (if we should sacrifice, do we help future generations more by implementing the proposed environmental policy or by doing something else?).

The economist's opportunity-cost rationale speaks to the question about description. If the goal is to improve the welfare of future generations, we should choose a policy that achieves the largest improvement for a given present cost. Consequently, we should compare the returns to the proposed environmental policy with those to other investments in order to see which are largest. Consider, for example, other investments with the same present-day costs as the environmental policy. If the discounted future benefits from these alternative policies are larger than the those from the environmental policy, we should consider implementing the alternative policies instead. We may be able to do more for future generations by subsidizing basic scientific and medical research or promoting education than by protecting the environment.

An obvious response would be to ask, "Why not invest in environmental protection *and* medical research?" This response brings us to the question about obligation—whether and how much to sacrifice. Unlike the question that asks us to describe and compare the benefits of one program to another, the obligation question asks us to contemplate our duties to future generations. As such, it fundamentally concerns ethical values rather than economic facts. Accordingly, equal standing is a more appropriate perspective from which to answer this question than is opportunity cost.

Proponents of the equal-standing principle have no problem with discounting for inflation or risk. But they find the pure-time-preference component of discounting to be morally controversial, even though the pure-time-preference discount rate is half the 6 percent discount rate drawn from today's markets. While a 19:1 ratio (present value to future value yielded by a 3 percent discount rate) is less philosophically forbidding than the 340:1 ratio (yielded by a 6 percent discount rate), it still is hard to reconcile with the equal-standing principle.

Violating "Hume's Law"

Separating environmental policy questions into questions about description and about obligation uncovers the root of much of the discounting controversy within economic circles and across disciplinary boundaries. This controversy is a consequence of trying to use facts about how people *do* discount to tell us how policymakers *should* discount. This attempt violates a maxim derived from eighteenth-century British philosopher David Hume, who asserted that facts alone cannot tell us what we should do. Any recommendation for what you, I, or society ought to do embodies some ethical principles as well as factual judgments. For example, to recommend policies if and only if their economic benefits exceed their costs would imply the ethical principle that increasing net economic benefits is the only worthy goal for society.

The fact that we *do* have time preferences may not tell us much about how we ought to regard future generations. Imagine a world where generations do not overlap. In this world, people are like long-lived tulips; every eighty years, a new batch comes to life after the previous batch disappears. Suppose the people in one of those generations happen not to care about any subsequent genera-

tions. They would then choose to exhaust resources and degrade the environment without regard for how these actions might lower the quality of life of the people who succeed them. The *fact* of this disregard, however, does not invalidate an ethical principle that people born far in the future deserve a good quality of life as much as people already living.

Using market discount rates to examine ethical questions has made the economics of discounting more complicated than it perhaps needs to be. For example, economists have long argued about whether to calculate pure-time-preference discount rates based on the returns that investors receive before they pay taxes or after they pay taxes and, if after, whether to include corporate income taxes or personal income taxes in the calculation. If pure time preference has only limited ethical relevance in determining how much we should discount, these issues become relatively unimportant.

Divergence between equal-standing and opportunity-cost discount rates would be less important if policies that always did the best from one perspective did the best from the other as well. Unfortunately, this does not always hold. A policy that generates benefits in the short run may have a higher discounted value in an opportunity-cost sense than a policy that produces benefits much later. If we use a lower discount rate—that is, one reflecting more equal standing—the policy with long-term benefits may come out on top. We might need to do more for future generations; moreover, we might be doing the wrong things now. At opportunity-cost discount rates, development of an urban park may be more beneficial than an equally costly plan to reduce greenhouse gas emissions by taxing gasoline. At low or zero discount rates, the gasoline tax may be the more beneficial policy.

Philosopher Mark Sagoff of the University of Maryland suggests that market discount rates may not be a good indicator of the ethical value that people, upon reflection, would place on protecting future generations. Accordingly, we might resolve the discounting issue by having the government set policy based on people's stated ethical views regarding how to weigh current lives and dollars against future lives and dollars. Through a telephone survey of 3,000 U.S. households, Maureen Cropper, Sema Aydede, and Paul R. Portney of RFF determined that the rate at which people apparently discount lives saved is comparable to after-tax returns in financial markets. For example, people discount lives a century from now at about 4 percent per year. Equal-standing advocates can draw scant comfort from such data, which might tell us how a democracy would react if it followed the public's pure time preferences but, according to Hume, don't tell us what the right time preferences are.

Ethically Justified Discounting

Reconciling discounting with ethics may seem impossible, but there is some hope. To say that present and future generations have equal standing in an ethical sense does not necessarily imply that they have the same claim on present resources, because the general level of wealth or well-being may be changing over time. If we follow the ideas of a recent Nobel Prize winner in economics, John Harsanyi of the University of California–Berkeley, we should sacrifice today for the benefit of future generations only if the average well-being of people in the future goes up by more than we lose on average today. If present trends continue, advances in technology and knowledge will make people better off in the future than we are today. In that case, more than a dollar of gains to them would be needed to make up for a dollar lost to us. Any future returns should then be discounted by this difference to ensure that future generations' gains in well-being exceed our losses. According to the view proposed by Harvard University philosopher John Rawls, we might not be justified in making *any*

sacrifice for future generations if they would be better off than we are now. If we expect future generations to be worse off than we are, however, Rawls' framework suggests that we should make present-day sacrifices.

More promising justifications for discounting come from critiques of the equal-standing idea itself. Philosophers such as Susan Wolf of Johns Hopkins University and Martha Nussbaum of Brown University have pointed out that to say that everyone has equal standing is to say that no one has special standing—including our families, friends, and fellow citizens. Insistence on equal standing denies the value that special interpersonal relationships hold for us and without which we could not be fully human. This argument may provide some support for asserting that generations closer to us should mean more to us than generations far in the future. (Thomas Schelling of the University of Maryland points out the irony of worrying so much about the welfare of future generations while doing so little to improve the welfare of many of the most destitute among us today.)

As long as resource scarcity makes trade-offs between the present generation and future generations inevitable, no consideration of environmental policies to benefit future generations should ignore economic opportunity cost. Ultimately, decisions to implement or not to implement such proposed policies will be the result of political processes, with all their virtues and imperfections. Justifications for the policies, which are tied in large measure to the degree of discounting, unavoidably involve ethical reflection and judgment. An appreciation of the necessary roles of both economics and ethics should clarify the nature of discounting and promote better understanding of our obligations toward future generations and how to meet them.

Suggested Reading

Arrow, K., and others. 1996. "Intertemporal Equity, Discounting, and Economic Efficiency." In J.P. Bruce, H. Lee, and E.F. Hates (eds.), *Climate Change 1995: Economic and Social Dimensions of Climate Change*. New York: Cambridge University Press.

Lind, R.C., and others. 1982. *Discounting for Time and Risk in Energy Policy*. Washington, DC: Resources for the Future (especially chapters by J. Stiglitz and A.K. Sen).

Parfit, D. 1984. *Reasons and Persons*. New York: Oxford University Press (especially Part Four, Future Generations).

Portney, P., and J. Weyant. 1999. *Discounting and Intergenerational Equity*. Washington, DC: Resources for the Future.

Rawls, J. 1971. *A Theory of Justice*. Cambridge: Harvard University Press (especially Chapter V, Distributive Shares).

Schelling, T.C. 1995. Intergenerational Discounting. *Energy Policy* 23(4/5): 395–401.

6

Time and Money
Discounting's Problematic Allure

Paul R. Portney

Although economists generally agree that it is essential to discount future benefits and costs at a positive rate, there is much less agreement about what that rate should be and how it should be applied in cases of very long-term projects—like the containment of global climate change— where there are significant matters of intergenerational equity.

Acting now to deal with problems whose consequences may not be felt for generations is obviously tricky business. The costs and benefits involved are hard to gauge, since they will be spread out over hundreds—perhaps thousands—of years. Still, waiting for the added certainty the future will bring is not always the best policy; sometimes we have to swallow hard and take preventative measures. Doing so, however, forces us to confront how much we are willing to sacrifice today for benefits that will be enjoyed later in our lives or in the lives of succeeding generations.

With that in mind, economists and other analysts make use of a technique called discounting to compare present with future costs and benefits on an equal basis. In its conventional use, streams of future benefits and costs are converted to present values through an appropriate discount rate. As long as the discount rate is positive, one dollar tomorrow is worth less than one dollar today. Take, for example, a project that twelve years from today will yield a return of $200,000. At an annual discount rate of 6 percent (reflecting, for instance, the cost of borrowing money this year versus next) that amount is now worth $100,000—all that an economist would advise investing.

To ponder whether and how to use discounting over much longer time spans, RFF and the Energy Modeling Forum at Stanford University invited the most influential thinkers on discounting to convene at RFF in November 1996. (During the energy crisis in the 1970s, RFF hosted a similar conference and invited some of the same participants, including Robert Lind.

Originally published in *Resources*, No. 136, Summer 1999.

Out of that meeting arose Lind's crystallization of a prescription for the proper discount rate, around which a consensus lasted for some time.) Each of these economists wrote an essay in response to a set of questions eliciting their opinions on how best to use discounting in decisionmaking for the far future. Their varied points of view are now available in book form.

A New Book

Though quite technical and no "primer," the recently released RFF book *Discounting and Intergenerational Equity* plainly shows agreement on some broad and basic points. The contributors speak with nearly one voice when they say it is appropriate— indeed essential—to discount future benefits and costs at some positive rate. Even those authors who favor a lower discount rate for the far (as opposed to near) future quite clearly believe that failing to discount would make for poor intergenerational decisionmaking. And even those few that could envision a zero or negative rate suggest such a case would be rare.

At the conference itself, the authors were nearly unanimous in recommending the use of a standard procedure for evaluating projects with timeframes of forty years or less. Within the scope of this relatively short period of time, they generally embraced discounting benefits and costs to make present-value comparisons. What's more, they tended to think the discount rate should reflect the opportunity cost of capital. Beyond the forty-year mark, however, discomfort set in, as the essays reveal.

Clouds in the Crystal Ball

To read the new RFF book is to get a sense of the unease among the best minds in the profession about the technical complexity and ethical ramifications of discounting far into the future. For one thing, there is no mistaking the very small present value of even very large costs and benefits if they will not be realized for hundreds of years.

Assume, for example, that the gross domestic product of the world will be $8 quadrillion in the year 2200 in current dollars. (This assumption is consistent with an annual growth rate of 3 percent from current world GDP over the next two hundred years.) Suppose next that we want to calculate the present value of that sum using the 7 percent discount rate that the Office of Management and Budget recommends for such purposes. The answer we get is a surprising $10 *billion*. In other words, it would not make sense for the world's present inhabitants to spend more than $10 billion today (or about $2 per person) on a measure that would prevent the loss of the entire GDP of the world two hundred years from now.

That conclusion may seem stunning. Yet the reason is clear enough: We could invest that same $10 billion at 7 percent today and have a sum more than sufficient to replace GDP two centuries ahead.

Still, what guarantee is there that the $10 billion invested would remain untouched during the intervening years? What if, instead, people living a century from now decided to dip into the fund to finance their own consumption? Those living two centuries from now would be left with neither the problem-mitigation project we eschewed nor the fund we created to make them whole.

Another difficulty that discounting the distant future presents is choosing between economic efficiency and distributional equity—and being able to tell the difference. Although the contributors to the RFF book are not among them, some people that are uncomfortable with the distributional consequences of climate change seem eager to tinker with the discount rate to make mitigation policies pass the efficiency test (when in fact they may not). There is no need to do so—efficiency is hardly the only criterion that matters in policy analysis. If, for example, it would be more efficient to reject a climate protection program, say, because it would be cheaper to invest the money in an interest-bearing asset, we might opt for the program, anyway, out of concern for the welfare of our descendants, especially if we doubt that compensation for now ignoring the problem will be available to future generations.

An Array of Approaches

Some of the book's contributors suggest using different discount rates depending on the period over which net present values are being tallied. This possibility of *nonconstant* discounting has surfaced in a growing number of studies, which show rather consistently that while individuals do appear to attach lower weights to distant benefits, they do not use a constant exponential discount rate. Rather, the longer the time period before effects are felt, the lower the implicit discount rate used.

Perhaps most surprisingly, at least three of the authors question the very utility of standard benefit-cost analysis for problems with significant intergenerational consequences. Thomas Schelling suggests, for instance, that we view the problem of climate change in much the same way that we try to decide the right amount of foreign aid to make available to poorer countries each year.

In an altogether different approach, Raymond Kopp and I suggest a mock referendum. The idea is to elicit from members of the present generation their willingness to pay to reduce both present and future risks associated with climate change. An aggregate willingness to pay would then be compared with the expected costs, say, of climate change mitigation. This approach would circumvent the need to estimate very long-term streams of benefits and costs, as well as the need to choose an appropriate discount rate. But it does present its own problems, such as how to frame and ask the questions to elicit honest responses.

A Fair Future

In the seventies, the energy crisis led people outside academia to think hard about the comparison of benefits and costs across time. Today it is the climate change debate that is largely responsible for reawakening interest in the subject, since it forces us to think about the legacy we may be leaving future generations. As soon as we begin to consider policies to affect the latter, up pops the concept of discounting. If a latter day consensus on how to use this tool is not in the offing, it is perhaps toward another kind of workability that the essays in the new RFF volume can begin to lead: by clarifying what exactly is being debated and why it is important.

Suggested Reading

Portney, Paul R. and John P. Weyant (eds.). 1999. *Discounting and Intergenerational Equity*. Washington, DC: Resources for the Future.

7

How Much Will People Pay for Longevity?

Alan J. Krupnick

Benefit-cost studies of programs to reduce air pollution typically show huge benefits, primarily from the estimated willingness-to-pay of individuals for reduced mortality risks. Some new research suggests that these estimates are sensitive to the magnitude of risk and to peoples' age and the state of their health. It appears that the "value of a statistical life" may be a good bit lower than that used in EPA benefit-cost studies and that the estimated benefits may depend in important ways on the health and age of the affected population.

Prolonging people's lives is arguably the most important outcome from improving air quality and drinking water. Together with morbidity improvements, these effects serve as the primary drivers for many of the major legislative mandates in the United States and Canada, such as the U.S. Clean Air Act and the Canadian Environmental Protection Act. The challenge before policymakers is to strike a balance between potential benefits in terms of lives prolonged (or, equivalently, death risks reduced) and the use of scarce resources to prolong them.

Striking this balance requires not only an estimate of the risk reductions related to reductions in pollution, but also an estimate of the public's preferences for obtaining this benefit, expressed in terms of their willingness to pay for it. Existing methods for determining the value of a statistical life (VSL)—a shorthand expression for the willingness to pay divided by the mortality risk reduction being experienced—have common shortcomings, according to our research. They tend to focus on the value adults in the prime of their life place on reducing their risk of dying, even though most of the people who benefit from environmental programs are older and/or may be suffering from chronic heart and lung diseases.

The existing methods also tend to focus only on immediate risk changes. When environmental programs reduce exposure to a carcinogen, the costs of doing so are often incurred in the present, whereas cancer-related mortality risks are reduced in the future, following a latency period. What is needed for an effective policy addressing pollutants with latent effects is an

Originally published in *Resources*, No. 142, Winter 2001.

estimate of how much people would pay now for a reduction in their risk of dying in the future.

In our research, we aimed to address these shortcomings by focusing on persons 40 to 75 years old to elicit their "willingness to pay" (WTP) for reductions in current and future risks of death. We wanted to determine the WTP for a reduction in death risk in an appropriate context for pollution, how WTP would vary with age, whether WTP would be influenced by current health status, and how latency would affect WTP.

Our findings yielded interesting and, in some cases, somewhat unexpected results. In general terms, what turns out to matter more than income or educational level in explaining people's WTP was their overall mental health and, specifically, whether or not they were specifically suffering from cancer. If they were in good mental health or had cancer, they were willing to pay more to see their risks reduced; with regard to cancer, respondents would pay substantially more (about 45%). At the same time, other expressions of physical health—and many were included in our survey—were not related to WTP.

Age does not influence WTP until age 70, according to our statistical findings. The 70–75 age group was willing to pay approximately one-third less than the average for a given reduction in annual mortality risk.

Our mean WTP estimates for a reduction in the risk of death over the next 10 years show that the value of a statistical like varies from approximately $1.2 million to $3.8 million (1999 C$), depending on the size of the risk value changed. These figures are 10% to 70% lower than Health Canada's age-adjusted VSL of $4.3 million (1999 C$), which was recently used in an analysis of proposed ambient air quality standards, and one-half (or less) the size of the $7.5 million (1999 C$) figure used by the U.S. Environmental Protection Agency (EPA).

Research Methods

The methods for developing empirical estimates of individual WTP for mortality risk reductions may be divided into two groups. Revealed preference studies primarily examine whether more risky jobs come with a higher wage. Stated preference studies rely on survey methods (termed contingent valuation methods) that pose realistic but hypothetical situations to individuals in which they can express their preferences in money terms for these complex effects.

Each approach to measuring WTP has its drawbacks. Revealed preference studies make untested assumptions about individuals' risk perceptions: that is, that risk perceptions correspond to objectively measured risks. Furthermore, it is often difficult to separate objective risk measures from other attributes of the job or product being examined. Stated preference studies are, in principle, capable of testing whether individuals correctly perceive mortality risks or changes in mortality risks.

However, these stated preference studies are not without their own pitfalls: respondents may not understand the risk changes they are asked to value, may not believe that the risk changes apply to themselves, and may lack experience in trading money for quantitative risk changes or lack the realization they are engaged in this activity. The result may be that WTP is found not to vary with the size of the risk change—an essential method of testing whether individuals correctly comprehend risk information that many existing studies omit.

Our approach was to devise and implement a contingent valuation study that would address these problems by:

- developing graphic depictions of risk and a series of education statements to enhance respondent comprehension;
- testing in several ways for respondent understanding of risk and other facets of the survey; and
- providing examples of comparable activities from everyday life, such as obtaining mammograms or colon cancer screening tests, to inform people about how they spend money to reduce death risks in their everyday lives.

Survey Sample Profile

The survey was administered to 930 people in

Figure 1. Use of grids to represent probabilities in mortality risk questionnaire

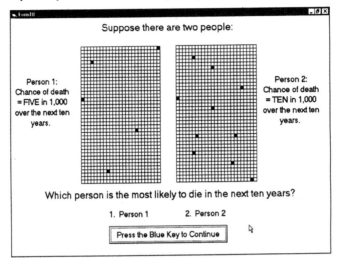

Hamilton, Ontario, in 1999, by a Canadian survey-research firm. Respondents were recruited by phone through random-digit dialing and asked to go to a facility in downtown Hamilton to participate in the survey. There, they worked on computers with simplified keypads, which were color-coded and specially labeled for use with the survey. Respondents moved through the survey at their own pace. Words on each screen appeared in a large font and there was a voice-over accompaniment.

The goals of the survey were to estimate what older people would pay for a reduction in their risk of dying and to examine the impact of health status on WTP. We sought a target population aged 40 years (the mean age of workers in the wage compensation studies) to 75 years and were able to assemble a sample of people that was very similar to the Ontario population in age, income, and the like.

The average age of the respondents was 54 years, with 31% of the sample above age 60, and 9% above age 70. Although 80% of the sample completed high school, only 20% had completed a university degree. The average household income in the sample was $54,000 (1999 C$). Most respondents rated their health as very good to excellent, although 41% reported some chronic res-

piratory or heart disease. The majority of respondents also described themselves as being in good mental health.

Survey Structure

Survey development is still more of an art than a science. Our survey instrument was developed over several years and is based on extensive one-on-one interviews, work with many focus groups, and even a 300-person pretest survey administered in Japan. The survey was divided into five parts. Part I introduced the project's sponsors—RFF, Health Canada, and McMaster University in Hamilton, Ontario—and elicited personal information about the respondent, including questions about the respondent's health as well as the health of immediate family members.

Part II introduced the subject to simple probability concepts through coin tosses and roulette wheels. The probabilities of dying and surviving over 10-year periods were then depicted using a 1,000-square grid. The respondent went through simple exercises to become acquainted with our method of representing the probability of dying. The respondent was then shown two 25 by 40

Update

The U.S. Supreme Court ruled against the contention by industry that cost-benefit analysis should be used to set the nation's air quality standards (NAAQS). However, this has not stopped the conduct and use of health-benefits analysis in designing air quality policy at EPA or in examining whether existing standards make sense. Indeed, if anything, values for reducing mortality risks to use in such analyses are even more in demand. For instance, the World Bank has initiated a major study for use in setting priorities in China that utilizes the survey described in this chapter. In addition, the OMB has again endorsed the monetization of reductions in mortality risk for use in cost-benefit analyses of U.S. regulations and has required that they be supplemented with other types of health values from indices used in medical practice (such as the Quality Adjusted Life Year). And the Energy Policy Act of 2005 mandates a major study of the external costs of the production and consumption of energy. Mortality risk valuation is likely to make up a large share of these costs.

Since this article was written, the survey described herein for Canada has been conducted in the United States, Britain, France, Italy, Japan, and South Korea–and is currently being administered in Shanghai and Chongqing, China. The finding in Canada that the elderly (those over 69) have statistically significantly lower willingness-to-pay than other adults (40 and above) is not as strong in most of these other countries. For instance, the U.S. "senior discount" is 20% (and not statistically significant) versus 30% in Canada. The tendency for people with chronic illnesses to be willing to pay more than healthy people is still in evidence for certain cases, but there is no apparent "mental health" effect.

From a policy perspective, our overarching conclusion is still that the VSLs being used in the U.S. are too large, whether for older, younger, healthy, or ill populations.

---

grids: one for person 1, with 5 red squares (representing death), and one for person 2, with 10 red squares (see figure 1 above).

The respondent was asked to indicate which person faces the higher risk. If the respondent picked person 1, he or she was provided with additional information about probabilities and the question was asked again. The respondent was then asked which person he or she would rather be. Individuals responding "person 2" (the person with the higher risk) were asked a follow-up question to verify this answer and were given the opportunity to change their answer if they wished. The baseline risk of death for a person of the respondent's age and gender was then presented numerically and graphically.

Part III presented the leading causes of death for someone of the respondent's age and gender. Common risk-mitigating behaviors were listed together with the quantitative risk reductions they achieve and a qualitative estimate of the costs associated with them ("inexpensive," "moderate," and

"expensive"). The purpose of this section is twofold. We wanted, first, to acquaint the respondent with the magnitude of risk changes delivered by common risk-reducing actions and products (for example, cancer screening tests and blood pressure medication) and, second, to remind the respondent that such actions have a cost, whether out-of-pocket or not.

Part IV elicited WTP by asking if respondents were willing to pay a given amount and then, depending on their answer, they were given a follow-up bid to accept or reject. In all, three sets of WTP questions were asked. Respondents were first asked if they were willing to pay for an abstract product that, when used and paid for over the next 10 years, would reduce their current risk of dying over the next 10-year period by 5 in 1,000; that is, by 5 in 10,000 annually. In the second WTP question, risks were reduced by 1 in 1,000; that is, by 1 in 10,000 annually. And in the third WTP question (to those 65 and under), risks were reduced by 5 in 1,000 again, but not until age 70, reminding respondents that they might not be alive to experience this benefit and asking them how likely it was they thought they would live to this age. The first and second questions were reversed for half the sample in order to test formally, and with separate samples, whether the larger risk change resulted in a larger WTP.

The product in question was defined in abstract terms—"a drug or a product not covered by health insurance"—because we found that more specificity resulted in many respondents rejecting the scenarios as not applicable to them. We also made it clear that the risk reductions would be obtained by use of a private good. In practice, most environmental programs reduce mortality risks for all persons in an exposed population—in other words, risk reductions are a public good. However, in order to factor out potential altruism on the part of respondents, it was necessary to focus only on private WTP. To the extent that it is appropriate to consider altruism—a complicated issue—our estimates are biased downward, but no more so than the existing estimates commonly used by EPA and others.

Part V included an extensive series of debriefing questions, followed by some final questions regarding education and household income. The debriefing questions were used to identify respondents who had trouble comprehending the survey or did not accept the risk reduction being valued. The computerized survey was then followed by a standard 36-question, pencil-and paper survey addressing the respondent's physical and mental health in detail and permitting the construction of standardized physical and mental health indexes for use in explaining why WTP varied across individuals.

Conclusion

One key measure of the success of a contingent valuation study like this one is that, when different groups of respondents are asked to value risk changes of different magnitudes, WTP increases with the size of the risk change. Our research shows that the size of the risk reduction has a strong influence on WTP. Mean WTP for an annual reduction in risk of death of 5 in 10,000 is about 1.5 times the

WTP for an annual risk reduction of 1 in 10,000. WTP, therefore, is sensitive to the size of the risk reduction but not strictly proportional to it (median WTP is closer to changing proportionally with risk). This lack of proportionality means that the VSL also varies with the size of the risk change, raising a question as to which VSL is appropriate in any given case.

Indeed, the overarching technical conclusion of our study is not only that the VSL may be lower than that in use for pollution- related benefit–cost analyses, but also that different VSLs may be appropriate in some circumstances regarding the age and health of affected populations. The lack of an effect of physical health status on WTP (with the possible exception of the presence of cancer) suggests that any potential proliferation of VSLs may be limited.

In terms of public policy, we would conclude that benefits of air pollution reductions, which do not have a cancer effect and affect primarily an older population, are being significantly overestimated in the United States and possibly in Canada, as well as in other countries that rely on the current literature or mimic U.S. practice.

Suggested Reading

Alberini, A., and A. Krupnick. 2002. Valuing Health Effects. In *International Yearbook of Environmental and Resource Economics 2001/2002*, edited by H. Folmer and T. Tietenberg. U.K.: Edward Elgar.

Alberini, A., M. Cropper, A. Krupnick, and N. Simon. 2004. Does the Value of a Statistical Life Vary with Age and Health Status? Evidence from the U.S. and Canada. *Journal of Environmental Economics and Management* 48(1): 769–92.

Hammit, J.K. 2000. Valuing Mortality Risk: Theory and Practice. *Environmental Science and Technology* 34 (8): 1396–1400.

Krupnick, A. 2003. Valuing Risks to Life and Health: Alternative Approaches to Valuing the Health Benefits of New Government Regulations. *Resources* 150 (Spring).

8

The Faustian Bargain
Risk, Ethics, and Nuclear Energy*

Allen V. Kneese

In this classic 1973 essay, Allen Kneese replied to a request from the Atomic Energy Commission for a statement concerning benefit-cost analysis of nuclear energy. Kneese explores both the contribution and the limitations of such analyses to highly risky, uncertain, and long-term issues. In the end, he argues that benefit-cost analysis "can supply useful inputs" but "cannot begin to provide a complete answer to questions with such far-reaching implications for society."

I am submitting this statement as a long-time student and practitioner of benefit-cost analysis, not as a specialist in nuclear energy. It is my belief that benefit-cost analysis cannot answer the most important policy questions associated with the desirability of developing a large-scale, fission-based economy. To expect it to do so is to ask it to bear a burden it cannot sustain. This is so because these questions are of a deep *ethical* character. Benefit-cost analyses certainly cannot solve such questions and may well obscure them.

These questions have to do with whether society should strike the Faustian bargain with atomic scientists and engineers, described by Alvin M. Weinberg in *Science*. If so unforgiving a technology as large-scale nuclear fission energy production is adopted, it will impose a burden of continuous monitoring and sophisticated management of a dangerous material, essentially forever. The penalty of not bearing this burden may be unparalleled disaster. This irreversible burden would be imposed even if nuclear fission were to be used only for a few decades, a mere instant in the pertinent time scales.

Clearly, there are some major advantages in using nuclear fission technology, else it would not have so many well-intentioned and intelligent advocates. Residual heat is produced to a greater extent by current nuclear generating plants than by fossil fuel-fired ones. But, otherwise, the environmental impact of routine operation of the nuclear fuel cycle, including burning the fuel in the reactor, can very likely be brought to a lower level than will be possible with fossil fuel-fired plants. This superior-

Subtitle added for *The RFF Reader, Second Edition*. Originally published as "The Faustian Bargain," in *Resources*, No. 44, September 1973.

ity may not, however, extend to some forms of other alternatives, such as solar and geothermal energy, which have received comparatively little research and development effort. Insofar as the usual market costs are concerned, there are few published estimates of the costs of various alternatives, and those which are available are afflicted with much uncertainty. In general, however, the costs of nuclear and fossil fuel energy (when residuals generation in the latter is controlled to a high degree) do not seem to be so greatly different. Early evidence suggests that other as yet undeveloped alternatives (such as hot rock geothermal energy) might be economically attractive.

Unfortunately, the advantages of fission are much more readily quantified in the format of a benefit-cost analysis than are the associated hazards. Therefore, there exists the danger that the benefits may seem more real. Furthermore, the conceptual basis of benefit-cost analysis requires that the redistributional effects of the action be, for one or another reason, inconsequential. Here we are speaking of hazards that may affect humanity many generations hence and equity questions that can neither be neglected as inconsequential nor evaluated on any known theoretical or empirical basis. This means that technical people, be they physicists or economists, cannot legitimately make the decision to generate such hazards. Our society confronts a moral problem of a great profundity; in my opinion, it is one of the most consequential that has ever faced mankind. In a democratic society the only legitimate means for making such a choice is through the mechanisms of representative government.

For this reason, during the short interval ahead while dependence on fission energy could still be kept within some bounds, I believe the Congress should make an open and explicit decision about this Faustian bargain. This would best be done after full national discussion at a level of seriousness and detail that the nature of the issue demands. An appropriate starting point could be hearings before a committee of Congress with a broad national policy responsibility. Technically oriented or specialized committees would not be suitable to this task. The Joint Economic Committee might be appropri-

ate. Another possibility would be for the Congress to appoint a select committee to consider this and other large ethical questions associated with developing technology. The newly established Office of Technology Assessment could be very useful to such a committee.

Much has been written about hazards associated with the production of fission energy. Until recently, most statements emanating from the scientific community were very reassuring on this matter. But several events in the past year or two have reopened the issue of hazards and revealed it as a real one. I think the pertinent hazards can usefully be divided into two categories—those associated with the actual operation of the fuel cycle for power production and those associated with the long-term storage of radioactive waste. I will discuss both briefly.

The recent failure of a small physical test of emergency core cooling equipment for the present generation of light-water reactors was an alarming event. This is in part because the failure casts doubt upon whether the system would function in the unlikely, but not impossible, event it would be called upon in an actual energy reactor. But it also illustrates the great difficulty of forecasting behavior of components in this complex technology where pertinent experimentation is always difficult and may sometimes be impossible. Other recent unscheduled events were the partial collapse of fuel rods in some reactors.

There have long been deep but suppressed doubts within the scientific community about the adequacy of reactor safety research vis-à-vis the strong emphasis on developing the technology and getting plants on the line. In recent months the Union of Concerned Scientists has called public attention to the hazards of nuclear fission and asked for a moratorium on the construction of new plants and stringent operating controls on existing ones. The division of opinion in the scientific community about a matter of such moment is deeply disturbing to an outsider.

No doubt there are some additional surprises ahead when other parts of the fuel cycle become more active, particularly in transportation of spent fuel elements and in fuel reprocessing facilities. As yet, there has been essentially no commercial experience in recycling the plutonium produced in nuclear reactors. Furthermore, it is my understanding that the inventory of plutonium in the breeder reactor fuel cycle will be several times greater than the inventory in the light-water reactor fuel cycle with plutonium recycle. Plutonium is one of the deadliest substances known to man. The inhalation of a millionth of a gram—the size of a grain of pollen—appears to be sufficient to cause lung cancer.

Although it is well known in the nuclear community, perhaps the general public is unaware of the magnitude of the disaster which would occur in the event of a severe accident at a nuclear facility. I am told that if an accident occurred at one of today's nuclear plants, resulting in the release of only five percent of only the more volatile fission products, the number of casualties could total between 1,000 and 10,000. The estimated range apparently could shift up or down by a factor of ten or so, depending on assumptions of population density and meteorological conditions.

With breeder reactors, the accidental release of plutonium may be of greater consequence than the release of the more volatile fission products. Plutonium is one of the most potent respiratory carcinogens in existence. In addition to a great variety of other radioactive substances, breeders will contain one, or more, tons of plutonium. While the fraction that could be released following a credible accident is extremely uncertain, it is clear that the release of only a small percentage of this inventory would be equivalent to the release of *all* the volatile fission products in one of today's nuclear plants. Once lost to the environment, the plutonium not ingested by people in the first few hours following an accident would be around to take its toll for generations to come—for tens of thousands of years. When one factors in the possibility of sabotage and warfare, where power plants are prime targets not just in the United States but also in less developed countries now striving to establish a nuclear industry, then there is almost no limit to the size of the catastrophe one can envisage.

It is argued that the probabilities of such disastrous events are so low that these events fall into the negligible risk category. Perhaps so, but do we really know this? Recent unexpected events raise doubts. How, for example, does one calculate the actions of a fanatical terrorist?

The use of plutonium as an article of commerce and the presence of large quantities of plutonium in the nuclear fuel cycles also worries a number of informed persons in another connection. Plutonium is readily used in the production of nuclear weapons, and governments, possibly even private parties, not now having access to such weapons might value it highly for this purpose. Although an illicit market has not yet been established, its value has been estimated to be comparable to that of heroin (around $5,000 per pound). A certain number of people may be tempted to take great risks to obtain it. AEC Commissioner Larsen, among others, has called attention to this possibility. Thus, a large-scale fission energy economy could inadvertently contribute to the proliferation of nuclear weapons. These might fall into the hands of countries with little to lose or of madmen, of whom we have seen several in high places within recent memory.

In his excellent article referred to above, Weinberg emphasized that part of the Faustian bargain is that to use fission technology safely, society must exercise great vigilance and the highest levels of quality control, continuously and *indefinitely*. As the fission energy economy grows, many plants will be built and operated in countries with comparatively low levels of technological competence and a greater propensity to take risks. A much larger amount of transportation of hazardous materials will probably occur, and safety will become the province of the sea captain as well as the scientist. Moreover, even in countries with higher levels of technological competence, continued success can lead to reduced vigilance. We should recall that we managed to incinerate three astronauts in a very straightforward accident in an extremely high technology operation where the utmost precautions were allegedly being taken.

Deeper moral questions also surround the storage of high-level radioactive wastes. Estimates of how long these waste materials must be isolated from the biosphere apparently contain major elements of uncertainty, but current ones seem to agree on "at least two hundred thousand years."

Favorable consideration has been given to the storage of these wastes in salt formations, and a site for experimental storage was selected at Lyons, Kansas. This particular site proved to be defective. Oil companies had drilled the area full of holes, and there had also been solution mining in the area which left behind an unknown residue of water. But comments of the Kansas Geological Survey raised far deeper and more general questions about the behavior of the pertinent formations under stress and the operations of geological forces on them. The ability of solid earth geophysics to predict for the time scales required proves very limited. Only now are geologists beginning to unravel the plate tectonic theory. Furthermore, there is the political factor. An increasingly informed and environmentally aware public is likely to resist the location of a permanent storage facility anywhere.

Because the site selected proved defective, and possibly in anticipation of political problems, primary emphasis is now being placed upon the design of surface storage facilities intended to last a hundred years or so, while the search for a permanent site continues. These surface storage sites would require continuous monitoring and management of a most sophisticated kind. A complete cooling system breakdown would soon prove disastrous and even greater tragedies can be imagined.

Just to get an idea of the scale of disaster that could take place, consider the following scenario. Political factors force the federal government to rely on a single above-ground storage site for all high-level radioactive waste accumulated through the year 2000. Some of the more obvious possibilities would be existing storage sites like Hanford or Savannah, which would seem to be likely military targets. A tactical nuclear weapon hits the site and vaporizes a large fraction of the contents of this storage area. The weapon could come from one of the principal nuclear powers, a lesser developed country with one or more nuclear power plants, or it might be crudely fabricated by a terrorist organization from black-market plutonium. I am told that the radiation fallout from such an event could exceed that from all past nuclear testing by a factor of 500 or so, with radiation doses exceeding the annual dose from natural background radiation by an order of magnitude. This would bring about a drastically unfavorable and long-lasting change in the environment of the majority of mankind. The exact magnitude of the disaster is uncertain. That massive numbers of deaths might result seems clear. Furthermore, by the year 2000, high-level wastes would have just begun to accumulate. Estimates for 2020 put them at about three times the 2000 figure.

Sometimes, analogies are used to suggest that the burden placed upon future generations by the "immortal" wastes is really nothing so very unusual. The Pyramids are cited as an instance where a very long-term commitment was made to the future and the dikes of Holland as one where continuous monitoring and maintenance are required indefinitely. These examples do not seem at all apt. They do not have the same quality of irreversibility as the problem at hand, and no major portions of humanity are dependent on them for their very existence. With sufficient effort the Pyramids could have been dismantled and the Pharaohs cremated if a changed doctrine so demanded. It is also worth recalling that most of the tombs were looted already in ancient times. In the 1950s the Dutch dikes were in fact breached by the North Sea. Tragic property losses, but no destruction of human life, ensued. Perhaps a more apt example of the scale of the Faustian bargain would be the irrigation system of ancient Persia. When Tamerlane destroyed it in the 14th century, a civilization ended.

None of these historical examples tell us much about the time scales pertinent here. One speaks of two hundred thousand years. Only a little more than one-hundredth of that time span has passed since the Parthenon was built. We know of no government whose life was more than an instant by comparison with the half-life of plutonium.

It seems clear that there are many factors here which a benefit-cost analysis can never capture in quantitative, commensurable terms. It also seems unrealistic to claim that the nuclear fuel cycle will not sometime, somewhere experience major unscheduled events. These could range in magnitude from local events, like the fire at the Rocky Mountain Arsenal, to an extreme disaster affecting most of mankind. Whether these hazards are worth incurring in view of the benefits achieved is what Alvin Weinberg has referred to as a transscientific question. As professional specialists we can try to provide pertinent information, but we cannot legitimately make the decision, and it should not be left in our hands.

One question I have not yet addressed is whether it is in fact not already too late. Have we already accumulated such a store of high-level waste that further additions would only increase the risks marginally? While the present waste (primarily from the military program plus the plutonium and highly enriched uranium contained in bombs and military stockpiles) is by no means insignificant, the answer to the question appears to be no. I am informed that the projected high-level waste to be accumulated from the civilian nuclear power program will contain more radioactivity than the military waste by 1980 or shortly thereafter. By 2020 the radioactivity in the military waste would represent only a small percentage of the total. Nevertheless, we are already faced with a substantial long-term waste storage problem. Development of a full-scale fission energy economy would add overwhelmingly to it. In any case, it is never too late to make a decision, only later.

What are the benefits?

The main benefit from near-term development of fission power is the avoidance of certain environmental impacts that would result from alternative energy sources. In addition, fission energy may have a slight cost edge, although this is somewhat controversial, especially in view of the low plant factors of the reactors actually in use. Far-reaching clean-up of the fuel cycle in the coal energy industry, including land reclamation, would require about a 20 percent cost increase over uncontrolled conditions for the large, new coal-fired plants. If this is done, fission plants would appear to have a clear cost edge, although by no means a spectacular one. The cost characteristics of the breeder that would follow the light-water reactors are very uncertain at this point. They appear, among other things, to still be quite contingent on design decisions having to do with safety. The dream of "power too cheap to meter" was exactly that.

Another near-term benefit is that fission plants will contribute to our supply during the energy "crisis" that lies ahead for the next decade or so. One should take note that this crisis was in part caused by delays in getting fission plants on the line. Also, there seems to be a severe limitation in using nuclear plants to deal with short-term phenomena. Their lead time is half again as long as fossil fuel plants—on the order of a decade.

The long-term advantage of fission is that once the breeder is developed we will have a nearly limitless, although not necessarily cheap, supply of energy. This is very important but it does not necessarily argue for a near-term introduction of a full-scale fission economy. Coal supplies are vast, at least adequate for a few hundred years, and we are beginning to learn more about how to cope with the "known devils" of coal. Oil shales and tar sands also are potentially very large sources of energy, although their exploitation will present problems. Geothermal and solar sources have hardly been considered but look promising. Scientists at the AEC's Los Alamos laboratory are optimistic that large geothermal sources can be developed at low cost from deep hot rocks—which are almost limitless in supply. This of course is very uncertain since the necessary technology has been only visualized. One of the potential benefits of solar energy is that its use does not heat the planet. In the long term this may be very important.

Fusion, of course, is the greatest long-term hope. Recently, leaders of the U.S. fusion research effort announced that a fusion demonstration reactor by the mid-1990s is now considered possible. Although there is a risk that the fusion option may never be achieved, its promise is so great that it merits a truly national research and development commitment.

A strategy that I feel merits sober, if not prayerful, consideration is to phase out the present set of fission reactors, put large amounts of resources into dealing with the environmental problems of fossil fuels, and price energy at its full social cost, which will help to limit demand growth. Possibly it would also turn out to be desirable to use a limited number of fission reactors to burn the present stocks of plutonium and thereby transform them into less hazardous substances. At the same time, the vast scientific resources that have developed around our fission program could be turned to work on fusion, deep geothermal, solar, and other large energy supply sources while continuing research on various types of breeders. It seems quite possible that this program would result in the displacement of fission as the preferred technology for electricity production within a few decades. Despite the extra costs we might have incurred, we would then have reduced the possibility of large-scale energy-associated nuclear disaster in our time and would be leaving a much smaller legacy of "permanent" hazard. On the other hand, we would probably have to suffer the presence of more short-lived undesirable substances in the environment in the near term.

This strategy might fail to turn up an abundant clean source of energy in the long term. In that event, we would still have fission at hand as a developed technological standby, and the ethical validity of using it would then perhaps appear in quite a different light.

We are concerned with issues of great moment. Benefit-cost analysis can supply useful inputs to the political process for making policy decisions, but it cannot begin to provide a complete answer, especially to questions with such far-reaching implications for society. The issues should be aired fully and completely before a committee of Congress having broad policy responsibilities. An explicit decision should then be made by the entire Congress as to whether the risks are worth the benefits.

Part 3

Environmental Regulation

Market-Based Approaches to Environmental Policy
A "Refresher" Course

Paul R. Portney

Market-based approaches to environmental regulation attempt to harness the economic incentives of the marketplace to achieve environmental objectives at relatively low cost and in ways that encourage innovation in pollution-control techniques. This approach encompasses a range of policy instruments, including taxes on pollution and the use of "cap-and-trade" systems.

Upon hearing the term "market-based approaches to (or economic incentives for) environmental protection," some people assume this means letting unfettered competition between unregulated private firms determine how clean our air or water will be, how much open space we will have, or how many fish stocks will be driven to collapse.

Nothing of the sort is intended. In fact, market-based approaches to environmental protection are a clever form of government regulation. They are premised on the recognition that while competitive markets are a wonderfully efficient means of deciding what types and quantities of consumer goods should be produced, they generally fail with respect to environmental quality, the provision of "public goods" like open space and common-property resources like fisheries. Every undergraduate and graduate economics textbook discusses this notion of "market failure," and the environment is always the first illustration that is used.

Given the very necessary government role in protecting the environment, the real question becomes how best to do this. Market-based approaches to environmental protection are premised on the idea that it is possible to confront private firms, individuals, and even other levels of government with the same kinds of incentives they face in markets for labor, capital, and raw materials—that is, prices that force them to economize. The rationale for market-based approaches, in other words, is to try to put the powerful advantages of markets to work in service to the environment.

Originally published in *Resources*, No. 151, Summer 2003.

Command-and-Control Era

To paint a quick picture of traditional regulation, consider the case of air and water pollution control. Prior to the early 1970s, the regulation of air and water pollution was almost exclusively the responsibility of state and local governments. In fact, the Clean Air Act amendments of 1970 and the Federal Water Pollution Control Act Amendments of 1972 marked the first really substantial federal involvement in environmental protection.

Under the Clean Air Act, the federal government (in the form of the then-new U.S. Environmental Protection Agency, or EPA) began specifying the pollution-control equipment that any new plant had to embody. In addition, EPA required local areas to formulate plans to reduce pollution from existing sources so that the air quality standards that EPA began issuing would be met. These plans typically required large, privately owned industrial facilities to reduce their pollution the most, and often required other sources to roll back their pollution by uniform amounts. Both new and old facilities had to apply for and receive operating permits from EPA that specified allowable emissions. In addition, the federal government also began limiting for the first time the tailpipe emissions of new cars rolling off the assembly lines of both domestic and foreign manufacturers. While the emerging water pollution regulations differed somewhat, at their heart, too, were a series of technological requirements for both newly constructed and existing plants, coupled with mandatory permits that specified allowable emissions.

Despite protests to the contrary, both programs have had significant successes, most notably in the case of the Clean Air Act. Since 1970, air quality around the United States has improved dramatically in almost every metropolitan area and for almost every air pollutant. For one notable example, airborne concentrations of lead, an especially insidious threat to health, were 93% lower in 2000 than they were in 1980.

Success under the Clean Water Act has been less dramatic, though quite obvious in many places. Rivers that 30 years ago had almost ceased to support aquatic life have seen fish strongly rebound (even if it is still inadvisable to eat the fish one catches in some places).

Despite these successes, by the late 1980s dissatisfaction with the technology-based standards approach had become rampant. First, by requiring sources of air and water pollution control to meet emissions standards keyed to a particular type of technology, many regulations had effectively "frozen" pollution control technology in place. No one had an incentive to invent a more effective and/or less expensive pollution control technology as long as some other technology had received EPA's blessing. Second, by requiring regulated firms to have specific types of pollution control in place, they were denied the flexibility to modify their production process or reformulate their product(s) in such a way as to reduce their emissions because they would still be required to use whatever technology was applicable. Finally, it was becoming clear that the technology-based command-and-control system was overly expensive. Study after study showed that it would be possible to meet the same environmental goals—either in terms of ambient air quality or in terms of emissions from affected sources—for much less money than the current approach was costing.

Cap and Trade vs. Pollution Taxes

There are two principal market-based approaches to environmental protection, both of which owe much of their popularity today to a small group of economists, most notably the late Allen Kneese of RFF. While mirror images of one another in many important respects, one market-based approach looks not unlike the current regulatory system while the other appears to be a more radical departure. The more familiar-looking approach to air or water pollution control would still be based on a system of required emissions permits. Under this approach—generally referred to as a "cap-and-trade" system—each pollution source is given an initial emissions limitation. It can elect to meet this limit any way it sees fit: rather than being required to install specific types of control technology, the source can reduce its pollution through energy conservation, product or process reformulation (including substitution of cleaner fuels), end-of-pipe pollution control, or any other means. Importantly, and not surprisingly, each source will elect to reduce its pollution using the least expensive approach available to it.

More surprisingly, a source has one additional option under the cap-and-trade system: it can elect to discharge *more* than it is required so long as it buys at least equivalent emissions reductions from one or more of the other sources of that pollutant. All that matters is that the total amount of emissions reductions that take place from all sources are equal to the initial cap established by EPA (or another regulatory authority). Those sources that will elect to make significant emissions reductions under this system are precisely those that can do so inexpensively; likewise, those that elect to buy emissions reductions from other sources rather than cut back themselves will be those that find it very expensive to reduce. (This is the analogue to Adam Smith's famous "invisible hand" that steers producers and consumers to the most efficient allocation of resources.) Moreover, all sources have a continuing incentive to reduce their pollution—the more a source's emissions fall short of its limitation, the more emissions permits it will have to sell to other sources.

The flip side of this approach is one in which *no* limits are placed on each ton of pollution that a source emits, but in which each ton is taxed. Pollution taxes are paid to the government, which is then free to use the revenues as it sees fit— to reduce other taxes, spend on pollution control R&D, reduce the national debt, etc. While appearing very different from the cap-and-trade approach, this system creates the very same set of incentives. That is, the firms that can reduce their pollution inexpensively will invest in doing so because each unit of pollution reduced is that much less paid in pollution taxes. Firms that find it very expensive to reduce their pollution will continue to discharge and pay the taxes; note, however, the strong and continuing incentive the latter have to find ways to cut their emissions—and the higher the taxes on

> ## Update
>
> After making extensive use of environmental taxes, Europe has now turned to the cap-and-trade approach as well. In January 2005, the European Union introduced the largest emissions trading program in the world. Its objective is to meet the European commitment under the Kyoto Protocol to limit emissions of carbon dioxide for the containment of global climate change. Under the new EU program, sources in designated sectors of the economy in all 25 member countries can trade CO_2 allowances (permits) across member-country borders.

pollution, the stronger that incentive. Also, both a cap-and-trade system and a pollution tax create the same incentive to reduce pollution that the wage rate creates for firms to minimize the amount of labor they use or that the interest rate has in disciplining firms' borrowing.

The cap-and-trade approach began to be implemented in a small-scale way in the late 1970s and early 1980s in both Democratic and Republican administrations. But the first really large-scale application of cap-and-trade—which resulted in the most significant environmental policy success since 1970—came in the 1990 amendments to the Clean Air Act. In order to reduce emissions of sulfur dioxide by 50% in the eastern half of the United States, an ambitious cap-and-trade system was created under which more than 100 large coal-fired power plants were given initial emissions reductions. These plants could meet their emissions reductions targets themselves, through any means they selected, including shifting from high- to low-sulfur coal. However, the affected plants were also given the ability to purchase excess emissions reductions generated by other plants that found it easy to reduce their sulfur dioxide.

This approach has resulted in reductions in sulfur dioxide emissions that have been both larger and faster than required by the law. Moreover, the annual savings to electricity ratepayers nationally (compared to the previous command-and-control approach) range from 50–80% and these savings amount to $1–6 billion annually, depending on whose estimates one wants to use. As a result of this success, cap-and-trade approaches are now being proposed for additional reductions of sulfur dioxide, nitrogen oxides, and mercury under the Bush administration's Clear Skies Initiative. They have also been put forward by former EPA Administrator Christie Todd Whitman for reducing water pollution in certain watersheds, by state and local governments seeking smog reductions, and by foreign governments exploring lower-cost approaches to a variety of environmental problems. The European Union has just announced that it will use a cap-and-trade system to control carbon dioxide as it struggles to comply with the terms of the Kyoto Protocol, which is still alive in Europe.

Uncertainties Created by Each System

Large-scale experiments with pollution taxes are harder to find in the United States. Under the 1987 Montreal Protocol to phase out worldwide use of chlorofluorocarbons (CFCs) and other ozone-depleting substances, a tax was levied on CFC production during the time mandatory phase-out was taking place, although this is clearly a hybrid system under which command-and-control regulation was augmented by a pollution tax. The evidence to date suggests that this hybrid approach is working well—CFC emissions have fallen and early evidence is that the stratospheric ozone "hole" has stopped growing.

Interestingly, perhaps the most ambitious application of pollution taxes is occurring not at the federal or even state level of government, but at the local level. Hundreds of communities around the United States have adopted "pay-as-you-throw" systems for household garbage collection. Rather than charge every household the same amount for refuse collection, these communities are charging households a fixed amount per bag of garbage collected at curbside. This has had the effect of reducing the amount of yard wastes that end up in municipal landfills (households are composting more) and

possibly even changing households' purchasing decisions toward products which come with less packaging.

Why have cap-and-trade policies flourished in comparison to pollution taxes in the United States? Perhaps most obviously, a system in which discharge permits are issued, but made saleable, looks rather like the regulatory system currently in place in the United States, with the added twist of marketability. Another reason has to do with the uncertainty each system creates. Specifically, under a cap-and-trade system, the total amount of pollution is firmly fixed—that is the purpose of the cap. What is uncertain are exactly where the emissions will occur (this depends upon who trades with whom), and how much an emissions permit (the right to emit one ton in a given year, say) will cost—the latter is determined in a competitive market.

Under a pollution tax, sources are allowed to discharge as much as they want, as long as they pay the per unit charge for each ton emitted. Thus, there is uncertainty about the total amount of pollution discharged (though we can be sure that the higher the tax, the lower the amount of pollution discharged). There is no uncertainty under the latter system about the maximum amount it will cost to reduce a ton of pollution, though, because that will not exceed the per-ton tax. The total amount of revenue raised by such a system is not predictable, because if sources can reduce their emissions less expensively than is believed to be the case, they will discharge less to avoid the tax. In years past, environmentalists objected to pollution taxes on the grounds that sources faced no pollution limits at all and could continue to pollute as long as they paid the corresponding taxes. Note, however, that this approach makes sources pay for every single unit of pollution that they discharge—unlike the command-and-control system in which firms are given considerable amounts of "free" emissions in the form of any discharges they may make so long as they are beneath their permitted levels.

The choice between cap-and-trade systems and pollution taxes rests in part on the pollutant in question. For pollutants like sulfur dioxide, CFCs, or carbon dioxide that mix equally in the atmosphere and that pose few or no local health effects, cap-and-trade works well because we are unconcerned about where emissions take place. On the other hand, if we are concerned that limiting emissions might impose too big a burden on the economy, the pollution tax approach is best because sources know that they will never have to pay more for a ton of pollution discharged than the tax. Effluent charges also raise revenue—not a trivial issue in many places, including developing countries.

One thing is for sure. Market-based approaches to environmental protection have become the default option in much of modern environmental policy, both in the United States and abroad. But it would be a mistake to claim that command-and-control regulation is dead. First, there are some cases where market-like solutions won't do the job. If an imminent, serious hazard to human health and the environment is discovered, an outright ban is likely to be the appropriate policy response. Second, some still prefer that companies be punished for their emissions by making them pay as much as possible to alleviate them. But this is premised on the misguided notion that firms pollute because they are malevolent, rather than because pollution is one consequence of making things that society demands. Moreover, such an approach really only punishes the customers, employees and shareholders of the firm, for they are the ones who will end up bearing the costs.

Suggested Reading

Burtraw, Dallas, and Karen Palmer. 2004. SO$_2$ Cap-and-Trade Program in the United States: A 'Living Legend' of Market Effectiveness. In *Choosing Environmental Policy: Comparing Instruments and Outcomes in the United States and Europe*, edited by W. Harrington et al. Washington, DC: Resources for the Future, 41–66.

Stavins, Robert N. 2000. Market-Based Environmental Policies. In *Public Policies for Environmental Protection*, edited by P. Portney and R. Stavins. Washington, DC: Resources for the Future, 31–76.

———. 2003. Experience with Market-Based Environmental Policy Instruments. In *Handbook of Environmental Economics*, Vol. 1, edited by K. Maler and J. Vincent. Amsterdam: Elsevier, 355–435.

10 Trading Cases
Five Examples of the Use of Markets in Environmental and Resource Management

James Boyd, Dallas Burtraw, Alan Krupnick, Virginia McConnell, Richard G. Newell, Karen Palmer, James N. Sanchirico, and Margaret Walls

Systems of tradable allowances (so-called cap-and-trade systems) are proving to be a very flexible, efficient, and popular instrument for environmental regulation. Such systems can achieve environmental objectives by placing limits or "caps" on damaging activities and, at the same time, provide incentives for doing this in cost-effective ways that encourage innovation in improved control technologies. The wide range of applications of this approach is just beginning to be appreciated.

Two decades ago, a "command-and-control approach" dominated environmental policy and regulations. Individual decisionmakers were told how to comply with an input or output standard, rather than being allowed to respond to market signals in the most economical way. Until well into the 1970s, few market-based instruments for environmental policy existed. Even then, examples were limited to an effluent charge program in Germany, some deposit-refund systems, and—depending on one's definition of a market-based instrument—performance bonds, which require potential polluters to demonstrate that they can compensate those damaged by their activities. In addition, a few markets were created to protect natural resources, such as transferable quotas for fishing in Canada and Iceland, and transferable development rights for land in the New Jersey Pine Barrens and a few other locations in the United States.

Created markets to protect the environment from pollution were conspicuously absent until 1975 when the U.S. EPA introduced "bubbles" and other approaches to relaxing economic growth restrictions for areas violating air quality standards under the Clean Air Act (CAA). This "bubble policy" allows a plan with multiple stacks to meet its emissions limits over all stacks at once rather than each stack separately. Everything changed in 1990 when Congress broke the logjam over legislation to limit acid rain by reducing sulfur dioxide (SO_2) emissions from power plants. Industry opposed the estimated huge cost of a command-and-control program, while environmentalists adamantly fought to bring plant emissions under tight control.

Originally published in *Environmental Science and Technology* 37 (June 1, 2003).

To end the controversy, Congress and the first Bush administration turned to tradable pollution permits. In the late 1960s, economists Crocker[1] and Dales[2] first introduced the idea of capping total emissions and allowing polluters to trade them. The idea had gained currency among economists because trading would be far less expensive than a command-and-control regulatory policy and could guarantee environmental improvements if a hard cap on total SO_2 emissions were set and backed up by continuous emissions monitoring. Further, the initial allocation of tradable permits could be designed to reduce political opposition by, for example, giving away more permits to better-performing utilities, as studies showed that the efficiency gains were independent of this allocation.

The resulting SO_2 allowance trading program (initiated as the Acid Rain program under Title IV of the CAA Amendments) turned out to be hugely successful and fostered confidence in this instrument, which has literally swept through government and academic circles around the world. This article reviews how tradable permits—now synonymous with created markets—are faring in five divergent areas. We describe how and why this strategy improves on command-and-control approaches, its remaining problems, how broadly trading programs have been implemented, and what the future may hold for them. We also present the debates that surround applying these approaches to water pollution, solid waste recycling, fisheries, and land use.

SO_2 allowance trading

The SO_2 allowance trading program's success stems from its simplicity, effective monitoring, definite penalties, and the opportunity for banking allowances.

Simplicity is the most important of these features. The program combines an aggregate cap on the annual allocation of emission allowances at all large, fossil fuel-fired electricity generation facilities with nearly unfettered opportunities to trade or bank allowances. The environmental community accepted this program because allowance trading was expected to reduce emissions by nearly 50%. Most believe that Congress would not have imposed the same amount of reductions if not for the prospective cost savings of trading.

Another key feature is the availability of continuous emission monitors (CEMs), which measure postcombustion levels of SO_2 and NO_x and estimate CO_2. Note that CEMs come at a substantial cost and may not have been required technically because engineering formulae can reliably predict SO_2 emission rates. However, mistrust between environmental advocates and the regulated community led to uncertainty about when and whether postcombustion controls would operate, so compliance assurance under trading required measurement of actual emissions. The technology to do so, as well as the computing capability to process large amounts of emissions data, emerged just in time to facilitate the program. The electronic Allowance Tracking System is transparent, so EPA determined that disclosing terms of contracts or prices or agency approval for trades was unnecessary. Hence, simplicity and transparency contributed to this program's overall success.

The certainty of penalties contributed to virtually 100% compliance. Noncompliance triggers a prespecified financial penalty plus surrender of a number of allowances for the subsequent year. Finally, banking contributed greatly to the cost savings of the program by providing a mechanism for firms to insure against adverse situations in fuel markets or in their own compliance activities. Banking also facilitates "buy-in" by some of the regulated parties because it endows them with an asset that only has value with a successful and stable program.

The program will cost just over $1 billion per year by 2010. Exogenous trends in the electricity, coal, and railroad industries would have contributed extensively to greater use of low-sulfur coal and SO_2 emission reductions even in the absence of the 1990 regulations. However, perhaps the most remarkable feature of the SO_2 trading program is that it successfully enabled the industry to fully capitalize on these advantageous trends, because of the flexibility inherent in allowance trading.

Compared to an approach that mandated a specific emission rate at every facility, the allowance trading program reduced program costs by an estimated 30–50%.[3] And compared to a prescriptive technology approach such as requiring scrubbers at a certain class of facilities, the savings are close to perhaps 200%.

Problems. Despite the success of the program and obvious cost savings, the program has not been perfectly efficient. The reason rests largely with the tendency of firms to look first to their own facilities to achieve compliance, relying on the market as a backup. State public utility commissions that establish cost-recovery rules for firms still operating where electricity prices are regulated exacerbate this risk-averse behavior. Ample anecdotes, statistical analyses,[4] and simulation studies show state regulators provided incentives for compliance strategies that raised program costs.

Many observers were also concerned that trading could worsen pollution in some areas. Indeed, the original design for the SO_2 trading program would have divided the nation into two separate trading zones in order to address this concern. However, policymakers recognized that a more constrained market would increase the program's cost, and they opted for a uniform national market. As it turns out, the geographic trading pattern added to the program's already dramatic benefits.[5]

The program also has been criticized because of the difficulty in changing the cap in response to new scientific and economic information. During the 1990s, costs to control SO_2 emissions from coal-fired power plants fell to less than half of the levels predicted at the time of the 1990 CAA, and new information suggested that the health benefits of reducing sulfates may be several-fold greater than anticipated. Consequently, the net benefits now exceed expectations.[6] However, the emission cap leaves regulators unable to adjust to new information, short of an act of Congress.[7]

Future. The perceived success of the SO_2 program and new information about the benefits of reducing air pollution have contributed to a new wave of legislative proposals aimed at further reducing SO_2 and NO_x emissions from power

plants. It reflects a remarkable consensus in the policy community that all of the proposals would expand the use of a cap-and-trade approach to achieve these emission reductions.

Most would also apply cap-and-trade to reduce mercury and CO_2 emissions. Trading mercury allowances is controversial because it is a hazardous air pollutant, and inclusion of CO_2 is controversial because the Bush administration's climate change proposal lacks mandatory CO_2 reductions. Two issues are already emerging in debates about these cap-and-trade proposals: How can the emissions cap adjust when new information emerges about benefits or costs? A cap that can change either automatically or easily in response to new information could address these issues.

And, especially, how do different approaches to distributing emission allowances affect the efficiency and cost distribution?[8]

Trading water emissions

The National Pollutant Discharge Elimination System (NPDES) program is the centerpiece of U.S. water quality regulations. It sets numerical and technological discharge standards that must be met if a source is to discharge waste into surface waters. Because regulators determine the technologies for control standards and because those standards are uniformly applied across sources, the current NPDES system is best thought of as a command-and-control approach to water quality regulation. Note that NPDES permits are only required for so-called point sources, which tend to be larger industrial, commercial, and public treatment facilities. Some large agricultural operations are also considered point sources, but nonpoint runoff from farms, roads, lawns, and most small pollution sources, such as septic tanks, are not directly regulated.

To encourage state experimentation with trading programs, EPA issued a water quality trading policy this year.[9] The policy is not a regulatory rule but does set objectives and guidelines for program design.

Why trading? The desire for effluent trading arises from significant disparity in pollutant control costs across sources of water pollution. This is particularly true for nonpoint sources because these controls are often significantly cheaper than those on already regulated point sources. Trading opens the door for those with more expensive controls (point sources) to pay those with cheaper controls (nonpoint sources) to reduce pollution. Trading can also generate a strong incentive to innovate, because more pollution control translates into salable credits.

Problems. To date, water quality trading in the United States exists primarily in the form of pilot programs. Technical challenges, such as the difficulty of modeling the transport of pollution in a watershed, are a barrier to such programs. In addition, the failure to regulate nonpoint sources and excessive reliance on end-of-pipe effluent standards, as opposed to ambient water quality standards, has hindered pilot programs and their expansion into wider trading initiatives.

By and large, nonpoint pollution sources are currently not regulated. This means there is no clear baseline against which to measure emissions reductions, nor is there an understanding of how to monitor, enforce, and evaluate these polluting activities. Even trading between regulated point sources is inhibited by a lack of explicit regulatory authority and by regulatory provisions that reduce incentives to trade. First, trades cannot be used to comply with technology-based standards (NPDES standards). This rule limits the scope of trading because compliance with those standards is typically the largest component of control costs. Second, sources must, in effect, "overcomply" to have credits to sell, but doing so presents two regulatory risks. First, the Clean Water Act (CWA) has antibacksliding provisions to ensure that, once achieved, water quality will be sustained. Overcomplying sources that return at a later date to more "normal" levels of compliance may run afoul of this provision. Second, overcompliance signals to the regulator that greater levels of control can be attained cost-effectively. This can lead to a future ratcheting down of the NPDES standard. In either case, the CWA provisions undermine the incentive to overcomply, and thereby trade.[10]

Update

One of the most recent and largest of the new trading systems is the new European Union Emissions Trading System for controlling CO_2 emissions under the Kyoto Protocol to contain global climate change. Under the system, specified sources of carbon emissions will be able to trade emissions allowances across the borders of all 25 EU member countries. More generally, emissions trading is being widely considered as a basic policy instrument for addressing the problem of global warming.

Ideally, trading programs allow decentralized decisionmaking. However, most "trades" to date are simply control reallocations grafted onto permits required for point sources. These so-called "offsets" transfer control from a point source to some other, cheaper source to reduce costs. In effect, point source permits are modified to allow a predetermined, purchased control activity as a substitute for on-site controls. They are also a mechanism to finance nonpoint source effluent reductions. But decisionmaking is not decentralized.

A cap-and-trade program is required for decentralized water quality trading. Here, the regulator establishes a cap on total releases from a defined set of sources, but allows the sources to meet the cap with whatever distribution of responsibilities emerges from the market. However, this structure builds on a more watershed-level, ambient quality-driven, regulatory approach.

Furthermore, for any trading policy to have political and scientific defensibility, trades must be seen as environmentally "equivalent". Unfortunately, a distinguishing characteristic of water quality problem is that the costs of pollution to society are highly dependent on the location of pollution sources in the physical, biological, and social landscape; type of pollutant; and timing of releases. When location and timing matter, individual environmental trades will not have uniform social and biological consequences. Of particular concern are trades that redistribute releases in a way that yields pollution "hot spots." Given the technical complexity and costs of ecological evaluation, determining precise equivalence is tough. The challenge for policy makers is to provide safeguards that ensure that overall environmental quality is maintained or enhanced while preserving the flexibility necessary for pollutant trading.

Future. If the water quality trading idea falters in the coming decade, it will not be for lack of interest or support. EPA appears to strongly support trading, at least as a regulatory aspiration. Indeed, a long-neglected part of the CWA, the Total Maximum Daily Load (TMDL) provisions, require states to identify waters that are not in compliance with water quality standards, establish cleanup priorities, and implement improvements. TMDL rules and watershed-level planning complement effluent trading by creating a mechanism to set total load caps based on ambient conditions and make initial allocations that could be traded.[11] A political commitment is required, however, to shift cleanup burdens to currently unregulated nonpoint sources; reform inflexibilities in the CWA; and finance the technical, scientific, and administrative innovations required by a flexible, watershed-level, multisource approach to water quality improvement.

Recycling credits

Solid waste policy, particularly since the mid-1980s, has focused on promoting recycling. States often set minimum recycling rate targets for communities or mandate residential curbside recycling programs. Recycling targets are also an integral part of so-called "extended producer responsibility" (EPR) laws that have gained popularity in Japan and throughout Europe. Such laws typically require producers to arrange to take back products from consumers and ensure that a certain percentage is recycled. EPR laws have been applied to packaging, batteries, electronic devices, and automobiles. The EPR concept has caught on to some extent in the United States, where it is typically referred to as "product stewardship."[12]

Why trading? Making producers responsible for recycling can help spur product designs that are more recyclable in the first place, thus reducing costs for meeting waste reduction targets. However, requiring all producers to do the same thing, as most EPR programs do, is inefficient. Overall, costs could be even lower if some flexibility were introduced, such as through a tradable recycling credit (TRC) program.

To make a TRC program work, policymakers would set a recycling rate target. It could be either an overall weight-based recycling target (such as 50% of the weight of some product or group of products) or a set of specific targets by component material. A recycling credit is generated when a pound of product or material is recycled. The parties responsible for meeting the recycling rate goal must show enough credits to meet their annual obligation. They can either ensure that enough of their own products have been recycled or purchase credits from others who have exceeded the minimum.

For example, recycling electronics depends on how easy they are to disassemble, the type and mix of materials they contain, whether or not individual materials, including plastics, are labeled, the presence of contaminants, and other factors. Regulators would have difficulty dictating exactly what product design changes would spur more recycling, but a TRC system has the potential to provide those incentives through the marketplace. More recycling means more earned credits, which a manufacturer can sell and thus reduce its compliance cost.

Problems. For the system to work, manufacturers must be easily matched with their own equipment at the end of its life. In fact, any system that promotes "design for environment" must have such a feature. In practice, this is only one of several issues that must be addressed. Should responsibility for collection of used equipment be left to the marketplace or spelled out in the program setup? How will hazardous materials such as lead and mercury be handled? How will durable products, like electronics, produced years ago figure into recycling rates?

Use of trading programs. In practice, trading mechanisms are rarely used to promote recycling.

The one prominent exception is in the United Kingdom where tradable credits, known as packaging waste recovery notes (PRNs), can certify compliance with the 1997 Packaging Regulations.[13] The Packaging Regulations set recycling rate targets for specific materials, with responsibility for meeting the targets shared among companies that manufacture packaging, produce raw materials, fill packaging, and sell products at retail. A compliance scheme is a third-party organization that, for a fee, will take on the recycling obligation for individual companies. Compliance schemes and individual companies can meet the recycling obligation by purchasing PRNs from certified reprocessors, which generate the PRNs whenever they recycle packaging.

The British system makes no attempt to track the manufacturer of the packaging that is ultimately recycled because the sheer volume and myriad of packaging on the market would make the process too complicated. As a result, the touted product design incentives may not even exist in the PRN system.

The system provides a means to bring down the overall cost of the program by generating another way for companies to meet their recycling obligations. Companies can either take responsibility for recycling their own products, join a compliance scheme that ensures obligations are met, or purchase PRNs in the marketplace. If the costs of compliance get too high, companies can switch to purchasing PRNs, which helps moderate the prices charged by compliance schemes. The recycling targets in the early years were rather low and were thus easily met—even exceeded—and at very low cost. Actual packaging recovery in 2001 fell slightly short of a new higher target. Trading PRNs has steadily grown over time, and PRN prices declined between 1998 and 2001; in 2001, the stricter recycling rate target drove PRN prices up slightly. Although it is too soon to fully evaluate the program, further analysis would provide useful information for designing tradable credit schemes for other products and in other countries.

Future. The European Union is finishing its Waste Electronic and Electrical Equipment (WEEE)

directive, which is expected to receive final approval in the spring of 2003. The WEEE directive mandates minimum collection and recycling targets for electronic and electrical equipment waste by 2005. In the United States, the National Electronics Products Stewardship Initiative, which includes representatives of industry, government, recyclers, retailers, and environmental groups, is trying to reach consensus around a national program to promote electronics recycling. Either of these initiatives could provide the next innovative tradable credit system that promotes both recycling and design for environment at low cost.

Individual transferable quotas

Most commercial marine fisheries are regulated under command-and-control regulations governing the sizes of vessels, types of nets, season lengths, and areas open to fishing.

Why trading? Such regulations fail to check the number of vessels and the level of fishing and encourage fishermen to work around those technological constraints mentioned in the previous paragraph. Without a sense of ownership over the fish until they are caught, fishermen race to catch as much as possible, as fast as possible. The historical record shows that the race will continue until fish stocks are depleted and the number and types of vessels in a fishery exceed its viable capacity.

Individual transferable quota (ITQ) systems are a promising means to correct this market failure and are analogous to other cap-and-trade programs, such as the U.S. SO2 allowance trading program. ITQs limit fishing operations by setting a total allowable catch (TAC), which is then allocated to fishing participants, typically on the basis of historical catch. Because fishermen have access to a guaranteed share of the TAC, ITQs significantly reduce the incentives to race to fish.

The benefits of reducing the race to fish include a longer season and a shift from maximizing quantity to improving quality. For example, fishing seasons for the Alaskan halibut fishery were reduced to two 24-hour openings by the early 1990s, but the season has lengthened more than

200 days since ITQ systems were introduced in 1994. The flexibility to time fishing trips when port prices are higher and the elimination of large supply gluts of fresh product have resulted in per pound price increases of more than 40%.[14]

When shares are transferable, the least efficient vessels will find it more profitable to sell their quota rather than fish for it. Over time, this will reduce excess capacity and result in the total catch caught at the lowest cost. Furthermore, fishermen gain a financial stake in the resource that creates an incentive for them to take into account the health of fish stock and how fishing impacts future catches.

Problems. The earliest market-based quota programs were established in Canada and Iceland in the late 1970s. As of 2000, ITQ programs were established for more than 75 fish and crustacean species, including four in the United States.[15] New Zealand is the world leader in ITQ programs, having established in 1986 the most comprehensive ITQ system, which currently includes more than 40 species.

Cap-and-trade programs in fisheries have many of the same issues found in other situations, such as initial allocation of rights, monitoring, and enforcement. Difficulties in the initial allocation process were evident in New Zealand where fishermen were able to choose two of the previous three years for calculating their average catch. The short time horizon and limited ability to choose resulted in 1400 out of 1800 fisherman lodging objections, and most received amended catch histories. In New Zealand, compliance and enforcement of ITQs follow a detailed set of reporting procedures that track the flow of fish from a vessel to a licensed fish receiver on land to export records, along with an at-sea surveillance program that includes on-board observers.

In theory, ITQ programs are analogous to other cap-and-trade programs, but there are some important differences. For example, controlling and forecasting emissions from a power plant is arguably easier than predicting the size and composition of a catch on any trip, especially in multispecies fisheries where populations cannot be directly targeted without incidental catch of other stocks. Some pro-grams have incorporated design features into their system to increase flexibility. In New Zealand, for example, fishermen have a set number of days after landing their catch to arrange quota on the various species in that catch. In addition to leasing or buying quota to cover their catch, quota owners have three options: pay a "price" to the government, which is set based on the nominal port price, to discourage discarding catch at sea and target those stocks without sufficient quota; enter into a non-monetary agreement to fish against another's quota; or surrender their catch to the government. Originally, quota owners could also carry forward or borrow (except for annual leases of quota) up to 10% from next year's quota, but this feature has been removed for administrative reasons.

Future. Although the assessments of these programs are generally positive, their future is unclear. For example, in the United States, a six-year moratorium on implementing new ITQ systems expired in September 2002, but policymakers continue to debate various design elements, such as if shares should be transferable and have limited duration, and whether shareholders must remain active in the fishery. Distributional concerns regarding the potential concentration and industrialization of the fishery are at the core of these design questions. Those opposed to quota management systems argue that implementing such a system will result in the loss of the small-scale fishermen, a claim that is analogous to preserving the family farm. In many cases, these concerns can be addressed in the design stage with caps on ownership levels and restrictions on transferability. (In the United States Alaskan halibut and sablefish ITQ system, sale or lease of the quota can only occur within vessel classes.) However, constraints may also limit the gains from trade.

Tradable development rights

Land-use zoning designates the density and allowable uses of every property. It is the most common way that a local government intervenes in land markets and is a major tool in controlling and directing development. Local governments also use zoning to preserve land from development for the public benefit, such as keeping farmland to maintain the rural character of a community and preserving forests to protect watersheds or ecological habitats.

However, common, low-density zoning does not ensure enough land will remain undeveloped. Landowners are still free to sell their land to developers, and often low-density zoning has simply resulted in spreading development out over even larger areas. Moreover, reducing the allowable housing density is often unpopular with landowners because it reduces the value of their land. So-called takings rules are designed to protect private property owners from this problem by preventing land from being taken by the government without compensation. Thus, political realities may result in too little land being protected. The economic problem is that the benefits from preserving certain undeveloped properties will be conferred on many in the community, but private landowners have no easy way to obtain compensation for these values.

Why trading? Can markets help to provide a solution? Transferable development rights (TDRs) are somewhat successful. Development rights are assigned to undeveloped land through zoning rules, and owners of the land can sell them. The market can be designed so that once development rights are sold from a parcel, that land is preserved in perpetuity. The overall amount of preservation or protection and its location can be targeted to appropriately reflect the public value of preserving such land. And, the pricing mechanism has the potential to direct the most valuable parcels eyed for development toward development, and the highest value agricultural parcels to stay in farming.

The development rights market can be established in many ways. Using baseline zoning to delineate each area, certain areas can be targeted for development and others for preservation. Developers must purchase TDRs for each unit of development, and the preservation areas, which are zoned low density, can sell development rights. In another version, the local government itself steps in and purchases development rights from landowners.

Problems. TDR markets intended to preserve land first appeared in the United States in the 1970s, but they became a popular land-use tool in the 1980s and 1990s. Close to 100 programs have been initiated to preserve farmland and environmentally sensitive areas since 1980.[16] Montgomery County, Maryland, initiated a TDR program in 1980 to preserve a large tract of rural land and is now credited with permanently preserving almost 30,000 acres of farmland in 23 years. Calvert County, Maryland, has had a successful program since the early 1980s, in which owners of 13,000 acres of prime farmland have sold development rights. In Nevada, the Tahoe Regional Planning Agency used a TDR system initiated in the late 1980s to protect water quality in Lake Tahoe.

TDR markets face several complex design and regulatory issues. Foremost, it is virtually impossible to put a hard "cap," or limit, on the amount of development during any given time period. Land in areas targeted for preservation can be zoned to very low-density uses, which reduces land values, but owners still have the option of developing the land even at the lower density.

Because there is no cap on development in any single year, both the supply and demand for development rights, as well as the resulting price of TDRs, will be critical for determining the amount of land that is preserved. Often, reducing the allowable zoning density in preservation areas creates a large supply of development rights, but the difficulty comes in establishing a demand for those rights. Local governments are not inclined to reduce zoning density in other areas to create demand, so they often allow developers to purchase development rights to build at higher density than existing rules permit. However, with little demand for high-density development in many communities—especially during the 1990s—and an oversupply of development rights, there have been few sales of TDRs. Such is the case in Montgomery County, where demand for rights has dropped and TDR prices are currently low.

Another complication is that the amount of development can actually rise from the sale of TDRs. In most programs, landowners in preservation areas receive development rights equal to what they could have built given existing zoning. Owners whose land has low development value are most likely to preserve their land and sell TDRs. Some of them, however, might never have sold their land for development, and yet they receive development rights that can be sold and used elsewhere. Thus, total development over time could be higher under many current program designs.[17] This may be, in part, why we see local areas designing programs in which the government buys development rights from landowners and retires them from use.

In addition, transaction costs may be high in TDR markets because many programs to date have not used central banking or clearinghouse methods to facilitate trades. Maryland's Calvert County has had a successful TDR program since the mid-1980s without providing banking. Instead, the local government has kept transaction costs low by providing information to potential sellers and buyers.[18] The county government has also become an active participant in the market in recent years by purchasing development rights to prevent development and stabilize the price of TDRs. Through such government price stabilization efforts, TDRs can become more like a price mechanism, in which local governments essentially promote permanent land preservation and set a higher price on new development. Prices of TDRs, then, reflect the societal benefits of preserving land.

Future. Transferable development rights will likely gain popularity as a land-use tool. TDR mechanisms are being considered for other land preservation uses, including hillsides, scenic views, wetlands, and wildlife habitat. An excess supply of rights and few transactions in many markets doesn't mean that TDRs can't work, but that the markets must be carefully designed to achieve community land use goals.

Suggested Reading

Burtraw, Dallas, and Karen Palmer. 2004. SO$_2$ Cap-and-Trade Program in the United States: A 'Living Legend' of Market Effectiveness. In

Choosing Environmental Policy: Comparing Instruments and Outcomes in the United States and Europe, edited by W. Harrington et al. Washington, DC: Resources for the Future, 41–66.

Stavins, Robert N. Market-Based Environmental Policies. 2000. In *Public Policies for Environmental Protection*, 2nd ed., edited by P. Portney and R. Stavins. Washington, DC: Resources for the Future, 31–76.

See also Chapter 23 in this volume: James Sanchirico and Richard Newell, "Catching Market Efficiencies: Quota-Based Fisheries Management," and Chapter 27 in this volume: Virginia McConnell, Margaret Walls, and Elizabeth Kopits, "A Market Approach to Land Preservation."

Notes

1. Crocker, T. D., The Structuring of Atmospheric Pollution Control Systems. In *The Economics of Air Pollution*, Wolozin, H., ed. (New York: W. W. Norton, 1966).

2. Dales, J. H., *Pollution, Property and Prices* (Toronto: University Press, 1968).

3. Carlson, C.; Burtraw, D.; Cropper, M.; Palmer, K. , *J. Pol. Econ.* 108 (2000): 1292–1326. Ellerman, A. D.; Joskow, P. L.; Schmalensee, R.; Montero, J.-P.; Bailey, E. M., *Markets for Clean Air,* (London: Cambridge University Press, 2000).

4. Arimura, T., *J. Environ. Econ. Manage.* 44 (2002): 271–289.

5. Burtraw, D.; Mansur, E. *Environ. Sci. Technol.*, 33 (1999): 3489–3494. Swift, B., *Environ. Rep.* 31 (2000): 954–959.

6. Burtraw, D.; Krupnick, A. J.; Mansur, E.; Austin, D.; Farrell, D., *Contemp. Econ. Pol.* 16 (1998): 379–400.

7. Zuckerman, B.; Weiner, S.L., Environmental Policymaking: A Workshop on Scientific Credibility, Risk and Regulation (Cambridge, MA: Massachusetts Institute of Technology Center for International Studies, 1998).

8. Bertraw, D.; Palmer, K.; Bharvirkas, R.; Paul, A., *The Effect of Allowance Allocation on the Cost of Carbon Emission Trading,* Report DP01-30 (Washington, DC: Resources for the Future, Aug. 2001), www.rff.org/disc_papers/PDF_files/0130.pdf.

9. *Water Quality Trading Policy* (Washington, DC: U.S. Environmental Protection Agency, U.S. Government Printing Office, January 13, 2003), www.epa.gov/owow/watershed/

trading/finalpolicy2003.pdf.

10. Stephenson, K.; Shabman, L.; Geyer, L., *The Environmental Lawyer* 5 (1999): 775–781.

11. Boyd, J., *The Duke Environmental Law and Policy Forum* 11 (2000): 39–87.

12. Palmer, K.; Walls, M., *The Product Stewardship Movement: Understanding Costs, Effectiveness, and the Role for Policy*, Resources for the Future Report (Washington, DC: Resources for the Future, November 2002), www.rff.org/reports/2002/htm.

13. Implementing Domestic Tradable Permits: Recent Developments and Future Challenges (Paris: Organization for Economic Cooperation and Development, August 2002), Chapter 6, www1.oecd.org/publications/e-book/9702121E.pdf.

14. Casey, K. E.; Wilen, J.; Dewees, C. *Mar. Resour. Econ.* 10 (1995): 211–230.

15. Newell, R.; Sanchirico, J. N.; Kerr, S., *Fishing Quota Markets*, Discussion Paper 02-20 (Washington, DC: Resources for the Future, August 2002).

16. Pruetz, R., Saved by Development: Preserving Environmental Areas, Farmland, and Historic Landmarks with Transfer of Development Rights (Burbank, CA: Arje Press, 1997).

17. Levinson, A., *Regional Science and Urban Economics* 27 (1997): 286–296.

18. McConnell, V.; Kopits, E.; Walls, M., *How Well Can Markets for Development Rights Work? Evaluating a Farmland Preservation Program*, Discussion Paper 03-08 (Washington, DC: Resources for the Future, March 2003), www.Rff.org/disc_papers/PDF_files/0308.pdf.

11

Economic Incentives versus Command and Control
What's the Best Approach for Solving Environmental Problems?

Winston Harrington and Richard D. Morgenstern

Environmental policies can rely more on direct regulation (the command-and-control or CAC approach) or, alternatively, on economic incentives for environmental protection. A study of their actual use on both sides of the Atlantic belies some of the more simplistic claims that have been made. Most polices, in fact, contain some elements of both approaches. Comparisons of the two approaches in terms of a set of criteria reveal that in practice there are some systematic differences. Economic incentives, for example, have been able to achieve greater cost savings. The good news is that in all the cases studied, there has been significant environmental improvement.

Now, decades after the first environmental laws were passed in the United States, policymakers face many choices when seeking to solve environmental problems. Will taxing polluters for their discharges be more effective than fining them for not meeting certain emission standards? Will a regulatory agency find it less costly to enforce a ban or oversee a system of tradable permits? Which strategy will reduce a pollutant the quickest?

Clearly, there are no "one-size-fits-all" answers. Many factors enter into the decision to favor either policies that lean more toward economic incentives (EI) or toward direct regulation, or what is commonly referred to as command-and-control (CAC) policy. Underlying determinants include a country's governmental and regulatory infrastructure, along with the nature of the environmental problem itself.

Even with these contextual factors to consider, we thought it would be useful to compare EI and CAC policies and their outcomes in a real-world setting. To do this, we looked at six environmental problems that the United States and at least one European country dealt with differently (see box on page 68.) For each problem, one approach was more of an EI measure, while the other relied more on CAC. For example, to reduce point-source industrial water pollution, the Netherlands implemented a system of fees for organic pollutants (EI), while the United States established a system of guidelines and permits (CAC). It turned out, in fact, that most policies had at least some elements of both approaches, but we categorized them as EI or CAC based on their dominant features.

Originally published in *Resources*, No. 152, Fall/Winter 2004.

We then asked researchers who had previously studied these policies on either side of the Atlantic to update or prepare new case studies. We analyzed the 12 case studies (two for each of the six environmental problems) against a list of hypotheses frequently made for or against EI and CAC, such as which instrument is more effective or imposes less administrative burden.

The Evolution of CAC and EI

Only recently has it been possible to find enough EI policies to carry out a project such as this. Until about 15 years ago the environmental policies actually chosen were heavily dominated by CAC approaches. In the United States, the 1970s saw a great volume of new federal regulation to promote environmental quality, none of which could be characterized as relying heavily upon economic incentives. Since then, however, there has been a remarkable surge of interest in EI approaches in environmental policy. Since the late 1980s, whenever new environmental policies are proposed, it is almost inevitable that economic incentive instruments will be considered and will receive a respectful hearing.

The reasons for the newfound popularity of EI policies are unclear. Perhaps it is due to the growth in awareness of economic incentive approaches among policymakers and policy analysts the 20 or so years between 1970 and 1990. In the 1970s these approaches were generally unfamiliar to those outside the economics profession. Another possibility is the emergence of tradable emission permits in the late 1970s. Before then, the main EI alternative to the regulatory policies being implemented was a per-unit tax on pollution (sometimes referred to as an effluent fee). By the 1980s the policy community was generally aware of a "quantity-based" EI alternative – tradable emission permits – that seemed to provide the same assurances of the achievement of environmental goals that were offered by CAC approaches.

A third possible cause is the widespread disappointment with outcomes of the CAC regulations adopted in the 1970s. The nearly limitless variety of American industries and industrial processes required the EPA to write very detailed and complex regulations, but despite these efforts, the Agency faced a raft of legal challenges. Regulatory complexity combined with litigiousness delayed the implementation of most regulations far beyond the schedules envisioned by Congress. In other words, much of the enthusiasm for EI could be attributed to disenchantment with CAC.

The Two Sides of the Pond

Initially, it is worth underscoring some differences between the United States and Europe that serve as a backdrop to policy decisions and implementation.

First, of course, we are comparing a single federal system in the United States with the many countries of the European Union (EU). Beginning in the late 1960s, environmental policymaking became centralized in the United States. In Europe, each country has adopted policies according to its own timetable, generally beginning in the late 1960s in the wealthiest nations and sweeping south and east to the former Soviet empire by the 1990s. Environmental policy in Europe is now a mix of country-specific and EU-wide measures, which these cases reflect.

Second, there are major differences between the United States and Europe in the extent of pre-regulatory studies undertaken. Because of the U.S. requirement on the Environmental Protection Agency and others to conduct a Regulatory Impact Analysis before taking action, substantially more information was available about the hoped-for benefits of U.S. policies. A further issue concerns the greater reliance on taxes for regulatory purposes in Europe compared to the United States. A number of European nations use such taxes—sometimes combined with incentive compatible rebate schemes—to achieve environmental objectives. In the United States, environmental taxes are virtually nonexistent.

Overall, however, and despite these various differences in approaches, we were not able to discern clear differences in regulatory outcomes across the Atlantic: in some cases one or more European

The Six Environmental Problems We Studied

In our analysis, we selected six environmental problems to serve as a "control" in order to compare EI and CAC approaches in trying to solve each one, which are summarized below. We paired a policy from the U.S. with one implemented in one or more European countries.

For clarity's sake, although almost all contain some blend of EI and CAC elements, those that are more closely associated with EI instruments are listed first:

1. SO_2 emissions from utility boilers: Permit market (U.S.) vs. sulfur emission standards (Germany)

2. NO_x emissions from utility boilers: Emission taxes (Sweden and France) vs. NO_x New Source Performance Standards (U.S.)

3. Industrial water pollution: Effluent fees (Netherlands) vs. Effluent Guidelines and National Pollutant Discharge Elimination System permits (U.S.)

4. Leaded gasoline: Marketable permits for leaded fuel production (U.S.) vs. mandatory lead phase-outs plus differential taxes to prevent misfueling (most European countries)

5. Chlorofluorocarbons (CFC): Permit market (U.S.) vs. mandatory phase-outs (other industrial countries)

6. Chlorinated solvents: Source regulation (U.S.) vs. three distinct policy approaches (Germany, Sweden, Norway)

---

nations acted sooner or more aggressively to address environmental problems while in other cases the United States acted sooner or more aggressively.

Testing the Hypotheses

Since the 1970s, when western countries began forming comprehensive environmental policies, there has been a good deal of speculation and disputation over the differences between EI and CAC instruments in practice. These discussions boil down to assertions or hypotheses about comparative advantages of each instrument. We compiled a list of the 12 most commonly stated hypotheses, recognizing that different observers might develop very different lists.

Below we discuss the five hypotheses that we consider most important in evaluating a policy instrument. For each, we state the hypothesis, review the original rationale in making it, and test whether the hypothesis holds up in light of one or more of our case studies.

EI instruments are more efficient than CAC instruments: that is, they result in a lower unit cost of abatement.

RATIONALE: It is commonly believed that EI instruments have an efficiency advantage over CAC instruments, although the case is not airtight. EI instruments are more cost-effective at achieving a given emission reduction. But to get from cost-effectiveness to efficiency requires additional assumptions, including that the system is one of perfect competition and that the emissions are not location-specific. A theoretical counter to this hypothesis is that a CAC instrument can be as efficient if the emission standard for each plant is chosen so that the marginal costs of abatement equal the marginal social costs of pollutant damage.

PERFORMANCE: The cases we analyzed show that EI is generally more efficient. For example, in looking at the U.S. program of marketable permits to lower SO_2 emissions, realized costs are only about one-half what was expected back in 1990 and about one-quarter of the estimated cost of various CAC standards. EI also achieved substantial cost savings in the elimination of CFCs and lead in gasoline, in part because of cost heterogeneities that could be exploited during the phase-down period. However, in instances where the regulations are so stringent that practically all-available abatement measures must be taken, there is little scope for choosing the most cost-effective ones, and EI instruments do not achieve significant cost savings over CAC. EI also enjoys little advantage if all plants face similar abatement costs. Both these conditions limited, for example, the efficiency losses of using CAC to reduce German SO_2 emissions.

The real advantages of EI instruments are generally realized over time, because they provide a continual incentive to reduce emissions, thus promoting new technology, and permit maximum flexibility in achieving emission reductions.

RATIONALE: The effects of CAC on technology are potentially complex. On the one hand, costly regulations provide a spur to find less costly ways of compliance. On the other, the requirement to install a specific technology conceivably discourages research, since discovering new ways to reduce emissions can lead to more stringent regulations. More stringent performance standards for new plants have the stated objective of promoting technology, but they can also have the pernicious effect of postponing retirements of older, dirtier plants and discouraging entry by outside firms.

PERFORMANCE: EI provides greater incentives than CAC for continuing innovation over time in most, but not all, cases studied. For example, the Swedish NO_x tax induced experimentation in boiler operations that led to substantial reductions in NO_x emissions. Because NO_x emissions from boilers are idiosyncratic, it was unknown beforehand what would work in each boiler. Therefore, achieving these reductions from CAC would not have been possible. Similarly, the U.S. SO_2 trading policy induced many nonpatentable boiler-specific innovations on utility boilers. Elsewhere, the Netherlands became a world leader in water purification technologies and its industries adopted more advanced, process-integrated measures to reduce pollutants.

Innovation also occurs under CAC, but the results are often different. For example, the lead phase down induced emissions reductions in all plants during the period when a CAC policy was employed, but when the policy allowed permit trading and banking, the reductions were concentrated in newer plants with longer expected lifetimes, where the improvements were most cost-effective. In the U.S's SO_2 policy, examination of patents suggests that in a CAC regime only cost-reducing innovations are encouraged, while under EI both cost-reducing and emission-reducing innovations are encouraged.

CAC policies achieve their objectives quicker and with greater certainty than EI policies.

RATIONALE: In the early 1970s, CAC was seen as the way to expedite compliance, even if the approach was not the least costly. It appeared then that EI instruments, particularly emission fees, would not achieve the same objectives.

PERFORMANCE: The evidence from the cases is mixed. Supporting the relative effectiveness of CAC is the U.S. effort to phase out the solvent trichloroethylene (TCE), in which EPA ultimately mandated limits. The EI aspects of the rule did not attract significant industry participation. In phasing out leaded gasoline in Europe, progress would have significantly slowed without mandating catalytic converters and maximum lead content in addition to tax differentials.

At the same time, several cases argue that EI policies are more effective. In the Dutch water case, for example, the influence of effluent fees on organic waste load reductions was prompt and large. Similarly, by eliciting industry cooperation, the trading and banking program probably achieved a much more rapid phase-down of lead in gasoline than would have been possible with a CAC program that industry would have opposed.

A final point on effectiveness is that two cases show that both approaches can result in significant environmental gains, but without careful design, can yield undesirable longer-term side effects. In the United States, NO_x emissions from coal-fired power plants were reduced, but the standards, which only affected new plants, caused firms to extend the life of older, more polluting plants to avoid the costs associated with newer ones. In Sweden, TCE users persuaded the public and authorities that complete implementation of a ban would cause them undue harm. They received numerous waivers and exceptions, thus undermining the authority of the environmental agency and perhaps emboldening other firms to oppose other regulations.

Regulated firms are more likely to oppose EI regulations than CAC because they fear they will face higher costs, despite the greater efficiency of EI instruments.

RATIONALE: Although EI instruments may have lower social costs overall, firms pay higher costs under EI than CAC. Under CAC, the argument goes, the polluting firm pays to abate pollution; under many EI instruments, the firm pays the cost of abatement plus a charge for the remaining pollution it discharges. The firm is better off only if the abatement cost is lower by an amount at least as great as the charge for the remaining emissions.

PERFORMANCE: Experience on both sides of the Atlantic suggests that no government has put this hypothesis to the test, which, in a way, is strong support for it. In nearly all cases, governments eliminate the burden of EI instruments by returning the revenues to the firms. For example, in France, revenues collected through NO_x discharge fees subsidized the firms' abatement investments, while in Sweden the charges were returned to the firms on the basis of the energy they produced. In the United States, where the EI instrument of choice is a tradable permit, the permits have always been given away rather than auctioned off.

CAC policies have higher administrative costs.

RATIONALE: Administrative costs are determined by the amount of interaction between the regulator and regulated source. Supporters of this hypothesis note that the complexity of setting and enforcing specific requirements is higher than implementing fee-based EI policies. In addition, fees for increased emissions tend to rise gradually, whereas with CAC, a line separates compliance from violation. The potentially high incremental cost at the point of violation gives regulated sources an incentive to defend themselves legally rather than accept sanctions, thus adding to the regulators' burden.

PERFORMANCE: The cases show no clear pattern. While the CAC-oriented Effluent Guidelines program in the United States imposed high administrative costs on EPA, so did the EI instruments of the lead phase-down program. Looking at SO_2 reduction, the EI-oriented U.S. trading program gained a reputation for low administrative costs, but the SO_2 reduction program in Germany does not show evidence of higher administrative costs than a comparable EI program. Overall, because the evidence on this hypothesis is mixed, we could not form a firm conclusion about whether policy outcomes supported or refuted it.

Apples and Oranges?

Questions of effectiveness and efficiency were at the core of the initial selection of policy instruments in the 1970s and 1980s. As these cases show, EI instruments appear to produce cost savings in pollution abatement, as well as innovations that reduce the overall cost. The concern that EI instruments are not as effective is not borne out in our analysis. However, the finding about EI's economic efficiency is tempered by evidence that polluting firms prefer a CAC instrument because of its perceived lower costs to them. In all but one of the case studies, the actual or potential revenues raised by EI instruments had to be reimbursed in some way to the firms. This, of course, means the revenues cannot be used for other purposes.

In the 1970s, almost all environmental policies relied on direct regulation, with very rare instances of EI instruments. Since the late 1980s, whenever a new policy is proposed, policymakers at least consider, and often select, an EI instrument. That said, almost all the policies that we studied are a blend of both, beginning as a CAC policy and then having EI elements added or substituted. In the 12 cases we studied, in fact, only a few (reduction of SO_2 emissions in Germany; TCE in Germany and Sweden) had no EI elements in their design. Moreover, we can report significant environmental results from the cases we studied. Averaged across all twelve, emissions fell by about two-thirds when compared to baseline estimates.

Most outcomes either met or exceeded policymakers' original expectations. This is encouraging news for those seeking environmental improvements in the future.

Suggested Reading

Harrington, W., R. Morgenstern, and T. Sterner (eds.). 2004. *Choosing Environmental Policy: Comparing Instruments and Outcomes in the United States and Europe*. Washington, DC: Resources for the Future.

12 Unleashing the Clean Water Act
The Promise and Challenge of the TMDL Approach to Water Quality

Jim Boyd

The Total Maximum Daily Load (TMDL) provisions of the Clean Air Act have taken on a new life that provides an ambient approach to water quality management instead of the earlier technology-based, end-of-pipe, control of point sources. The challenges of the approach are formidable, but it focuses efforts on what we all ultimately care about: the cleanliness of the nation's bodies of water.

The environmental movement's greatest intellectual triumph is the now-common understanding that environmental conditions are the end-product of complex interactions between a variety of physical, biological, and social systems. Environmental policy itself is growing toward a more holistic, and complex, approach to the diagnosis and resolution of environmental problems; however, this growth will not come without difficulty.

Nowhere are the promise and challenge of holistic policy-making better illustrated than in changing approaches to water quality regulation. Last August, the U.S. Environmental Protection Agency (EPA) proposed new rules to invigorate the largely dormant Total Maximum Daily Load (TMDL) provisions of the Clean Water Act (CWA). The final rule, due this summer, will have immediate implications for water quality monitoring and analysis.

Over the next two decades, the rules will put in motion significant, state-led changes in the regulation of pollutant sources. Instead of the technology-based, end-of-pipe approach to controlling point sources that has characterized water quality enforcement to date, the TMDL program promises an "ambient" approach to water monitoring and standards. Rather than focus on releases from known sources of pollution, regulations will increasingly address the overall quality of waterbodies. In a nutshell, the TMDL approach is to monitor lake, river, and estuarine water quality; identify the nature and location of polluted waters; trace pollutants to their sources; and impose controls adequate to guarantee the health of various waterbodies.

Originally published in *Resources*, No. 139, Spring 2000.

Implicit in the TMDL approach is a focus on the causes and effects of pollution throughout a watershed. More explicitly, the TMDL program will seek the identification of any and all sources of pollution. It also will focus on what we all ultimately care about the most—the cleanliness of our waterbodies. It all sounds sensible and straightforward, but in fact it is a radical, untried departure from current practice.

CWA regulation over the last 25 years has yielded significant water quality improvements. Nevertheless, the current approach is somewhat limited due to its focus on point sources, the most easily identifiable and rectifiable pollution sources. Point sources are typically large factories or municipal sewage treatment plants. The fact that they were responsible for a significant fraction of water quality problems in the past and were easy to identify justified this narrow approach. But the low-hanging fruit of low-cost, high-volume point source controls has been harvested. Today, significant water quality improvement requires the expansion of controls to nonpoint sources.

While industrial and municipal point sources will no doubt continue to be vivid symbols of the nation's water pollution problems, this image is increasingly inappropriate. Water pollution from agricultural, commercial, and urban sources—called nonpoint pollution—while harder to caricature, should be the focus of our dissatisfaction. Hundreds of thousands of river miles and millions of lake acres remain impaired on account of it. Because nonpoint sources are a primary cause of those impairments, TMDLs will change the politics, economics, and implementation of water quality regulation. Along the way, water quality will increasingly be seen as interdependent with other spheres of concern, notably air quality and land use programs. Moreover, the tools required to understand the fate of pollutants, assign responsibility, and monitor compliance within watersheds will tax regulators' technical and financial resources. Such is the price of holistic policymaking.

The Nonpoint Source Challenge

Water quality improvements over the last 25 years have resulted primarily from point source controls. Future improvements must come principally from nonpoint source controls. Today, agricultural runoff, in the form of pesticides, fertilizer, and animal wastes, is the single largest contributor to the impairment of rivers and lakes. Logging and construction activities, many of them on federal lands, are a significant source of sediment contamination, as runoff carries fine-grained soils from roads and construction sites into lakes and streams.

In urban and suburban areas, watershed degradation is closely tied to increased population density and residential and commercial development. In such areas, the relatively impermeable nature of the groundcover leads to rapid, unfiltered runoff to rivers, lakes and oceans from roadways and parking lots, chemically treated lawns, and commercial establishments. Increased attention is also being given to atmospheric deposition, where pollutants from airborne dust and industrial and commercial air emissions are absorbed by surface waters or precipitated via rainfall.

One of the reasons why nonpoint sources are such a significant problem is that they present serious implementation, monitoring, and enforcement challenges. Nevertheless, the water quality problems they cause can no longer be ignored. In this context, it is not surprising that political and legal pressures are being applied to the EPA, and in turn to the states, to make something of the regulatory potential contained in the CWA's TMDL provisions.

The Changing Politics of Water Quality

The seeds of this shift in regulatory emphasis have been in place since the CWA's passage in 1972. The act contains provisions that call for enforcement to be driven by ambient water quality rather than end-of-pipe controls and for states to identify waters for which the point source controls elsewhere in the act "are not stringent enough to implement any water quality standard applicable to such waters." States must prioritize any waters so identified, based on analysis of use and severity of degradation, and establish total maximum daily pollutant loads sufficient to bring the waters into compliance.

During the first two decades of CWA enforcement, the states as well as the federal government largely ignored the TMDL provisions. But the failure of most states to attain water quality goals and the federal government's desire to bring more sources into the regulated sphere has led to a reexamination of latent enforcement power in the CWA. The TMDL provisions are important because they require statewide assessments and public documentation of water quality problems, and they appear, at least in principle, to imply that states must allocate pollutant load reductions to sources not currently covered by load restrictions. Since the bulk of current impairment is caused by nonpoint sources, any state seeking further load reductions (at least on a cost-benefit basis) will be led directly to nonpoint sources. In this way, the shift to ambient monitoring and standards almost necessarily leads to a greater emphasis on nonpoint sources.

EPA's authority to implement the new TMDL rules is being challenged by a variety of agricultural interests on the grounds that authorizing legislation is required, given ambiguities in the CWA. The agency's opponents argue that the CWA covers only waters impaired by point sources, thus leaving EPA without authority to promulgate such rules. The agency contends that the CWA provides it with ample authority to step in and issue nonpoint controls if there is evidence of ongoing impairments and inadequate state responses to them. These issues must be resolved in the coming years. The scope of federal authority will be challenged, as will state efforts to assign responsibility for load reductions under their own statutes. (In fact, the legal scrum is already well underway. A recent RFF report, "The New Face of the Clean Water Act: A Critical Review of the EPA's Proposed TMDL Rules," explores these issues in more detail.) But despite bumps in the road, movement toward a system of regulation that addresses nonpoint sources and that views water quality as a watershedwide issue is inevitable.

Any enforcement of the CWA's TMDL provisions will alter the politics of load reduction. The need to meet *in situ* water quality standards sets up a state-by-state confrontation between well-organized industrial interests (which can claim to have already paid their pollution control dues) and organized agricultural, commercial forestry, and municipal interests who resist the "expansion" of CWA-driven requirements to their hard-to-solve nonpoint problems. Absent nonpoint controls, point sources can reasonably expect to be held responsible for load reductions on TMDL-impaired waterbodies. This scenario is obviously of great concern to current point source permit holders.

Technical Hurdles Posed by Watershed-Level Regulation

A striking feature of this political and legislative history is that we have been down this road before. Statutory approaches that predate the CWA, such as the Water Quality Act of 1965, also called for ambient water quality standards and state-driven implementation plans (two fundamental features of the TMDL approach). The failure of these earlier approaches to water quality regulation is a cautionary tale. Water quality-driven standards and controls present a variety of daunting challenges. Looking to the future, these challenges loom large.

The first step in the TMDL process is the listing of a waterbody as impaired. Impairment is established in reference to criteria set by the states; the criteria describe the standards, data to be used, and relevant guidelines necessary to ensure the quality of data analysis. These monitoring, classification, and notification requirements are the first administrative and technical challenge for states. While this is the least taxing of the exercises set in motion by the TMDL process, it is worth noting that many states have had difficulty in meeting even these preliminary requirements.

With knowledge of impairments, states must put forward defensible plans for source reductions to bring the affected waterbody into attainment. This kind of exercise is fraught with technical diffi-

Update

In 2000, EPA released a proposed rule governing state development of TMDL plans for impaired waters. That draft rule provided the impetus for this article and the issues it discusses. However, the rule was subsequently withdrawn and a replacement rule has not been issued. Nonetheless, the ambient-focused TMDL program continues to grow into the nation's principle water quality management program. EPA has issued new guidelines on matters such as monitoring and assessment of water conditions and the essential elements of an acceptable TMDL plan. In response, the states have moved aggressively to list waters as impaired and to develop TMDL load limits as required under the Clean Water Act and by their own state legislation on water quality management. These federal and state actions have, at times, been motivated by court rulings on governmental responsibilities for implementing the TMDL program. Thus, the TMDL program continues to grow, even without the rule that motivated this article, and the issues and challenges that confront these efforts are as applicable to the current situation as they were at the time this article was written.

culties. Analysis of loadings and the effect of load reductions requires some form of watershedwide modeling that captures transport processes (such as infiltration and runoff), groundwater and surface water interactions, pollutant accumulation and decay, and in-stream mixing. In the case of nonpoint source loads, the science is relatively undeveloped due to the complexity of the interacting systems involved. Knowledge of the relationship between control practices and loadings is particularly poor.

Because of the wide range of pollutant sources, pathways, and factors that affect loadings, source contributions will rarely if ever be known with certainty. Instead, the regulator must rely on models that attempt to capture the factors that affect the transport, deposition, and ultimate fate of pollutants in the waterbody. Models will also be required to predict how changes in land use brought on by economic growth will add to future loadings.

The modeling techniques and data required for TMDL implementation will contribute significantly to the costs of implementation. Some simplicity and cost savings will undoubtedly be possible as states become more practiced in TMDL development (and as more resources are devoted to the development of data and models for use in this kind of program.) However, the degree to which data sources and modeling techniques can be standardized is limited. Each listed water segment is, in some sense, unique because of its hydrology, transport pathways, pollutant sources, and so forth. TMDL development will invariably involve site-specific analysis.

A lack of scientific certainty will not by itself legally hobble TMDL plans, since certainty is not a prerequisite for program implementation. Uncertainty does place a premium, however, on administrative procedures that provide the greatest possible level of scientific credibility to standards, models, and data collection. Pollutant sources, unhappy with their designation, will undoubtedly seek relief from TMDL controls by challenging a state's modeling tools, water quality criteria, and data collection procedures. Accordingly, the technical details of state TMDL programs should be subject to ongoing notice and comment procedures and evaluation by expert panels. This is likely to be a source of both significant up-front and long-run program costs. Credibility, transparency, and enforceability are particularly paramount if flexible environmental controls, such as effluent trading, are to be realistically contemplated.

The Scope of Interactions

The TMDL program's ambitions are all the more notable when their interactions with other areas of environmental law and regulation are considered. Consider the potential impact of TMDLs on air quality regulation. Air deposition is a major source of water pollution; a prime example is nitrogen oxides deposition to the Chesapeake Bay. Air deposition links water quality in one state with air emissions in another. While the implications of this linkage have not been fully contemplated, it does create the distinct possibility of jurisdictional conflict both across state borders and within EPA program offices. (See *Resources* 137, "A Dilemma Downwind" for more on the inter-jurisdictional implications of clean air policy.)

In addition, the TMDL rules will increasingly highlight the artificial distinction between water quality and quantity issues, particularly in the West. Water quantity decisions, which are controlled primarily by state law, often have a direct impact on water quality: changes in stream flow affect the transport of pollutants; the amount of water taken or returned to a waterbody may significantly affect the dilution of pollutants; and water supply often determines the suitability of a waterbody as habitat for fish or other species. Because of these interactions, water quantity decisions (relating to irrigation, damming, reservoir management, basin-to-basin trades, and the like) may affect a water's TMDL status. Accordingly, TMDLs will in some cases constrain water transfers involving impaired waterbodies.

Despite the challenges it presents, the TMDL approach clearly demonstrates movement toward a welcome, mature phase of water quality regulation. The key feature of EPA's proposed TMDL rules is that they are motivated by, and address, water quality issues created by the widest range of sources. The holistic, watershed-level analysis required by the TMDL process will inevitably identify a larger sphere of often-unregulated discharge sources. For these reasons alone, the TMDL program is likely to promote significant, desirable changes in the implementation of water quality regulation.

13

Penny-Wise and Pound-Fuelish?
New Car Mileage Standards in the United States

Paul R. Portney

Corporate Average Fuel Economy (CAFE) Standards are clearly inferior in economic terms to the use of a hefty tax on gasoline to induce greater fuel economy. But in the absence of such taxes, the case against CAFE standards is less compelling. Nevertheless, there are some modifications that would make the program more flexible and efficient.

Our seemingly endless debate about energy policy in the United States has been especially sharp since May 2001 when the Bush administration announced its new national energy policy. If anything, that debate has been much sharper still since the terrorist attacks of September 11, 2001, reminded us of the perils of using as much oil as we do in the United States.

Most of us remember the tiresome beer commercial in which seemingly normal people debated whether a particular brand was better because it "tasted great" or was "less filling." At the risk of only some exaggeration, we have our own version of this debate over domestic energy policy, with half the protagonists shouting "produce more" while their opponents shout "use less." The former look especially fondly at the Arctic National Wildlife Refuge (ANWR) as a possible source of additional oil, while the latter focus on improved fuel economy standards for new cars as the way to slake America's unquenchable thirst for petroleum. Both sides suffered at least temporary losses when the Senate—in the space of a few short days—recently rejected efforts to open ANWR for oil exploration and to tighten the Corporate Average Fuel Economy (or CAFE) standards for all new light-duty vehicles produced in the United States.

Last year, I had the pleasure of chairing a committee assembled by the National Research Council (the study arm of the National Academy of Sciences) to examine the past and possible future effects of the CAFE standards (hereafter referred to as the CAFE Committee). The committee's final report, Effectiveness and Impact of Corporate Average Fuel Economy (CAFE)

Originally published in *Resources*, No. 147, Spring 2002.

Standards, was published last summer. Accordingly, I watched the debate over fuel economy standards quite closely. Here I report on the findings of that study and offer some suggestions on the way readers might think about the CAFE program. Let's begin with a bit of history regarding the fuel economy standards and what we know (and don't know) about their early effects.

Looking Back

Because of several disruptions in world oil markets during the 1970s, the price of oil went from less than $20/barrel in 1970 to more than $80/barrel in 1981 (converted to year 2000 dollars). Even before the end of that decade, Congress passed legislation requiring all new passenger cars and light-duty trucks (in other words, pickup trucks, minivans, and the now-ubiquitous sport utility vehicles, or SUVs) to meet federal mileage standards. Cars were required by Congress directly to meet a fleet average of 27.5 miles per gallon (mpg) by 1985, and the National Highway Traffic Safety Administration mandated that light-duty truck fleets were to average no less than 20.7 mpg. Since new cars were averaging only about 16 mpg in 1977, the year before the CAFE requirements begin to ramp up, and new trucks about 13 mpg, these required increases were quite significant.

What effects did the new standards have? Perhaps surprisingly, this is a harder question to answer than one might think. The principal confounding factor is that the price of gasoline had been going up since well before the CAFE standards were established. This created a strong demand on the part of new car buyers for more fuel-efficient cars, as well as an incentive for automakers to produce them. The CAFE Committee found that these two forces working together—higher gasoline prices and federally mandated fuel economy standards—resulted in a greater than 50% improvement in new car and light-duty truck fuel economy between 1978 and 1985. As a result, the country enjoyed significant reductions in oil consumption and also emissions of carbon dioxide, a greenhouse gas.

Update

Since 2002, there has been no movement whatsoever with respect to the fuel economy standards pertaining to passenger cars, despite the fact that Congress enacted major energy legislation in 2005. This means that the standard to be met by passenger cars is unchanged since the late 1970s.

However, in 2003, and again in 2005, new, tighter fuel economy standards were proposed for Light Duty Trucks (a category encompassing minivans, pickup trucks, and Sport Utility vehicles). If the more stringent standards for LDTs proposed in August 2005 go into effect by 2010, as expected, by that time the LDT fleet should be averaging 23.5 miles to the gallon—a 14 percent improvement over the 20.7 mpg standard that existed in 2003. The new regulations proposed in 2005 also established a new and more complicated system in which the LDT fleet was divided into six different segments, based on the size of the vehicles. The larger the vehicle class, the less stringent the fuel economy standard that size class has to meet.

While these standards are a step in the right direction, the tighter standards that will apply to LDTs in the future fall far short of the improvements recommended by the 2001 National Academy of Sciences Committee on Corporate Average Fuel Economy.

In fact, the CAFE Committee estimated that by the year 2000, improved fuel economy was reducing oil consumption by 2.8 million barrels per day (or about 14% of the current total) and reducing annual emissions of carbon in the United States by about 100 million metric tons (or 6% of current annual emissions). The committee could not determine how much of these improvements were due to the price effect (which subsided rather dramatically beginning in 1981 when oil prices began their fall back to about $20/barrel in year 2000 prices) and how much was due to the effects of the CAFE standards. Since 1981, it is highly likely that fuel economy remained where it did solely because of the federal standards.

There is another, less happy consequence to the rapid improvement in fuel economy between 1978 and 1985, however. Because automakers were being forced both by consumer demand for more fuel-efficient cars (for a time, at least) and by government regulations, they had little choice as to the way they could improve fuel economy so rapidly. The result was an almost decade-long cohort of new cars and light-duty trucks that were smaller and lighter than their predecessors. According to all but two dissenting members of the CAFE Committee, the rapid downsizing and "downweighting" of new vehicles that began in 1978 was responsible by 1993 for about 2,000 more fatalities annually than would have been observed had vehicles remained as large as they were prior to 1978. As we shall soon see, this does not necessarily mean that further enhanced fuel economy must come at the cost of highway safety, but the rapid improvements of the late 1970s and early 1980s most likely did.

Looking Ahead

Given the improvements of the past, why the continuing concern about future fuel economy? Despite the fact that both passenger car and light-duty truck fleets continue to meet their respective standards, the average fuel economy of the combined new car fleet has declined about 8% since 1986. "How can this be?" you might reasonably ask. Actually, the answer is quite simple, as Figure 1 illustrates. In 1975, when the law establishing the CAFE program was passed, light-duty trucks (once again, this category comprises pickups, minivans, and SUVs) accounted for about 2 million of the 10 million total vehicles sold that year in the United States. By 2001, however, light-duty truck sales accounted for 51% of the 17 million-plus new vehicles sold. Since these light-duty trucks are only required to average 20.7 mpg, as opposed to 27.5 mpg for passenger cars, their growing share of all

Figure 1. Historic vehicle sales

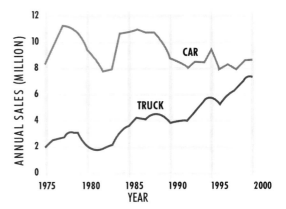

Note: Trucks include vehicles under 8,500 pound gross vehicle weight (GVW) that are not classified as passenger cars.

Source: Friedman et al. (2001)

new-vehicle sales is gradually pulling down the combined new vehicle fuel economy average. Along with robust growth in the number of miles that all cars are being driven, this shift in the new-car mix is a major reason why oil use and imports are growing steadily.

What can and should be done about this? The former is much easier to answer than the latter. Regarding possible future fuel economy improvements, the CAFE Committee thoroughly investigated the technological potential for short-, medium- and long-term gains. According to the committee report, "Technologies exist that, if applied to passenger cars and light-duty trucks, would significantly reduce fuel consumption within 15 years."

I cannot even begin here to identify all the technologies the committee considered, but they include mostly things that are already in limited use for some parts of the new vehicle fleet rather than technologies for which dramatic breakthroughs are required. Examples are such things as variable valve timing, intake valve throttling, variable-compression ratio engines, continuously variable transmissions, friction reductions, 42-volt electrical

Table 1.

Vehicle Class	Base mpg	Enhanced mpg (% Improvement)	Purchase Price Increase ($)	Lifetime Fuel Savings ($)
CARS				
Subcompact	31.3	35.1 (12)	502	694
Compact	30.1	34.3 (14)	561	788
Midsize	27.1	32.6 (20)	791	1,140
Large	24.8	31.4 (27)	985	1,494
LIGHT TRUCKS				
Small SUVs	24.1	30.0 (25)	959	1,460
Mid SUVs	21.0	28.0 (33)	1,254	2,057
Large SUVs	17.2	24.5 (42)	1,629	2,910
Mini Vans	23.0	29.7 (29)	1,079	1,703
Small Pickups	23.2	29.9 (29)	1,067	1,688
Large Pickups	18.5	25.5 (38)	1,450	2,531

systems, and reduced aerodynamic drag and rolling resistance.

So what if the technological potential exists for fuel economy improvements? It is almost always possible to do better technologically than we are currently doing—whether from an automotive, computing, medical, or agricultural standpoint. The really important questions are how much will these improvements cost and what benefits will we derive from them?

The committee provided at least some information along these lines. Beginning with technologies that could improve fuel economy rather inexpensively, and moving successively to those that could do so but at greater expense, the committee first sketched out what economists would recognize as a marginal cost curve for fuel economy improvement. This was done on the assumption that the automakers would have at least 10 and as many as 15 years to make these changes—an extraordinarily important assumption, as we shall later see. By combining these cost estimates with estimates of the discounted value of the fuel that would be saved, the committee summarized part of its work in a table like that in Table 1.

As the table indicates, through the application of the technologies the committee identified, it would be possible in 10 to 15 years to improve the fuel economy of a mid-sized passenger car (for example, a Buick Regal, C-class Mercedes, or Honda Accord) from the current mpg average of 27.1 to 32.6 (a gain of 20%). This would add an estimated $791 to the purchase price of the car but would be more than offset by the $1,140 in discounted (at 12%) fuel economy savings over the assumed 14-year life of the car. Additional fuel-saving technologies could be applied, but according to the committee these technologies would add more to the purchase price of the car than they would save in discounted fuel costs. The larger the car, the greater the savings: the fuel economy of a mid-size SUV (such as a Ford Explorer or a Toyota Highlander) could be improved from its current 21 mpg to 28 mpg (33%). This would add $1,254 to its purchase price but would result in more than $2,000 in discounted fuel savings over its lifetime.

One question immediately arises: would these estimated improvements in fuel economy adversely affect safety? No, according to the committee. In fact, the committee found that neither passenger safety nor vehicle performance (acceleration and towing capacity, for example) would suffer when measured against today's standards so long as the technologies the committee identified were introduced throughout the fleets. The committee even figured into its calculations a slight increase in the weight of vehicles because of safety requirements likely to be imposed over the next 15 years. (It is possible, even likely, however, that performance would suffer in comparison to what it might be in 10 to 15 years were automakers not required to improve fuel economy.)

Thinking More Deeply

Does all this mean that it's a good idea to impose more stringent fuel economy standards on automakers? Possibly, but not necessarily. First, one could argue, most people already know full well they could get better fuel economy by purchasing a different car. After all, no one buys a large SUV thinking it will stretch his or her gasoline dollar. Rather, at gasoline prices that typically range between $1.25 and $1.75 per gallon, there simply isn't very great demand among the American public for "fuel-sippers." Although I take strong issue with several of the arguments put forward by automakers during the recent Senate debate on CAFE, they are dead right on at least one count. CAFÉ standards require them to produce more fuel-efficient cars than large segments of the public appear to want—at least at current gasoline prices.

Second, if the government does require better new-car fuel economy, or if automakers provide it voluntarily, then the cost of driving a given distance falls (you'll use less gas per mile driven). This means the number of miles traveled will increase—about 1 to 2% for each 10% reduction in the cost of driving, according to research. This "rebound" effect—and its possible contribution to air pollution, increased congestion, and accident risks—has to be factored into CAFÉ policymaking.

Third, if people are much more sensitive to the upfront cost of buying a new car than to the fuel savings they will enjoy over its life, tighter CAFE standards could slow down the retirement of older vehicles on the road. ("We can't afford a new car, so we'd better keep ol' Bessie for a while.") We have observed this effect (called "new source bias") in decisions regarding the construction of new coal-fired power plants, certainly (see the article by Gruenspecht and Stavins in this issue), and it could keep gas-guzzlers on the road longer than we expect.

Fourth and finally, suppose CAFE standards are made more stringent. Although the CAFE Committee argued that this need not adversely affect safety or performance so long as automakers adopt the technologies identified by the committee, there certainly would be no requirement that they do so. If they chose to meet tighter standards by, once again, making cars smaller and lighter, drivers and passengers could be put at greater accident risk. Of course, consumer insistence on vehicle safety could force automakers down the technological route to enhanced fuel economy.

Given these possible shortcomings, CAFE standards must be weighed against the benefits of improving fuel economy. It is clearly worth something to reduce emissions of carbon dioxide and there are benefits as well to lessening our dependence on oil and, hence, our vulnerability to oil price shocks.

Suppose that a ton of carbon reduced is valued at $50, the figure used by the CAFE Committee (admittedly at the high end of the current range of estimated benefits of carbon abatement). Suppose further that the external benefits of each barrel of reduced oil consumption are valued at $5 (again, at the high end of estimated values). Together, these are equivalent to a $0.25 premium on the price of a gallon of gasoline. For this premium to be larger, either additional benefits of fuel economy improvements have to be identified or larger values justified for carbon reduction and/or oil consumption reductions.

A Bottom Line

By far, the hardest question for any policy analyst to answer is this one: What would you do if the decision were yours to make? First, recognize that CAFE standards are distinctly inferior to higher gasoline taxes (and thus prices) as a way of dealing with both climate change and oil market externalities, a key finding in the CAFE Committee report. Higher gasoline prices would motivate new car buyers to demand better fuel economy; accordingly, automakers would be more willing to produce such vehicles since the demand would be there. Much more importantly, higher gasoline prices would also create an incentive for those driving the 200 million plus vehicles already on the road in the United States to drive less, carpool (or take public transport) more, and keep their cars in better tune. By working only on the new-car margin, CAFE is an incredibly slow way to deal with climate change and oil consumption. Thus, in my world of worlds, I would gradually increase gasoline taxes (along with taxes on all other carbon-based fuels), while rebating the tax revenues to the public by reducing other taxes so as not to exert drag on the economy.

But what if our elected officials continue to lack the wisdom or, more likely, the will to increase the taxes on gasoline and other carbon-based fuels? Is the CAFE program an acceptable, second-best alternative? Yes, I reluctantly conclude, but only if it is modified in ways the committee recommended.

I would support gradual increases in the required fuel economy targets automakers face, beginning in model year 2007 and extending through 2017. By that time, the passenger car fleet ought to be averaging 35 mpg and the light-duty truck fleet, 28 mpg. However, manufacturers whose fleets fall short of these targets must be able to purchase fuel economy "credits" from companies whose cars or light-duty trucks exceed the goals.

There is no reason why an automaker wishing to specialize in heavy-duty pickups or large SUVs should have to produce smaller vehicles to offset its fleet impact so long as it can pay another manufacturer to make "gasoline misers." Moreover, if fuel economy improvements are harder to come by technologically than the CAFÉ Committee believed (so that safety might be compromised), the government should offer to sell extra fuel economy credits to automakers at some predetermined price—a "safety valve," if you will, to ensure that the fuel economy program does not become more expensive than it should.

There are no easy calls regarding fuel economy. Now you have mine.

14 Is Gasoline Undertaxed in the United States?

Ian W.H. Parry

Even though there would be substantial environmental and public safety benefits from higher taxes on gasoline, the United States continues to have the lowest gasoline taxes among the industrialized nations for a number of reasons.

Gasoline taxes vary dramatically across different countries. While the United Kingdom has a gasoline tax equivalent to $2.80 per gallon, the highest among industrial countries, the United States has the lowest tax of 40¢ per gallon (18¢ federal tax and on average about 22¢ state tax) (see Figure 1). It is commonly thought that Europeans have a greater tolerance for high fuel taxes than Americans, as they have shorter distances to travel and better access to public transport, although the fuel tax protests in Britain in September 2000 suggested that the political limits of such taxes might have been reached.

A number of arguments are made for implementing high gasoline taxes. By discouraging driving and fuel combustion, gasoline taxes help to reduce local air pollution, carbon dioxide emissions (a greenhouse gas), traffic congestion, traffic accidents, and oil dependency. Taxing gasoline is one way of forcing people to take into account the social costs of these problems when deciding how much, and what type of vehicle, to drive.

Gasoline taxes also provide a source of government revenues. In Britain, gasoline tax revenues are several times highway spending, and the Labour government has argued that if gasoline taxes are reduced, schools and hospitals will have to close. But this argument is somewhat misleading as the revenues could always be made up through other sources, such as income taxes. The real issue is what level of gasoline taxation might be justified when account is taken of the full social costs of driving, and the appropriate balance between gasoline taxes and other taxes in raising revenues for the government.

Originally published in *Resources*, No. 148, Summer 2002.

Environmental Effects

Gasoline combustion causes local air pollution, notably smog and carbon monoxide. This pollution can reduce visibility, but its main harm is to human health. For example, poor air quality can exacerbate respiratory problems and lead to premature mortality. Economists have assessed the damages caused by air pollution using epidemiological evidence on the link between air quality and human health, and studies estimating people's willingness to pay to reduce risks of adverse health effects. Damage estimates have fallen over the last 20 years or so as the vehicle fleet has become cleaner, at least partly in response to emissions- per-mile regulations that are imposed on new vehicles. According to a recent study by Kenneth Small and Camilla Kazimi (University of California–Irvine), pollution damages are around 2¢ per mile (after updating to 2000), or about 40¢ per gallon, though there is still much uncertainty over these estimates.

Economists have also attempted to assess the potential damage from carbon emissions, such as the economic damage to world agriculture from future climate change and the costs of protecting valuable coastal regions against rising sea levels. These estimates are highly speculative; for example, it is difficult to value the ecological impacts of climate change, to allow for the small possibility of catastrophic climate change from instabilities within the climate system, and there is much controversy over the appropriate discount rate to use for converting future damages into current dollars. A typical estimate from the literature is around $25 per ton of carbon, which translates into only 6¢ per gallon, though some studies obtain much higher numbers. These preliminary figures suggest that advocates of higher transportation taxes to reduce carbon emissions may be on weak ground. Their efforts might be better spent focusing on reducing combustion of other fossil fuels, particularly coal.

Traffic Congestion and Accidents

Raising gasoline taxes nationwide is not well suited to addressing traffic congestion, which is specific to

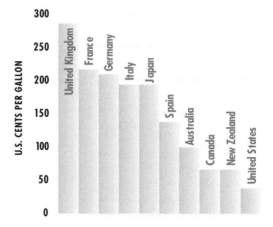

Figure 1. Gasoline excise taxes in different countries

Source: Parry and Small 2002.

certain roads in urban centers at particular times of day. Gasoline taxes do encourage people to use public transportation and to car-pool, but they also penalize driving on uncongested roads, such as in rural areas or urban centers on weekends. A much better way to alleviate congestion on, say, I-66 in the northern Virginia suburbs of metropolitan Washington, DC, would be to charge people to drive on the road at peak periods. (This could be done electronically by deducting from a pre-installed credit card on the windshield). Time-of-day pricing would be more effective at reducing congestion than gasoline taxes as it would encourage people to use busy roads at off-peak periods and look for alternative routes. With regard to accidents, it would be better to tax a driver's annual mileage, taking account of vehicle type and personal characteristics (such as experience and prior crash history). And raising the penalties for drunk drivers would be a more direct way to reduce alcohol-related crashes.

Nonetheless, peak-period fees and mileage related taxes have not been widely implemented in the United States (though there have been limited experiments with congestion pricing in California and Texas). Gasoline taxes might be the next-best response for curbing congestion and accidents, so it is still appropriate to consider congestion and accident benefits in an overall assessment of gasoline taxes. For congestion, this would require estimating marginal congestion costs averaged across both urban and rural roads and peak and off-peak travel periods, for the whole United States. Congestion costs are measured by the extra time it takes to drive under congested conditions compared with free-flowing traffic, multiplied by the monetary value of travel time (usually taken to be about half the market wage). Based on the available evidence, Kenneth Small and I concluded that the best estimate for the "averaged" congestion cost is about 3.5¢ per mile, or 70¢ per gallon.

The cost to society from traffic accidents largely depends on human fatalities and injuries—in the United States around 40,000 people are killed on the roads each year. Other costs include traffic hold-ups and property damage. The costs of fatalities and injuries include not only economic costs (such as medical expenses) but also the personal or "quality-of-life" costs; economists usually measure people's willingness to pay for improved safety at the equivalent of several million dollars per fatality avoided.

It is very tricky to decide what portion of accident costs individuals might take into account in their driving decisions, and what portion they might not consider. It seems plausible that people will take into account the injury risks to themselves (and other occupants of their vehicles), and some of the property damage if they anticipate higher insurance premiums in the event of a crash. Most likely, they will not take into account the costs of traffic delays or the injury risk to pedestrians.

Whether one person's driving raises the accident risk to other drivers is not clear: the frequency of collisions rises with more traffic, but, if people drive more slowly or more carefully in heavier traffic, a given accident will be less deadly. Based on earlier work by Mark Delucchi (UC–Davis) and the U.S. Federal Highway Administration, Kenneth Small and I put the average accident externality at around 3¢ per mile (60¢ per gallon) for the United States, although there is a wide range of other plausible estimates.

Adding up the tally of the costs of driving that individuals do not take into account gives a total, so far, of $1.76 per gallon, which is a substantial amount (though the uncertainty and controversy surrounding damage estimates should be borne in mind). However, a major problem with the gasoline tax is that it taxes fuel rather than distance traveled. Over the long haul, it is estimated that roughly 60% of the tax-induced reduction in gasoline will be from improvements in fuel efficiency (people buying smaller cars, retiring older, fuel-inefficient cars more often, etc.); only about 40% will come from people driving less. For each gallon of gasoline reduced by fuel taxes, the accident and congestion benefits are only 40% as large as they would be if all of the reduction in fuel were due to reduced driving and none to improvements in fuel efficiency. Since imposing fuel taxes also creates economic costs by altering travel behavior, this reduction in benefits results in the optimal tax being smaller by a similar portion. Adjusting the above figures reduces the appropriate gasoline tax to less than a dollar.

Gasoline Taxes as Part of the Overall Tax System

Gasoline taxes also provide revenues for the government and this raises the issue of whether the ability to provide revenues constitutes a reason to set higher levels of taxation than warranted on the grounds of pollution, congestion, and accidents. Leaving aside other considerations, whether it is better to finance some of the government's budget through gasoline taxes or not depends on the economic costs of gasoline taxes compared with other taxes, such as income taxes. Income taxes lead to economic costs or "excess burdens" because they distort the overall level of employment in the economy; for example, by reducing take-home pay they reduce the labor force participation rate of married women. Gasoline taxes cause economic costs by changing travel behavior and raising transportation costs for businesses.

Economists usually find that it is less costly to raise revenue from taxes with very broad bases, such as income taxes, than narrowly focused taxes on specific products that are easy to avoid by spending on other products. However, exceptions to this rule are products whose demand is relatively insensitive to price, which is the case for gasoline. Consequently, some level of taxation might be appropriate, in excess of that justified for curbing pollution, congestion, and accidents. Kenneth Small and I calculated this component of the optimal tax at roughly 20¢ per gallon for the United States.[1]

Taking account of both the revenue-raising benefits and the benefits from reduced fuel consumption, the optimal gasoline tax is just over $1 per gallon for the United States, according to our estimates, though the estimate is sensitive to different assumptions and may change over time. This is 2.5 times the current U.S. gasoline tax, but still less than half of the current tax rate in the United Kingdom.

It is often thought that low-income groups bear a disproportionate burden of gasoline taxes, although the actual evidence on this is more mixed. For example, a study by James Poterba (MIT) found that spending on gasoline as a share of total household expenditure was only slightly lower for the richest 10% of households than for the poorest 10%, and the gasoline expenditure share was highest for middle-income groups. In my view, gasoline taxes should mainly be evaluated by weighing their overall economic benefits and costs. Distributional concerns are much better addressed through altering the progressivity of the income tax system, or providing a safety net through the benefit system.

Other Arguments For Gasoline Taxes

The September 2001 terrorist attack, the recent debate in the House and Senate over competing energy bills, and fears about a resurgent Organization of Petroleum Exporting Countries (OPEC) have heightened concerns about U.S. dependency on imported oil, particularly from the Middle East. Oil dependency makes the U.S. economy vulnerable to volatile oil prices that might result from deliberate exercise of market power or changes in world market conditions. However, the

United States will always be vulnerable to oil price fluctuations—regardless of how much it imports from the Middle East—because the domestic price of oil is determined by the world price. The only way to reduce the economic disruptions from volatile world oil prices is to reduce the oil intensity of the U.S. economy.

In principle, an economic argument can be made for some level of gasoline taxation on the grounds of macroeconomic vulnerability, although a broader oil tax would be a more appropriate policy. Changes in oil prices can impose costs elsewhere in the economy that are not taken into account by energy producers and consumers. For example, it is costly in the short term for other firms to adjust their capital stocks in response to sharp, and unforeseen, swings in oil prices. A recent report on fuel economy standards by the National Research Council put the economic cost of our dependency on foreign oil at less than 12¢ per gallon.

Some people argue that we need to maintain a larger military in order to ensure uninterrupted oil supplies from the Middle East. The amount of extra military expenditure is tricky to assess; for example some Middle East military spending is to protect the security of Israel. Moreover, most research, for example a study by Douglas Bohi and Michael Toman at RFF, suggests that we may not need to spend more to ensure a continued supply of imports because it would be hard for Middle East countries to prevent other countries they supply from redirecting oil to the United States. Even if they were successful in stopping such shipments to the United States, the sharp increase in world oil prices would provide strong incentives for other OPEC members to defect and sell to the United States, or for non-OPEC suppliers such as Canada, Russia, and Mexico, to increase production.[2]

Another consequence of motor vehicle driving is the wear and tear on the road network that must be repaired at taxpayers' expense. However most of the damage is caused by heavy trucks rather than autos: road damage increases exponentially with a vehicle's axle weight so that a truck weighing 10 times as much as a car does one thousand times the damage. This problem really calls for a tax on diesel fuel rather than gasoline or, better still, a tax on truck mileage adjusted for axle weight.

There is a range of other unintended consequences associated with the production, transportation, and use of gasoline, including the costs of oil spills, leakage from storage tanks at refineries, and disposal of nonrecycled cars. However, the costs involved tend to be small in magnitude relative to those of congestion, accidents, and pollution, and they are better dealt with by policies other than higher gasoline taxes.

The Politics of Tax Reform

Congress has ducked the issue of whether to raise the Corporate Average Fuel Economy (CAFE) standards, which impose minimum miles-per-gallon requirements on sales of new vehicles, and left the National Highway Traffic Safety Administration to review and recommend new standards. The argument for tightening the CAFE standards is to address some of the factors discussed above, particularly greenhouse gases and oil dependency. However a much more effective policy to address these two concerns would be to raise the federal gasoline tax. Higher gasoline taxes would reduce fuel consumption not only by encouraging the development of more fuel-efficient vehicles, but also by encouraging people to drive cars rather than sport-utility vehicles and minivans, to buy new (more fuel-efficient) vehicles more often, and to reduce the overall amount of mileage. Indeed, tighter CAFE standards would lower the cost per mile of driving, and could worsen some of the other problems discussed above, such as traffic congestion.

However, substantially higher taxes on motorists are not on the political radar screen for the foreseeable future; despite a major effort in 1993, the Clinton administration was able to raise the federal gasoline tax by only 4¢ per gallon, and the rate has remained unchanged since then. Nonetheless, there is considerable scope for reducing the social costs of driving, without increasing the overall burden of taxation to motorists as a group. For example, opposition to local peak-period pricing schemes—which are much more effective at reducing congestion than gasoline taxes—might be dampened somewhat by returning revenues collected to motorists in the form of lower state gasoline taxes.

Suggested Reading

Litman, Todd. 2005. Fuel Taxes: Increasing Fuel Taxes and Fees. In *TDM Encyclopedia*, Victoria Transport Policy Institute. Available at: www.vtpi.org/tdm/tdm17.htm.

Parry, Ian W.H., and Kenneth A. Small. 2002. Does Britain or the United States Have the Right Gasoline Tax? Discussion Paper 02-12. Washington, DC: Resources for the Future.

Porter, Richard. 1999. *Economics at the Wheel: The Costs of Cars and Drivers*. New York: Academic Press.

Notes

1. This calculation assumes that extra gasoline tax revenues would substitute for income tax revenues, leaving total public spending the same. If extra revenues instead financed more highway spending, the optimal gasoline tax could be higher or lower, depending on whether the social benefits of the extra spending were larger or smaller than the social benefits of lower income taxes.

2. It could also be argued that the United States could raise gasoline taxes to take advantage of its market power in world oil markets. Effectively some of the burden of the tax increase would be borne by Middle East suppliers in lower world oil prices. This argument calls for an oil import tariff rather than a gasoline tax; however, it might do more harm than good by provoking retaliatory tariffs against the United States.

15

Pay as You Slow
Road Pricing to Reduce Traffic Congestion

Ian W. H. Parry and Elena Safirova

As U.S. cities become increasingly plagued by traffic congestion and gridlock, road pricing is emerging as a promising approach to provide relief. Recent advances in electronic payment technology make the pricing of roads feasible and compelling.

Congestion pricing is being recognized increasingly as the only truly effective way to alleviate the ever-worsening traffic gridlock in the nation's cities, and the time for its implementation has arrived with recent developments in electronic payment technology. The federal government could encourage state and municipal authorities to introduce road-pricing mechanisms by removing a variety of legal obstacles to converting high-occupancy vehicle (HOV) lanes into high-occupancy/toll (HOT) lanes and, more generally, by helping to pay start-up costs of road-pricing initiatives.

The Case for Congestion Pricing

Traffic congestion imposes substantial costs on society. According to recent studies by the Texas Transportation Institute, travel delays and the resulting extra fuel combustion now cost the nation about $70 billion each year. The average time per year an urban motorist loses to congestion during peak hours has grown from 16 hours in 1982 to 62 hours in 2000. Although detailed methodologies used to compute travel delays and to monetize them are not unanimously accepted by all researchers, those numbers provide an idea of the magnitude of the problem. And congestion is likely to become even worse in upcoming years, with continued growth in vehicle ownership and the demand for driving. Meanwhile, environmental con-

Originally published in *New Approaches on Energy and the Environment: Policy Advice for the President* (Washington, DC: Resources for the Future, 2004).

straints, neighborhood opposition, and budgetary limitations are making it ever more difficult to build new roads to accommodate increasing demand.

Typically, it takes only a modest reduction in the number of drivers to unclog a congested road and get the traffic moving faster. Charging people for driving on busy roads at peak periods is the best way to achieve this; such charges encourage some people to drive earlier or later to avoid the rush hour peak, to take less congested routes into town, to car pool, to use mass transit, or to reduce the number of trips, such as by working at home or by combining several errands into one trip. Other policy approaches are far less effective at reducing traffic congestion. Increased subsidies for mass transit may help lure some people away from driving on busy roads. But this policy can be partially self-defeating: if roads become less congested at peak period because more people are using transit, this attracts onto the roads some people who were not previously driving at peak period because of high congestion. In short, the roads may just fill up with traffic again; this is not the case under road pricing, however, as the charges discourage people from getting back into the car as congestion falls. The same phenomenon tends to undermine other approaches that do not raise the cost of driving, such as expanding cycle access or promoting telecommuting. And higher fuel taxes, which raise the costs of all driving, whether it is in urban or rural areas or occurs during peak or off-peak periods, are an extremely blunt way to reduce traffic jams; before the recent introduction of road pricing, driving in central London was not much faster than walking, despite gasoline taxes seven times as large as those in the United States.

Forms of Road Pricing

Over the last two decades, a considerable amount of money has been invested in adding high-occupancy vehicle (HOV) lanes to urban freeways to try to induce more carpooling; more than 2,000 lane miles were added at a cost of nearly $9 billion. The results have not been encouraging. Nationwide, the share of carpooling in work trips

actually fell between 1990 and 2000, from 13.4 to 11.2 percent, while the share of single-occupant vehicles in work trips increased from 72.7 to 75.7 percent, with slightly declining shares of transit and nonmotorized trips making up the balance.

HOV lanes, at least those currently with traffic flows well below those on parallel, unrestricted lanes, result in underuse of scarce road capacity at peak period. Converting them into HOT lanes by allowing their use by single-occupant vehicles in exchange for a fee while continuing to permit high-occupancy vehicles to use the lanes for free, would benefit many motorists. Those who value the travel time savings enough to pay the fee would benefit, while those who continue to use unpriced lanes may benefit from reduced congestion as some drivers switch to the premium lane. Carpoolers who were already using the HOV lane may be slightly worse off as more vehicles join premium lanes, but ideally the fees would be variable and set at levels to maintain free-flowing traffic, even at the height of rush hour. Tolls would be deducted electronically from accounts linked to transponders as vehicles pass under overhead meters; carpoolers would pass by a manned booth where the vehicle occupancy is briefly checked.

Over time, urban centers could develop a network of linked HOT lanes giving drivers access to any part of the region without major holdups, and allowing local authorities to provide express bus service throughout the region, thereby reducing the need for constructing costly express light rail systems.

To date, only three examples of freeways with HOT lanes exist in the United States: one in Los Angeles, another in San Diego, and a third in Houston. However, serious efforts are under way on HOT lane projects in many other urban centers with severe congestion problems. And many motorists are used to paying tolls that were initially designed to pay for the costs of road construction; it is conceivable that these tolls could be made to vary with the level of congestion.

A number of popular objections to road pricing have been raised, but none of those objections really seem to hold much water for HOT lane proposals. One objection is that motorists are opposed to paying for something that they previously used for free. However, under the HOT lane scheme, drivers will not be forced to pay tolls, as they can always use the parallel, unpriced freeway lanes.

For the same reason, it is not true that low-income families, who are least able to afford new taxes, will be driven off the roads; they actually may benefit from reduced congestion on unpriced roads. And, in fact, evidence suggests that it is not the rich who exclusively use premium lanes; in California, people of all income levels use HOT lanes when saving time is important to them. It therefore seems unnecessary to give discounts for low-income drivers using HOT lanes (as required by the recent House bill H.R. 3550), as that undermines the effectiveness of HOT lanes in providing free-flowing traffic.

Another impediment is simply unfamiliarity with the concept of road pricing and skepticism about its effectiveness. For this reason, it makes sense to introduce pricing incrementally, increasing the number of priced lane segments as their success in alleviating congestion becomes evident to the general public. California's two HOT lanes, which have been operating for several years, have demonstrated the ability of variable electronic pricing to maintain free-flowing traffic, and surveys in California now show widespread public acceptance of the HOT lane concept. And despite many predictions that it was doomed to fail, the introduction of road pricing in central London has reduced congestion delays by around 30 percent.

Other forms of road pricing exist as well. Outside of the United States, a number of area (cordon) pricing schemes have successfully reduced congestion in city centers, including London, Rome, Trondheim in Norway, and Singapore. Area pricing makes sense when, as in many old European cities, the central business district is dominated by a maze of narrow, winding streets, making pricing of individual roads impractical. In contrast, congestion in many U.S. cities is concentrated on wide highways and large arterial roads feeding into the center, and in this case, pricing of individual highways and arterials makes more sense. But area pricing still could play a useful role

Update

Since we wrote this article, the new SAFETEA-LU bill reauthorizing transportation spending has been passed. This bill also renewed the Value Pricing Pilot Program and created a new Express Lanes Demonstration Program (ELDP). The ELDP will establish 15 demonstration projects for expressway tolls on different facilities; eligible facilities include currently tolled roads, roads that presently have HOV lanes, roads that are modified to provide new tolled lane capacity, or un-tolled roads, where new, tolled capacity is added. Excess toll revenue may be used for highway and transit projects, provided that the tolled facility is adequately operated and maintained, and toll collection must be completely automated.

However, the bill did not modify states' interstate highway tolling authority granted under two pilot programs created in the 1991 and 1998 highway bills. To date, other than conversion of HOV lanes to HOT lanes, no state has successfully imposed tolls on existing interstates under either of these programs.

in areas such as Manhattan, where downtown streets are severely congested and people can get around by transit or walking if they are unwilling to pay area fees.

Other schemes might involve pricing of particular elements of the transportation infrastructure, including time-varying tolls on bridges. Such tolls currently exist on bridges over the Hudson River, connecting New Jersey and New York City, and on bridges in Lee County, Florida. Other applications might include fees for access to national parks that currently are congested during peak visiting hours.

These alternative-pricing schemes likely would be met with more political opposition than HOT lane conversions, as motorists are left with no option but to pay if they wish to keep using the same roads. They also are more vulnerable to the criticism that poor people might be forced off the roads. Nonetheless, toll revenues might be used in ways to help the poor, such as by spending it on projects to extend transit access to low-income neighborhoods.

Recommendations

Although decisions about urban road pricing are ultimately the responsibility of metropolitan planning organizations, any of a number of initiatives could be taken at the federal level to jump-start its implementation.

- *Remove the ban on the imposition of tolls on interstate highways.* Most road networks in metropolitan areas include portions of the interstate highway system, and to the extent that these roads are congested, they need to be priced if a locality is to deal effectively with regionwide traffic jams. Although in principle the Transportation Equity Act for the 21st Century (TEA-21) allows tolling on highway segments as part of the pilot program, to date no state has successfully applied for this authority. The pilot program requires the local authority to show that funds from the state's apportionment and allocations would never be sufficient to pay for maintenance and improvements of the road in question over time, which is very difficult to demonstrate. Alternatively, states can impose tolls on interstate highways if they pay back the federal government for funds already invested in the highway in question. The TEA-21 should be amended to allow local tolling on interstates, without these highly cumbersome restrictions.
- *Allow all HOV lanes to be converted to HOT lanes.* Again, provisions in TEA-21 prevent localities from converting HOV lanes to HOT lanes unless those lanes are in the Value Pricing Pilot Program (VPPP); most HOV lanes currently are excluded from the VPPP, as it is limited to only 15 congestion-pricing projects in the entire nation. The TEA-21 should be amended to permit local authorities to convert any HOV lanes to HOT lanes, whether or not they are currently covered by the VPPP.

- *Federal aid for start-up costs of road-pricing initiatives.* Introducing pricing on existing roads involves various set-up costs, including costs of installing monitoring technologies and barriers to separate priced and unpriced lanes. And the creation of HOT lane networks in metropolitan areas would require the construction of many additional lanes to link up the existing fragmented systems of HOV lanes, implying a substantial amount of new investment (adding one lane mile costs on average $4 million in right-of-way purchase, labor, and material costs). Toll revenues could cover many of these costs; a study by the Reason Foundation finds that about two-thirds of the necessary investment costs for constructing fully integrated HOT lane networks for the nation's eight most crowded urban centers could be funded by using toll revenues to finance tax-exempt bonds. But extra funding through the federal aid transportation program, such as the VPPP, could help to kick-start road-pricing initiatives.
- *Introduce legislation to address privacy issues.* Another impediment to the implementation of congestion pricing is the uncertain legal basis for electronic toll collection. Many drivers will be reluctant to have transponders in their vehicles until specific legislation has been enacted establishing for what purposes information collected on driving habits can and cannot be used. This concern will become more pressing with increasing use of Global Positioning Systems (GPS) to make electronic payments.
- *Establish a national standard for electronic-tolling technology.* At present, different technologies exist for electronic tolling, including the EZ-Pass, which dominates northeastern states, and the Fastrak, established on the West Coast. Incompatibility between these two systems imposes an additional burden on long-distance road users, particularly trucks, and for a segment of tourist travelers, as vehicles need to be fitted with more than one transponder, and different transponders might interfere with each other. Legislation to establish a national standard for electronic-tolling technology, before

road pricing becomes more prevalent in U.S. cities, would help avoid unnecessary duplication of technology installation costs and facilitate a nationally integrated pricing network.

If the government were to adopt these types of initiatives, we might at last begin to reverse the trend of ever-increasing urban congestion.

Suggested Reading

Litman, Todd. 2005. Road Pricing, Congestion Pricing, Value Pricing, Toll Roads, and HOT Lanes. In *TDM Encyclopedia*. Victoria Policy Institute. Available at www.vtpi.org/tdm/tdm35.htm.

Poole, Robert W., and C. Kenneth Orski. HOT Networks: A New Plan for Congestion Relief and Better Transit. Los Angeles, CA: Reason Foundation. Available at: www.rppi.org/ps305.pdf.

Safirova, Elena et al. 2004. Welfare and Distributional Effects of HOT Lanes and Other Road Pricing Policies in Metropolitan Washington DC. In *Road Pricing: Theory and Practice, Research in Transportation Economics* 9, edited by G. Santos. Elsevier, 179–206. Available at: www.rff.org/Documents/ RFF-DP-03-57.pdf.

16 Cleaner Air, Cleaner Water
One Can Lead to the Other in the Chesapeake Bay

A cross-media approach to environmental management reveals that the control of one form of pollution can often have beneficial effects on other dimensions of environmental quality. In this case, the reduction of nitrogen-oxide emissions into the atmosphere not only cleans up the air, but it results in cleaner water.

Save The Bay. For years, bumper stickers have carried the simple, urgent slogan on the back ends of cars in states that surround the Chesapeake. And for years governments in these jurisdictions have been responding to the call to do something about deteriorating water quality in the 200-mile-long arm of the Atlantic. Of course the task is far from simple. Tracking pollutants to their many sources and then finding ways to combat them is a major undertaking.

Merely identifying the sources of pollution has produced surprises. Agricultural runoff and municipal water treatment were fingered long ago as major culprits in the nitrogen buildup that chokes out aquatic life in the Bay. Only quite recently, however, did researchers discover that airborne nitrogen-oxide emissions—from utilities that generate electricity and from cars and trucks on the highways—can do the same kind of damage.

Now that researchers know about the connection between the Chesapeake region's air and water, however, they have begun to take a cross-media approach, which not only adds to the complications of analysis but improves the environmental outcome and reduces the price paid to achieve it. Analysts are thus on the lookout for "two-fers"—like a law that mandates cleaner air but whose implementation leads also to cleaner water.

RFF has completed a study this fall that substantiates just such an instance of ancillary benefits and the news is good for the Chesapeake Bay. RFF estimates that the huge body of water will benefit substantially from the large reduction in nitrogen-

Originally published in *Resources*, No. 133, Fall 1998.

Figure 1. Chesapeake Bay watershed and airshed

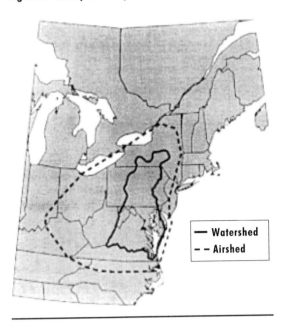

— Watershed
-- Airshed

oxide emissions from utilities and other large sources that EPA has proposed for the Eastern United States under the Clean Air Act, plus expected reductions from mobile sources.

"The Bay obtains a bonus," Senior Fellow Alan J. Krupnick says of EPA's latest effort to curb NO_x emissions because they are a precursor to smog. EPA's more stringent emissions standards would reduce airborne nitrogen compounds (nitrates) that reach the Bay by at least 26 percent, the RFF study shows. But RFF projected even larger reductions if the EPA program were structured differently.

Program design matters, in other words, and is a "key message" of the analysis, Krupnick emphasizes. The Chesapeake Bay community and others with a stake in NO_x emissions reductions should "not be indifferent," he says, to the features of a NO_x trading program.

The RFF team estimated that the cost of complying with the new EPA program would be 40 percent cheaper than the cost all sources incur now to meet their obligations under the Clean Air Act. But RFF projected even larger savings if EPA extended the trading program to all the different sources of NO_x emissions—not just utilities. Savings could be achieved by shifting some of the burden of NO_x abatement away from, say, electric utilities, and onto automobiles. At least as the RFF study turned out, the lower cost would be accompanied by fewer nitrate loadings to the Bay. The model showed a reduction in nitrates of more than 10 percent over what they would have been if only utilities could play the trading game.

Even greater cost savings (half the cost of command and control) could be had from an ozone exposure reduction program—one that targeted NO_x emissions reductions geographically, concentrating cleanup in the Midwest and New York. But in that case the Bay would fare worse than under command and control. Thus not every tack that EPA might take to reduce ozone would benefit the Chesapeake as much as any other, nor do cost savings and NO_x emissions reductions always go hand in hand. The crucial calculation for the Bay's health hinges on where the NO_x is reduced.

These findings are gleaned from the first of a two-part analysis on which RFF researchers are working with colleagues Paul Guthrie and Brian Morton. The study is sponsored by EPA's Office of Policy, Planning, and Evaluation and the agency's Chesapeake Bay Program.

Suggested Reading

For the report on which this article is based, see RFF Discussion Paper 98-46, "The Chesapeake Bay and the Control of NO_x Emissions: A Policy Analysis."

Part 4

Environmental Accounting and Statistics

17 Accounting for the Environment

Spencer Banzhaf

Our ability to design effective policies for environmental management depends critically upon our access to accurate information about environmental quality and trends. The lack of information has created serious problems for environmental policymaking. There is a strong case for an independent agency with the sole function of collecting and disseminating data on the environment.

Recently, Rep. Doug Ose (R-CA) proposed legislation (the "Department of Environmental Protection Act") that would elevate the U.S. Environmental Protection Agency (EPA) to a cabinet department and create within it a Bureau of Environmental Statistics (or BES). While cabinet status for EPA may have symbolic or organizational advantages, the creation of a BES could prove to be the most meaningful portion of the bill—and an important development for future environmental policymaking.

The Ose bill would authorize the proposed BES to collect, compile, analyze, and publish "a comprehensive set of environmental quality and related public health, economic, and statistical data for determining environmental quality . . . including assessing ambient conditions and trends."

Why do we need another bureaucratic agency collecting statistics? The overarching reason is that we simply do not have an adequate understanding of the state of our environment. In many cases, the network of monitors measuring environmental quality is insufficient in geographic scope. For example, in many cases our knowledge of national air quality is based on a few monitors per state; our knowledge of water quality is even weaker. The measures we do have typically focus on potential problem areas—a sensible approach from the standpoint of enforcement, but not for surveying the overall state of things. Accordingly, we must make inferences about overall quality from observations at these trouble spots. The consequence is a biased understanding of environmental quality.

Of course, this easy answer begs the further question of

Originally published in *Resources*, No. 151, Summer 2003.

why we need a better understanding of the state of our environment. There are several good reasons.

First, we have a natural desire to understand broad trends that affect our society and its welfare. Indeed, it is for this reason that we first began to collect many of our national economic statistics, including the familiar measures of gross domestic product (GDP) and inflation. Yet from the origins of GDP accounting, in A.C. Pigou's seminal *Wealth and Welfare* (1912), it was acknowledged that GDP is only a proxy and not a perfect measure of welfare because it omits many important components that do not pass through markets. Even then, the environment was acknowledged to be one of the important omissions. Since that time, we have invested enormous resources in improving measures of the market components of national well-being, but we have not proportionately broadened that effort to other components, like the environment. It is time to do so.

Second, our ability to design effective policies to balance environmental quality with other objectives, or to attain environmental objectives in the most efficient and effective manner, is hampered by inadequate information. As professional social scientists, we at RFF would probably always want more data to analyze. But the knowledge gap is more significant than a mere shortage of beans for bean counters. It manifests itself in every stage of policy design and evaluation.

Looking in the rearview mirror, in many cases we do not know whether existing policies have been effective, making it difficult to assess what remains to be done. Looking forward, we often find that the playbook of strategies with which one might attack environmental problems is limited by lack of information. Sometimes, the lack of information creates practical problems for implementing and enforcing a strategy. For example, it is difficult to imagine a serious effort to manage the total maximum daily load of pollutants into our nation's watersheds, as EPA has proposed, without more complete data about pollution loadings and their sources. At other times, the lack of information makes it difficult to anticipate the effects of a policy, creating political uncertainties. For example, the

Update

The Ose bill, which would have created a Bureau of Environmental Statistics, died in Congress. But there is continuing interest in establishing an agency to coordinate the collection and dissemination of basic environmental information.

cap-and-trade system, proven to be a highly cost-effective way to reduce air pollution nationally, may allow remaining pollution to concentrate in particular areas. Without a more thorough monitoring network, it is impossible to know whether these so-called hot spots are a serious problem. The consequence is hesitation in further use of this potentially effective policy instrument.

A third reason we should want better environmental statistics is that many expensive environmental regulations, with serious consequences for businesses and local economies, are triggered by incomplete information. A prominent example is compliance with air quality standards. Counties and regions that fail to meet these standards risk loss of federal highway dollars, bans on industrial expansion, and mandatory installation of expensive pollution-abatement equipment. Compliance is often based on readings from a small number of monitors. A fair question is whether some communities have been singled out while others have escaped detection. Moreover, although readings from only one monitor may push a portion of a county over a pollution threshold, reestablishing a clean slate once air quality has improved is much more difficult. Recent research by Michael Greenstone of the University of Chicago has shown that many counties remain in official non-compliance even though readings from the available monitors have shown compliance for many years. The Catch-22 is that a county must prove compliance throughout its jurisdiction even if the monitoring network is inadequate to shed light on all areas.

Creating a BES would also facilitate "one-source shopping" for members of Congress, agency administrators, and the public, who currently must navigate a maze of agencies to construct a picture of the nation's environment. In addition, an independent BES might lend more credibility—a sense of objectivity—to our environmental statistics, giving the public a commonly accepted set of facts from which to debate policy, much as the Bureau of Labor Statistics and the Bureau of Economic Analysis have done for economic statistics.

Lessons Learned from the CPI

Indeed, our experience with economic statistics teaches us a number of lessons for a BES. First, statistics can be politically controversial. Although widely accepted now, some economic statistics were the focus of past controversy. During World War II, for example, industrial wages were linked to changes in the U.S. Consumer Price Index (CPI). At the same time, the CPI began to move out of synch with the popular perception of price changes, recording much lower inflation rates than people experienced in their everyday lives, largely because it missed quality deterioration in the goods selling at modestly increasing prices: eggs were smaller, housing rental payments no longer included maintenance, tires wore out sooner, and so forth. The result was political uproar, with protests on the home front from organized labor. In the end, a lengthy review process, with representatives from labor, industry, government, and academic economists, resolved the issue.

Although environmental statistics will probably never hit people's pocketbooks as directly as did the CPI, they can get caught in the crossfire between business and environmental groups. Building in a regular external review process would help keep the peace during such moments. Crises aside, external reviews would ensure that a BES is balanced and objective, in both fact and perception, and help improve its quality over time.

Indeed, the regular external reviews of the CPI have raised points that would be of value to a future BES. Some are academic questions about sampling and analyzing data and could be addressed within the agency. Others may require congressional action

from the beginning, such as the need for data sharing. In our economic statistics, there is substantial overlap between information collected for the U.S. Census (housed within the Department of Commerce), unemployment statistics and the CPI (collected by the Bureau of Labor Statistics), and the GDP (collected by the Bureau of Economic Analysis). To address this concern, Congress recently passed the Confidential Information Protection and Statistical Efficiency Act, which allows the three agencies to share data and even coordinate their data collection.

Similar data-sharing issues would arise for environmental statistics. Currently, environmental statistics are collected not only by EPA but also by the Departments of Agriculture, Interior, Energy, and Defense. Even some of the economic statistics collected by the Census Bureau and other agencies would overlap in a complete picture of environmental statistics. Coordination across these agencies—and in some cases consolidating tasks into the new agency—would be essential for producing the best product without duplication of effort.

An additional insight gained from looking back on our experience is that economic statistics now play a much larger role in our economy and in economic planning than originally envisioned. Most generally, they have been used as a scorecard for the nation's well-being, a basis for leaders to set broad policy priorities (stop inflation, spur growth), and a basis for the public to assess its leaders. At a more detailed level, they now fit routinely into the Federal Reserve's fine-tuning of the economy. Finally, through indexing of wages and pensions, tax brackets, and so on, the CPI automatically adjusts many of the levers in the economic machine.

One could imagine environmental statistics playing each of these roles. First, despite their current weaknesses, environmental statistics already help us keep score of our domestic welfare. Second, they increasingly could be used to adjust policies. Initially, environmental statistics may serve as early warning signals for problems approaching on the horizon (or all-clear signals for problems overcome). Later, as the data develop and policies evolve to take advantage of them, they may even be used in fine-tuning. For example, on theoretical drawing boards, economists have already designed mechanisms that, based on regularly collected data, would dynamically adjust caps for pollution levels or annual fish catches. The only thing missing is the data with which to make such mechanisms possible.

A final lesson learned is that high-quality statistics cannot be collected on the cheap. We currently spend a combined $722 million annually on data collection for the U.S. Census (excluding special expenditures for the decennial census), the Bureau of Labor Statistics, and the Bureau of Economic Analysis, and more than $4 billion each year for statistical collection and analysis throughout the federal agencies. Over the past three years, these budgets have increased at annual rates of approximately 6.5% and 9.7%, respectively. Nevertheless, these efforts are widely considered to be well worth the cost.

By comparison, the current budget of $168 million for environmental statistics seems small. Consider that in 1987—the last year for which comprehensive data are available!—the annual private cost of pollution control was estimated to be $135 billion, and that government spends $500 million a year for environmental enforcement. With approximately 2% of our GDP at stake in these expenditures, and the welfare of many people, a top-notch set of environmental statistics seems long overdue.

Suggested Reading

Costs of the statistical programs of the federal government are tracked by the Office of Management and Budget. See *Statistical Programs of the United States Government* (www.fedstats.gov/policy).

Greenstone, Michael. 2003. Did the Clean Air Act Amendments Cause the Remarkable Decline in Sulfur Dioxide Concentrations? Working paper, University of Chicago, Department of Economics.

Pollution abatement costs were reported in U.S. Environmental Protection Agency, *Environmental Investments: The Cost of a Clean Environment*, EPA-230-11-90-083, Nov. 1990. More recent, but less comprehensive statistics are collected by the U.S. Census Bureau; see *Pollution Abatement Costs and Expenditures: 1999,* Nov. 2002 (available at www.census.gov/prod/2002pubs/ma200-99.pdf).

18 Greening the GDP
Is It Desirable? Is It Feasible?

Joel Darmstadter

Many environmentalists and some mainstream economists have voiced dismay over the limitations of Gross Domestic Product (GDP) as a measure of national economic performance. They claim that such a measure must be expanded to encompass the depletion of resources and other environmental impacts of economic activities. Extending the accounts in this way, however, raises a whole set of challenging conceptual and measurement issues.

In recent years, environmental activists, as well as a number of mainstream economists, have voiced dismay about the limitations of the gross domestic product (GDP) and related social-accounting aggregates as reliable measures of national economic performance and thereby as a legitimate basis for important policy decisions. This criticism has focused in particular on the extent to which measured GDP fails to reflect two important phenomena: the depletion of natural resources as well as damage to the ambient environment.

Why, critics ask, do we make allowances in our national accounts for the depreciation of structures and industrial equipment but not for the depletion of petroleum lifted from reservoirs? Why include damage to and losses from physical capital but not for the deterioration of an urban airshed? Hence, the call for an adjusted, "green" GDP—or more precisely, NNP (net national product)— that would rectify these measurement weaknesses. It is worth noting, however, that, being primarily a measure of the country's output of marketed goods and services, GDP has long been recognized to have certain—virtually unavoidable—flaws. GDP is no guarantor of human happiness or, by itself, an entirely reliable key to human welfare. For example, the fact that two countries with comparable levels of per capita GDP can have strikingly different degrees of inequality raises profound ethical issues regarding human well-being. It is also the case that certain nonmarket activities, such as household work by family members or crops grown and consumed on farms, understate national output. (Attempts to "impute"

Originally published in *Resources*, No. 139, Spring 2000.

A Terminological Clarification

The issue of incorporating green accounts in the nation's national-income-and-product accounts (NIPA) can be clarified by reference to a few basic relationships. More than 60 years ago, Prof. John Hicks, who later won the Nobel prize in economics, pointed out that a rising level of gross national product (GNP) (the difference between GNP and GDP need not detain us here) does not ensure that new investment in a country's private and public infrastructure compensates for the depreciation of such physical capital. In other words, GNP or GDP could continue growing (at least for a while) even while the physical capital, on which future prosperity depended, was wearing out. Hence, a precondition for at least maintaining prevailing levels of economic activity was constancy in the value of net national product (NNP), which equals gross output minus depreciation or, as NIPA labels it, "capital consumption allowances."

Critics of conventional NIPA measurement practices recognize the constant NNP condition as a necessary, but insufficient, basis for sustained levels of economic activity, since it fails to account for changes in the stock of environmental and natural resource assets. It is at this point where the seemingly dry question of NIPA measurement conventions links up with the deeper, more emotionally charged issue of society's prospects for a sustainable future.

market values to some of these activities have progressed both here and abroad.) And when persons voluntarily opt for leisure in preference to paid work, they most likely enjoy increased welfare while contributing to diminished market output.

The economics profession has hardly shrugged off such vexing conceptual and measurement problems. A landmark 1973 paper by Yale economists William Nordhaus and James Tobin sought to compare recorded output with a range of indicators designed to capture trends reflecting assumed changes in human welfare. Their preferred measure of economic welfare (MEW) per capita showed a long-term growth rate markedly below that of per capita NNP. At the same time, they observed that "progress indicated by conventional national accounts is not just a myth that evaporates when a welfare-oriented measure is substituted." And that judgement, I believe, remains valid today; one cannot lightly dismiss the extent to which existing national account aggregates, with all their defects, correlate well with a number of important indicators reflecting quality of life.

In spite of this long-standing awareness of such measurement issues, and attempts to grapple with them here and internationally, it is fair to say that the somewhat unique dilemma posed by use of natural resources and environmental "services" is of much

more recent origin—at least with respect to quantification. The most ambitious effort to address these issues was a proposed multiyear initiative by the U.S. Department of Commerce's Bureau of Economic Analysis (BEA), which was launched with an initial, highly tentative set of findings (limited to selected mineral commodities) issued in 1994.

For reasons that have never been made entirely clear, this effort became politically unwelcome in Congress, which quickly enjoined BEA from pursuing its long-range plan, pending an independent assessment of the possibilities and problems associated with such social accounting reforms. (See Joy E. Hecht, "Environmental Accounting: Where Are We Now, Where Are We Heading," *Resources* 135, Spring 1999.) A distinguished NRC panel, chaired by Nordhaus and entrusted with that assessment task, has now produced the result of its deliberations, *Nature's Numbers: Expanding the National Economic Accounts to Include the Environment* (National Academy of Sciences Press, 1999).

How significant is the NRC report as an analysis of, and brief for, a green GDP and related improvements in the nation's national-income-and-product account (NIPA) system? (See the box on this page for a description of basic GDP terminology.) In my judgment, the volume deserves to be viewed as an outstanding contribution to this com-

Pollution Control Expenditures and GDP

A tangential issue in the green accounting debate has to do with those environmental control or improvement expenditures, such as utility scrubbers, that do enter the GDP accounts, and constitute about 2% of the GDP. It is legitimate to question whether or not such investments adequately offset the value of the damage they are meant to avert. To the extent that they do not, conventionally measured net national product (NNP) would exceed a "true" estimate, provided one can ascribe a market-like equivalent to such damage. Some people may judge such damage to be beyond quantitative reckoning, believing their welfare to have been "incalculably" diminished. But as pointed out earlier in this essay, GDP, whether conventionally measured or subject to a green adjustment, cannot purport to reflect all aspects of changes in human welfare.

A largely irrelevant issue that sometimes arises in discussions of GDP and its shortcomings is a challenge to the inclusion of items like pollution-abatement spending in the first place. The challenge rests on the contention that, like other so-called "defensive" expenditures, such as dental checkups and oil changes, it does not add to material well-being but merely keeps it from getting worse. If some ombudsman of people's utility functions could establish the boundary between defensive and nondefensive outlays, perhaps this conundrum could be resolved. But don't hold your breath!

plex subject, one that should give combatants in this frequently passionate debate pause for some thoughtful reflection.

While evaluation of BEA's accounting explorations constitutes a significant part of the panel's report, *Nature's Numbers* also provides an up-to-date review of both the conceptual underpinnings of resource and environmental accounts as well as the methodological and empirical challenges in their estimation.

The three principal elements of environmental and resource accounts are nonrenewable assets, renewables, and environmental quality. Of these, it is the first whose estimation would appear to be the most tractable and whose exclusion from conventional accounting systems is least logical and excusable. After all, numerous activities involving subsoil mineral assets are already reflected in market transactions. For example, revenues from crude oil production, and investment in and depreciation of tangible physical assets, like drilling rigs and gathering lines, are part of the existing account structure. Why not, therefore, include the depletion of (or accretion to) the oil in the ground, the availability of which may be vital to sustaining the economic well-being of future generations?

Challenge of Choosing Appropriate Tools

As the NRC report observes, however, valuing changes in subsoil assets presents formidable challenges, even though it is a lot less complicated than the diminution of national wealth attributable to, say, the value of impaired visibility due to power-plant emissions. The major problem relates to the array of choices regarding the appropriate price to apply to the additions to, subtractions from, or revaluation of resources in the ground. Another challenge has to do with the choice of quantities by which such prices are multiplied. After all, while there may be a consensus as to the 10- or 20-year "on-the-shelf" inventory of proved oil or gas reserves, there may be much disagreement and uncertainty as to the extent of resources beyond proved reserves and the technological conditions under which assets shift from one to the other category.

No wonder that when it comes to green accounting, ranges of possible values—rather than the point estimates one finds in the conventional GDP accounts—are an inherent necessity in resource and environmental accounts. Thus, the report takes note of BEA's estimate of the value of subsoil mineral additions in 1987; these vary

Update

Despite the strong support from a committee of the National Research Council, Congress barred further federal research on environmental and resource accounting. Perhaps it was a legislative way of "killing the messenger"—of saying "we don't want to hear any bad news (or, for that matter, even evidence of progress) on environmental degradation or resource depletion." No one doubts that the topic abounds with elusive issues on both the conceptual level and empirical front. Yet numerous other governments, international institutions, and academic scholars, to their credit, are continuing to pursue serious studies in this field.

between 0.4% and 1.4% of that year's GDP.

Undoubtedly, the range of estimates designed to measure the value of changes in environmental quality would be wider still. Again, in the case of reduced visibility, measurement techniques that attempt to express such change in terms of imputed market values—for example, by using contingent valuation, hedonic pricing, and travel-cost methodologies—are now much more robust than several decades ago. They are far from universally accepted, however, and surely warrant emphatic caveats as one contemplates incorporating the estimates in the standard NIPA system. (See the box on this page regarding GDP treatment of pollution control expenditures.) Moreover, there is bound to be an irreducible set of negative environmental and social impacts that, while clearly adverse to social welfare, is not amenable to the dollar metric used to measure economic activity.

To the NRC committee, these problems are neither surprising nor a reason to throw in the towel on further development of resource and environmental accounting systems. On the contrary, the panel appears wholeheartedly to support the benefits to society of efforts to gauge the extent—however approximate—to which conventionally measured

GDP is either a serviceable or misleading proxy of overall economic and environmental health.

At the same time, the NRC committee made clear its view that this more rounded picture should take the form of periodic "satellite" accounts rather than being commingled with the GDP accounts, as presently calculated. Especially for shorter-term policy guidance, the existing GDP account structure must remain the system of choice. Perhaps in time, some components of resource and environmental accounts can be so integrated, just as—following a period of lengthy analysis and vetting—price indices have been modified to reflect emerging trends and new insights into technological change and new product development. But even under the best of circumstances, it is unrealistic to suppose that the existing annual time-series measures of economic performance can ever be augmented to track resource and environmental trends with the same frequency. The NRC committee states its views quite unambiguously:

> [E]xtending the [NIPA accounts] to include assets and production activities associated with natural resources and the environment is an important goal. Environmental and natural resource accounts would provide useful data on resource trends and help governments, businesses, and individuals better plan their economic activities and investments. The rationale for augmented accounts is solidly grounded in mainstream economic analysis ...[however,] environmental accounts must not come at the expense of maintaining and improving the current core national accounts, which are a precious national asset."

Perhaps the best way to encapsulate the value of the NRC study is to say that, for some time to come, the conceptual and philosophical aspects of accounting reforms have been firmly laid down and are not likely to be significantly enhanced by further scholarly discourse. The direction should now shift toward quantification (where feasible) and to Congress, that body's charge for an even-handed exploration of the difficult issues at hand having been conscientiously and admirably met.

Suggested Reading

Hecht, Joy E. 2005. *National Environmental Accounting: Bridging the Gap Between Ecology and Economics*. Washington, DC: Resources for the Future.

Nordhaus, William D., and Edward C. Kokkelenberg (eds.). 1999. *Nature's Numbers: Expanding the National Economic Accounts to Include the Environment*. Washington, DC: National Research Council/National Academy Press.

Part 5

Environmental Federalism

19

Environmental Federalism

Robert M. Schwab

For environmental problems that are highly localized, economics suggests that it makes more sense to set standards for environmental quality that reflect local conditions rather than uniform national standards.

Nearly all major federal environmental legislation divides the responsibility for controlling pollution between the federal government and the states. The exact division of responsibilities varies substantially from statute to statute.

Consider the Clean Air Act. It requires the federal government to set air quality standards for each of six pollutants and to establish performance standards for new stationary sources of pollution (such as factories and utilities) and mobile sources (such as automobiles). States are required to develop implementation plans, which must be approved by the federal government, to reach these goals; states also share the responsibility for enforcement. In sharp contrast, under the Clean Water Act the individual states are directed to set water quality standards. We thus have uniform national standards for air quality but state-specific standards for water quality.

Which approach is better? Should we move toward greater centralization, thus giving more responsibility for controlling pollution to the federal government? Or should we encourage further decentralization and allow state and local governments a greater voice in environmental policy?

This question is part of a much broader issue. The debate over the proper division of governmental responsibilities is long-standing. It was a key issue in the framing of the Constitution in the 1780s; it lies at the heart of the debate over the Reagan administration's call for a "new federalism" in the 1980s. Many of the questions about environmental federalism should be familiar to those who have thought about other issues in federalism.

Originally published in *Resources*, No. 92, Summer 1988.

Decentralized Policymaking

The case in favor of decentralized decisionmaking is simple, yet powerful and persuasive. For the moment, let's set aside the question of who should set policy and look at the principles we would like to see embedded in policymaking decisions. "Optimal" environmental policy—from an economic standpoint, at least—requires us to continue to reduce pollution in all parts of the country up to the point that the cost of further reducing pollution (what is known as marginal abatement cost) equals the benefits from that reduction (that is, marginal social damage). Focusing only on the economics of the issue, if it costs society $100 to reduce emissions by one additional ton but the benefits from doing so were valued at $200, we would be wise to cut back on pollution further; if, on the other hand, benefits were valued at only $50, we have set environmental standards too high. If we continue with this line of argument, the logic behind the economic principle that the optimal standard is such that marginal abatement cost equals marginal social damage becomes clear.

While it is true that we should apply the same *principle* in every region, we should not set the same environmental *standard* everywhere unless the costs and benefits of pollution control are the same in each region. There are excellent reasons to believe that this is not the case. For example, different industries are clustered in different regions, and it is much more costly for firms in some industries to reduce emissions than others. Thus differences in regional industry mix imply differences in regional abatement costs. Differences in meteorology, topography, and land-use patterns also play important roles in abatement costs. Few would disagree with the proposition that it will be much more costly for Los Angeles to reach any given level of air quality than most other cities; by one estimate, the Los Angeles area would have to reduce gasoline consumption by 82 percent in order to meet the air quality standards set under the Clean Air Act.

Similar considerations arise on the benefit side of the equation. Benefits, after all, are to some extent subjective. It could well be the case that the

Update

The debate over environmental federalism has intensified since this paper was published in 1988. This is true not only in the United States. In the emerging European Union, there is much controversy over the independent roles of the member countries in environmental policymaking and more centralized measures that "harmonize" policies in Europe. Likewise in the United States, there is a continuing discussion of the respective roles of the EPA and the states in the design and enforcement of environmental policy. The principles of environmental federalism provide insights into issues like the arsenic rule (see chapter 20), but they cannot explain recent state and local efforts to control carbon dioxide emissions to address global warming.

residents of some states would be willing to pay $200 for a slightly cleaner environment, but that residents of others would be willing to pay only $50. Population size also plays a role. Everyone in an area suffers the damages pollution causes. Therefore, everything else being equal, the greater the population in an area, the greater the benefit from a cleaner environment.

Ongoing research sponsored by RFF and funded by the National Science Foundation points to the validity of this argument. Paul R. Portney, Albert M. McGartland, and Wallace E. Oates have assembled data on damages and abatement costs for total suspended particulates (TSP) for Baltimore and St. Louis. Their preliminary estimate is that the optimal standard for TSP concentrations for Baltimore is nearly 50 percent less stringent than the optimal standard for St. Louis. That is, if we were to equate the costs and benefits of reducing TSP in both cities, the permissible level of pollution would need to be nearly twice as high in Baltimore.

How should we divide the responsibility for setting environmental policy in order to reach this optimal outcome (again recalling that we are only focusing on economic issues)? Federal legislation is rarely sensitive to regional differences in costs and benefits. Typically, federal legislation implies uniform standards; for example, under the Clean Air Act, the federal government sets uniform maximum allowable levels of pollution. Therefore, vesting the federal government with the authority to set environmental policy is unlikely to lead to an optimal outcome.

This line of reasoning should sound familiar to anyone who has been involved in the debate over fiscal federalism. It would make little sense to provide the same menu of public services in every community, or for the federal government to try to tailor policies to meet the local circumstances in each community. Thus we allow state and local governments to decide how much to spend on education, when trash is to be collected, and which beats police officers will patrol. The logic leading to this conclusion also suggests that state and local governments may be in a better position than the federal government to choose the correct level of environmental quality.

Possible Problems

Many people would accept the basic argument in favor of decentralization, but would continue to urge a strong role for the federal government in environmental matters nonetheless. They would contend that while the outcome under federal uniform standards is unlikely to be optimal, it is better than the outcome that would emerge if we were to allow state and local governments to set environmental policy.

Though it might not be immediately obvious, most of the arguments against decentralization turn on a single principle that has a strong basis in economic logic. If there are costs or benefits associated with the environmental policies adopted by one region that are borne by the residents of other regions, then it is unlikely that decentralized decisionmaking will lead to an optimal outcome. We can apply this principle in a range of circumstances.

The clearest cases are those where pollution generated in one state crosses political boundaries into another. For instance, if Illinois firms are forced to reduce pollution, we would expect the citizens of Illinois (and, hopefully, Illinois policymakers) to take into account the benefits from a cleaner environment that Illinois citizens would realize. Thus, Illinois would require its firms to reduce pollution to the point that marginal abatement cost equaled marginal damage suffered by Illinois residents. Under such a decentralized approach, we would not expect Illinois citizens to necessarily take into account the benefits Indiana residents receive. An optimal policy would require Illinoisans to go further and continue to reduce pollution until marginal abatement cost equaled the sum of damages to both Illinois and Indiana residents.

Clearly, in many cases this argument against decentralization is quite persuasive. The problems in coordinating efforts by Delaware, Virginia, Maryland, the District of Columbia, and Pennsylvania to clean up the Chesapeake Bay are well documented (and the Chesapeake is often offered as one of the more successful cooperative undertakings). In some cases, whether or not a pollutant is local is a policy issue; damages from sulfur dioxide emissions can be contained in a small area or dispersed over a much larger area if firms are required to use tall stacks. Some pollutants are international, acid rain being a good example, as the strained relations between Canada and the United States can attest. Some are truly global, carbon dioxide's impact on the climate (often called the greenhouse effect) and the impact of chlorofluorocarbons on the ozone shield being extreme examples of environmental issues that cannot be addressed by states or even nations. In such cases, the benefits from controlling pollution extend well beyond traditional political borders.

Many pollutants, however, cause damages over a much more limited area. Lead, carbon monoxide, and TSP are good examples. In many cases, the control of hazardous waste sites and the establishment of drinking water standards also fall into this category.

Similar issues arise repeatedly when assessing fiscal federalism. Suppose the residents of Community A build parks which are used not only by residents of A but by those of Community B as well (an example of a "fiscal spillover"). Under these conditions, we would expect that Community A had built too few parks. The standard policy solution in this case would be a grant from the federal government to Community A to encourage it to build additional parks.

We can analyze a second argument against decentralization in this framework. People throughout the country derive benefits from the quality of the environment in unique national sites such as national parks and wilderness areas; the residents of New York care about clean air in Yellowstone, as do the residents of Wyoming and Idaho. In many ways this issue is identical to the interjurisdictional pollution problem discussed above, though here the damages from pollution spill over state boundaries even though the pollution itself does not. Again, optimal policy almost certainly requires some federal role.

As a third argument, there are sometimes gains from uniformity and coordination. (Anyone who doubts this should try playing a tape made on a Beta videocassette recorder on a VHS machine or using a piece of software written for an Apple computer on an IBM.) In the context of environmental policy, it would be very expensive to demand that Detroit make cars that meet different standards in Montana, New Jersey, and so on. Clearly there are some benefits from uniform standards; automobiles are probably the most important example. But even in the case of automobiles, the benefits to be gained from a policy on uniformity do not rule out the possibility that the "correct" number of standards may be a happy medium between the current number—two (the Clean Air Act allows California to set its own standards for auto emissions)—and fifty, that is, one that applies to each state.

Interjurisdictional Competition

A fourth argument against allowing lower levels of government to set environmental policy focuses on the effects of interjurisdictional competition. The fear is that in their eagerness to attract new business and jobs, local officials would set environmental standards that are too lax. Thus, so this argument

goes, the federal government must set environmental policy in order to "save the states from themselves." Such a stance is consistent with a more general view that nearly all forms of competition among jurisdictions are destructive. The Advisory Commission on Intergovernmental Relations (the focal point for interest in federalism issues within the federal government) for years argued that state and local governments should be encouraged to cooperate with one another.

In the last several years, however, there has emerged a strong sense that competition among governments can be beneficial, just as competition among firms can be beneficial. According to this view, competition forces local government to make decisions that are in the best interest of their constituents. Thus competition stops local officials from overspending; if they were to try to set public budgets that were too high, firms would vote with their feet and move to competing communities.

The following simple example captures the spirit of this argument in the context of environmental policy. Suppose Illinois were about to set a standard for some pollutant. Who would bear the costs associated with this new policy? Since Illinois competes against other states to attract and hold firms, clearly the burden cannot fall on Illinois firms; they have the option to move to another state. Therefore the cost of reducing pollution must fall on Illinois residents in the form of fewer jobs and lower wages.

Under these circumstances, what level of environmental quality should Illinois policymakers set? They might reason as follows. Illinois residents will pay the costs of reducing pollution. Therefore if Illinois policymakers act in the best interest of their constituents, they should set a standard such that the costs of reducing pollution further are just balanced by the benefits. If they set a less stringent standard, Illinois residents would be willing to sacrifice additional jobs and wages in return for a cleaner environment; if they set a more stringent standard, the benefits from less pollution would not be worth the cost. This is exactly what optimal environmental policy requires: marginal abatement cost should equal marginal social damage.

Undoubtedly the real world is much more complicated than the above situation. Firms cannot instantaneously move from Illinois to Hawaii; Illinois firms may have quite a bit of influence with Illinois lawmakers; the costs of pollution control will be concentrated among workers in certain industries while the benefits of a cleaner environment are diffuse; residents of other states will realize some of the benefits if Illinois firms generate less pollution. But this illustration does make an important point. Competition among jurisdictions implies that the costs of government would be borne largely by those who realize the benefits from those policies and will therefore encourage efficient public decisions.

Children and Grandchildren

Finally, as some people have asked, if we vest state and local governments with the authority to set environmental policy, will they properly take into account the interests of future generations? Will communities be inclined to follow policies that yield benefits now (such as more jobs) but would result in substantial damages in the future (such as increased exposure to long-lived pollutants like hazardous wastes)? Problems of this kind would seem to be more acute at the state and local levels than at the federal level. We would presumably all want the federal government to take into account the interests of our children since it is likely that they will live somewhere in this country. But at least some of our children will live in communities elsewhere, and therefore their environmental heritage will depend on the decisions of others. Geographic mobility, so this argument goes, could result in myopic local decisions and suboptimal environmental quality for future generations.

This line of reasoning is not altogether persuasive. There is substantial evidence that many of the factors that determine the quality of life in a community are reflected in community property values—that is, people are willing to pay more to live in a community with good schools, a low crime rate, and a clean environment. Suppose a community were considering allowing a new firm to locate

in that community, and suppose further that this firm would cause $100 worth of environmental damages in the future. If environmental quality is reflected in property values, the value of homes in this community would fall by (the present value of) $100 if the firm were allowed to enter. At least in this simple case, current generations could not escape the future costs and benefits of their environmental decisions. Here again, the example ignores some important issues; in particular, it assumes that future residents are aware of and can evaluate the damages from long-lived pollution. But it does suggest that some of the concerns about decentralized policymaking need to be reevaluated.

Sensible Solutions

These considerations are not meant to suggest that we should abolish the U.S. Environmental Protection Agency and ask state and local governments alone to protect the environment. But I would argue that we should seriously consider moving some policymaking responsibility to lower levels of government, while still leaving the federal government with an important role in environmental issues. There are some strong arguments in favor of decentralization. An environmental policy that is correct in one region is unlikely to be correct in all. Federal regulation is rarely sensitive to these differences; instead, it often implies a single uniform policy in all regions.

It is important not to minimize the problems associated with this proposal. In particular, the spillover problem is difficult; the damages from some pollutants do not respect state boundaries. But sensible solutions may be available. Federal matching grants to state and local governments are often used to deal with the problem of fiscal spillovers. Perhaps similar policies (for instance, grants from the federal government to a state in return for an agreement to control pollution more tightly) would be effective in the interjurisdictional pollution problem. Further, we might envision a regulatory structure where states (and possibly lower levels of government) assume the responsibility for controlling some pollutants while the federal government retains the responsibility for others. It is not clear that we can solve all of the problems in environmental federalism, but the issues are sufficiently important that they deserve careful attention.

Suggested Reading

Congressional Budget Office. 1997. Federalism and Environmental Protection: Case Studies for Drinking Water and Ground-Level Ozone.

Oates, Wallace E. 2002. A Reconsideration of Environmental Federalism. In *Recent Advances in Environmental Economics*, edited by J. List and A. de Zeeuw. Cheltenham, U.K.: Edward Elgar, 1–32.

Vogle, David et al. Environmental Federalism in the European Union and the United States. Available online at http://www.tilburguniversity.nl/globus/activities/conference/papers/vogel.pdf.

The Arsenic Rule
A Case for Decentralized Standard Setting?

Wallace E. Oates

The costs of complying with a uniform national standard for arsenic concentrations in drinking water involves minuscule costs per household in some areas and enormous costs in others. An economic approach to the problem makes a compelling case for allowing variations in the standard across different jurisdictions that reflect these cost differentials. This could be done by allowing jurisdictions to decide on their own standard.

In the waning days of the Clinton administration, the U.S. Environmental Protection Agency (EPA) issued a new standard for the permissible level of arsenic in U.S. drinking water. The new arsenic rule reduced allowable arsenic concentrations by 80% from 50 parts per billion (ppb) to 10 ppb. A short time later, the new Bush administration put the revised standard on hold, citing the need for further scientific evaluation. But following a contentious period of debate, EPA Administrator Christine Whitman announced in October 2001 that the new arsenic standard would indeed be 10 ppb, as set in January.

The controversy has centered largely on the original EPA benefit-cost study. The study supporting the new measure presents, in fact, a close call. The estimated benefits are somewhat less than the costs for the benchmark case, but the government has argued that there are sufficient "intangible" benefits to make the measure worthwhile. Some subsequent studies, in contrast, find the EPA analysis far too optimistic and argue that the new measure comes nowhere near passing the benefit-cost test.

A close examination of the nature of the arsenic problem suggests, in my view, a quite different perspective on the whole matter. I will argue that rather than setting any uniform national standards, there is a persuasive, if provocative, case for decentralized standard setting. My proposal is that each water district in the United States be empowered to choose its own arsenic standard.

The basis for this proposal is twofold. First, the arsenic standard is a very close, real-world approximation of what

Originally published in *Resources*, No. 147, Spring 2002.

economists call a "local public good." The issue here is long-term exposure to a contaminant with certain carcinogenic risks. But the population at risk is restricted to regular users of the local water system—that is, the residents of the particular water district. Drinking water quality is a service shared by a well-defined local group of people (at least for most contaminants).

Second, there are striking variations in the cost of treatment across U.S. water districts. Treatment of drinking water is an activity that exhibits enormous economies of scale. Table 1 reports the cost per household of achieving the target of 10 ppb for water districts of different sizes. This target can be attained very inexpensively in large water districts—at less than $1 per annum per household in the largest class of districts. But, as the table shows, implementing this standard is a very dear proposition for residents of small districts; its cost per year can exceed $300 per household!

The move from 50 ppb to 10 ppb brings an estimated tiny reduction in risk. EPA estimates that the tighter standard may save approximately 20 to 30 statistical lives per year nationwide. But this is subject to a large dose of scientific uncertainty— some claim that a reasonable confidence interval will actually encompass zero lives saved.

The point here is that it may well be worth an extra $1 per year per household for such a small risk reduction as this. But it seems highly unlikely that this can justify an expenditure of more than $300. Indeed, such a sum could effect far greater reductions in risk if used for other public (or private) health measures, such as increased frequency of mammograms, colon screening, or a host of other measures. In short, the arsenic rule is a case where a uniform national standard seems highly inappropriate; one size simply doesn't fit all in this case.

Large versus Small Water Districts

It is interesting that the pattern of existing arsenic concentrations, in fact, reflects the cost differentials in Table 1. Most large water districts already meet the standard of 10 ppb. Of the 54,000 community

Table 1. Average Annual Cost Per Household for an Arsenic Standard of 10 PPB by System Size

System Size	Cost per HH
<100	$327
101-500	163
501-1,000	71
1,001-3,300	58
3,301-10,000	38
10,001-50,000	32
50,001-100,000	25
100,001-1 million	21
>1 million	0.86
All categories	32

water systems in the United States, about 95% are already in compliance with the proposed new standard. Of the systems that will have to introduce more stringent treatment procedures, 97% are small systems that serve fewer than 10,000 people each. The new measure would thus impact primarily small districts, precisely those for whom the new standard is most expensive and likely not worthwhile.

EPA is well aware of the costliness of this measure for small water districts. In fact, there is provision under the Safe Drinking Water Act for "exemptions" from the standard due to "compelling" factors that may include the inability of a particular district to meet the cost of complying with the standard. However, the term exemption is a little misleading here; it is not an exemption from meeting the standard but rather the granting of an extension of the period over which the district must come into compliance.

Whitman has indicated the agency's commitment to working with small districts to assist them in various ways (including grants and loans) to meet the new standard. But this really avoids facing up to what I see as the real issue here: the standard simply is inappropriate for small water districts. The nature of the problem suggests that the standard should be tailored to the circumstances of individual districts. And these circumstances might well reflect not just differences in costs, but also differences in preferences across various communities.

Update

Instead of allowing varying standards for systems of different sizes, EPA has reaffirmed that all systems must comply with the 10 ppb standard. However, EPA has authorized the states to allow small systems some additional time to come into compliance and is making available grants and loans to support treatment upgrades.

The best way to accommodate such variation is to allow districts to determine their own standards.

Let me offer a vision of how decentralized standard setting might work. EPA would play a critical role in providing basic information and guidance both for the risks associated with different arsenic standards and for the costs of treatment. The agency would, in a sense, provide a menu of choices to individual water districts. The districts themselves would then, either through their own elected officials or through a referendum if they wish, select their own standards for the arsenic concentration. In this way, both the large differences in treatment costs and any differences in preferences across localities would manifest themselves in local choices. The outcome would be a range of standards across districts, tailored to the particular conditions of each.

Best Left to Experts?

This is admittedly a tricky and contentious issue. Some believe that the setting of public-health standards should be left to the experts. This issue is not quite the same thing, they would argue, as a decision on whether or not to repave a local road. And yet, we give plenty of responsibility to decentralized levels of government. And I am not persuaded that the outcomes are generally inferior to uniform national standards set at the central level. The experts, in my proposal, still have a crucial role to play: providing basic information and guidance. The key point here is that a uniform standard for a local public good is not the economically right answer: it involves a waste of valuable resources. We can do better, sometimes much better, with programs that are responsive to local settings and conditions.

The arsenic rule, of course, is not the only candidate for decentralized standard setting. In fact various other pollutants of drinking water present similar opportunities for individualized standards that are responsive to local conditions. And this may well apply to certain other dimensions of highly localized environmental quality. But the arsenic rule presents an intriguing case that could be used as an experiment or initial foray into this kind of localized environmental decisionmaking.

Suggested Reading

Burnett, Jason K., and Robert W. Hahn. 2001. A Costly Benefit: Economic Analysis Does Not Support EPA's New Arsenic Rule. *Regulation*, Fall, 44–49.

Sunstein, Cass R. 2001. *The Arithmetic of Arsenic*. AEI-Brookings Joint Center for Regulatory Studies Working Paper 01-10, August.

U.S. Environmental Protection Agency, Office of Water website on the arsenic rule: http://www.epa.gov/safewater/arsenic.html (accessed October 4, 2005).

U.S. Environmental Protection Agency Science Advisory Board. 2001. *Arsenic Rule Benefits Analysis*. EPA-SAB-EC-01-008, August.

Wilson, Richard. 2001. Underestimating Arsenic's Risk: The Latest Science Supports Tighter Standards. *Regulation* (Fall): 50–53.

21

The Interstate Transport of Air Pollution
A Regulatory Dilemma

Alan J. Krupnick and Jhih-Shyang Shih

Environmental regulation becomes yet more complicated when pollution created in one jurisdiction flows into another. The management of air quality in the U.S. is bedeviled by precisely this problem. The case, for example, of ground-level ozone is one where regulatory efforts to reduce ozone concentrations in one state are undermined by inflows of pollutants from upwind states. Such transboundary pollution presents some fundamental challenges to the basic design of regulatory programs. One promising approach is the creation of regional air-management partnerships (RAMPS) in which all the states in a particular air-shed could co-ordinate policies to meet national air quality standards. Alternatively, the job can be lodged with the federal government.

Although air quality generally has improved over the last three decades, some areas, including a big portion of the eastern United States, have a poor record on attaining ambient air quality standards, and the number of counties in nonattainment has increased with EPA's new tighter ambient air quality standards. For ozone, 474 counties now violate the new 8-hour standard, up from 125 under the old standards. For PM2.5, 209 counties violate this new standard, many of these violating the ozone standard as well. And, following requirements to continually reassess the appropriate stringency of these standards, expectations are building that they will be further tightened in the future.

To help meet the new standards, the federal government has been issuing a number of important new national and regional regulations, including the Clean Air Interstate Rule, which mandates an even tighter sulfur dioxide (SO_2) and nitrogen oxide (NO_x) cap-and-trade program; the NO_x SIP Call; the NO_x Trading Program; and several rules that reduce particulate and other emissions from off-road and diesel engines.

Even so, it is unlikely that these rules will be enough to meet the new standards in some nonattainment areas, which raises the questions of how best to decide on and implement further control measures. In particular, continuing to follow the Clean Air Act's state implementation plan (SIP) process has

Adapted from A. Krupnick and J. Shih. A New Approach to Air Quality Management. In R. Morgenstern and P. Portney, eds. *New Approaches On Energy and the Environment: Policy Advice for the President* (Washington, DC: Resources for the Future, 2004).

some fundamental problems. One of the main ones is that the process is local while the air quality problem is regional.

The SIP process provides each state with the responsibility for developing an EPA-approved plan for reaching the National Ambient Air Quality Standard (NAAQS) by the mandated deadline. For most areas, this involves developing an emissions inventory, implementing a host of emissions reductions measures to apply at the local level, and conducting air quality modeling to show that the measures are enough to meet the standards.

This process is likely to be inefficient to meet stringent air quality standards, because scientific evidence shows that air pollutants, such as ozone, fine particulates, and their precursors, can be transported long distances across state boundaries. For example, recent research at RFF and Georgia Tech has shown that, on average, local emissions account for only 23 percent of local ozone and fine-particulate concentrations (Table 1). Thus even if one state reduced its ozone, ozone and its precursors still would be transported across its borders from other states. Because the pollution control efforts of upwind states affect the air quality in the downwind states, one state cannot tackle the ozone problem by itself; it will take multiple states' efforts to solve the problem.

In spite of these findings, under the current regulatory system and SIP design, an individual nonattainment state is required to submit only its own SIP for approval; hence the upwind state is not likely to account for the beneficial impact of its pollution control policy on its downwind neighbor, nor the fact that it may be cheaper to reduce emissions in an upwind area than in the area itself. However, cap-and-trade programs can "automatically" address this problem to some extent.

Previous Attempts to Address the Issue

Regional approaches to solve air quality problems are not new. The Clean Air Act (CAA) provides states with the authority to sue for problems of "overwhelming transport" and gives EPA the authority to set up organizations to address air pollution problems spanning multiple states or tribes. Specifically, petitions under Section 126 of the CAA, the formation of the Ozone Transport Assessment Group (OTAG), and the creation of the Ozone Transport Commission (OTC) are good examples of attempts—albeit incomplete ones—to address this issue.

Lawsuits are a clumsy way to implement policy, however, so this approach should not be relied on to reach attainment. Lawsuits cannot address how states will meet their own attainment problems or those of the region on a cost-effective basis.

OTAG was a partnership among EPA, the Environmental Council of the States, and various industry and environmental groups, which assessed the long-range transport of ozone and its precursors. Although it was successful as a means of information exchange among the states, it lacked regulatory authority and failed to recommend measures to address the problem.

Finally, OTC, a multistate organization comprising government leaders and environmental officials from 12 northeastern and mid-Atlantic states, has developed cap-and-trade strategies to help states in the region attain and maintain the NAAQS for ozone, and do so cost-effectively. However, recent research has shown that the states covered by OTC are mostly the areas violating air quality standards, in other words, downwind states. By limiting the universe of allowable trades within the Ozone Transport Region, the trading restrictions make it difficult for sources to find trading partners; the smaller the trading region, the smaller the potential for lowering the cost of meeting a given cap.

Possible Solutions

A couple of very different approaches are possible for dealing with the long-range transport issues. Regional air management partnerships (RAMPs) could be developed via expansion of existing regional institutions or establishment of new ones, or most parts of the SIP process could be eliminated —except for assuring that local emissions do not increase—and responsibility for attainment shifted primarily to the federal government.

Table 1. Summary of State Contributions (%) to Local Air Quality Per Unit Emissions Reduction (July Episode, Area-weighted)

Receptor state	1-hour ozone Reduced point source NO_x	8-hour ozone Reduced point source NO_x	Reduced point source NO_x	Episode average 24-hour PM 2.5 Reduced point and area source SO_2
AL	36	34	35	36
DE + MD	21	20	12	14
GA	24	23	24	27
IL	13	10	11	19
IN	19	17	18	22
KY	25	25	22	23
MA + CT + RI	16	20	19	21
MI	12	13	16	15
MO	12	14	21	12
NC	22	21	19	22
NJ	21	21	10	15
NY	14	13	16	20
OH	15	15	6	15
PA	16	15	17	21
SC	32	32	33	42
TN	41	39	38	35
VA	24	24	21	16
WI	67	67	69	40
WV	17	17	10	25
Average	23	23	22	23

Note: Table shows local contributions to reductions in one-hour and eight-hour daily maximum ozone from unit reductions in point-source NO_x, and reductions in 24-hour PM2.5 concentrations from unit reductions in point-source NO_x and reductions in point- and area-source SO_2 emissions. For example, if each state reduced point-source NO_x emissions by one ton, then 36 percent of the resulting ozone reductions in Alabama would be attributed to reductions of NO_x within Alabama.

Source: Shih et al. 2004. *Source-Receptor Relationships for Ozone and Fine Particulates in the Eastern U.S.* RFF Discussion Paper 04-25. Washington, DC: Resources for the Future.

Building Regional Air Management Partnerships (RAMPS)

Economists generally argue that the geographic reach of institutions regulating environmental pollution should be coincident with the geographic reach of the pollution. In the case of ozone and fine particulates, such an institution would need to be composed of all the states and tribes in an airshed working collectively to address these regional air pollution problems. A collective process could provide the states an opportunity to jointly consider the issues and to explore strategies that are not only cost-effective, but also equitable for every state within the entire airshed.

To pursue this option, a new institutional mechanism to address the regional air quality management issue could be established. The RAMPs concept is consistent with the recommendation made by the Subcommittee for Ozone, Particulate Matter, and Regional Haze Implementation Programs and in a 2004 National Research Council report, *Air Quality Management in the United States.* RAMPs could be implemented by either expanding the current OTC or creating a new regional organization to address the long-range transport problem.

The CAA gives EPA authority to establish transport commissions and air quality control regions. EPA would be able to establish RAMPs using this statutory authority, relying on its general rulemaking authority to provide direction and schedules. RAMPs could act as a forum for information sharing, reaching agreement and developing recommendations on how to solve regional air pollution problems. The new regional organization could provide technical support and assessment, and create areas of influence (AOI) and areas of violation (AOV) using air quality modeling and tracer experiments. Institutional mechanisms also could be structured to support the development and implementation of incentive- and market-based approaches to managing regional pollution problems, including developing positive incentives for upwind areas to reduce precursor emissions, such as emissions trading and air pollution funds. The organization also may endow areas of violation with some power to compel actions from areas of influence. Under RAMPs, states would retain primacy, subject to EPA oversight and Federal Implementation Plan (FIP) authority, to the greatest extent consistent with air quality and equity goals, with responsibility assigned at the lowest level of government practicable.

Assuming that the preferred approach to attaining the new NAAQS involves establishing regional emissions cap-and-trading programs, RAMPs could oversee the orderly transfer of emissions credits between jurisdictions, including developing protocols for tracking, verifying, recording, and otherwise overseeing the conditions of interstate and other interjurisdictional emissions reduction credit transactions. RAMPs also could be responsible for reviewing and approving each state's regulations into its SIP for attaining the NAAQS for ground-level ozone.

For this idea to work with the current regulatory system, state SIPs would need to incorporate policies consistent with their RAMPs. EPA could compel states to do this by rejecting the SIPs of all states in the airshed unless they had satisfactory measures enforcing this incorporation. Unless RAMPs have some authority, this arrangement is unlikely to be efficient, because there is no guarantee that a state will agree to bear more costs than other states.

Federal Responsibility with a Diminished SIP Process

A much more radical option would be to dismantle much of the SIP process, except for assuring no net increase in local emissions, and transfer authority for meeting standards to the federal government. This option has several advantages over the first approach. First, federal measures for reducing emissions, such as regional trading options and fuel quality regulations, are almost universally accepted as both the most effective and the most cost-effective. Conversely, state and local measures, such as transportation control measures, have performed poorly. Second, the alternative of establishing RAMPs is problematic, because our federal system does not favor conferring authority to regional, as opposed to state, institutions.

Third, eliminating much of the SIP process reduces costs and contentiousness. This overly bureaucratic and legalistic process draws attention and resources away from the more germane issue of ensuring progress toward the goal of meeting the NAAQS. Furthermore, the enormous modeling uncertainties in demonstrating attainment of the ambient standards could be eliminated and transferred instead to the federal government, which has the expertise, money, and economies of scale to do a better job. And given the large role for regional transport, the federal or regional level is the appropriate one for modeling and demonstration of attainment.

Fourth, forcing localities to demonstrate that their emissions are not increasing is a much more tractable and measurable task than requiring them to demonstrate attainment. Localities likely would still keep their inspection and maintenance programs in place, for instance, and could be required to show that the program was being well implemented. Fifth, the public's attention could turn to the government body that actually has the power to address the problem—the federal government—rather than the localities, which are largely helpless to further reduce concentrations.

The main drawback to this approach stems from its radical departure from the CAA. Congress probably would need to act. Also, many stakeholder groups would be skeptical of such a change at first, with perhaps the greatest concerns being about delay, backsliding on local emissions reductions, and the process by which transportation planning and air quality planning are required to be coordinated. This is why the nondegradation requirement would be so important.

Summary

Altering the geographic approach to managing air quality in this country is needed to match the realities of long-range transport of air pollution. It makes little sense to have localities responsible for attaining air quality standards when on average about 75 percent of the problem is not of their own making. Either this responsibility should shift to new regional air quality management institutions, backed by EPA, with ultimate responsibilities still lodged within the states, or it should be lodged at the federal level, with a dramatic reduction in state responsibilities. A presidential commission could be set up to consider these two quite different options for improving the management of our nation's air quality.

Suggested Reading

For a study of interstate water pollution problems in the U.S., see Hilary Simon. 2005. Transboundary Spillovers and Decentralization of Environmental Policies. *Journal of Environmental Economics and Management* 50 (July): 82–101.

National Research Council. 2004. *Air Quality Management in the United States.* Washington, DC: National Academy Press.

22 State Innovation for Environmental Improvements
Experimental Federalism

Winston Harrington, Karen L. Palmer, and Margaret Walls

One of the advantages of a federal system with a heavy reliance on decentralized decisionmaking is the opportunity for widespread experimentation and learning through innovative policies in individual jurisdictions—so-called laboratory federalism. The federal government can encourage such experimentation with grant programs that provide incentives to state and local governments to design and implement new approaches to public policy.

One of the virtues of the U.S. federal system of government is the ability to allow for geographic differences in policies and experimentation across state lines. But not enough experimentation takes place, because its costs are borne locally while the benefits spread across the country. More and possibly better designed experiments would very likely emerge if states and local governments could compete for federal grants, much like researchers compete for National Science Foundation funds to implement creative new policies, ones that all other states could learn from and emulate—or avoid. A particularly useful focus for this grants process would be economic incentive-based policy instruments; two environmental applications would be congestion pricing experiments for roads and incentive-based instruments to encourage product stewardship and recycling. The possibility of doing policy experiments is one of the greatest advantages of a federal system, but despite their value, practical barriers prevent many experiments from being implemented.

The Value of Policy Experiments

Policymaking is often a leap in the dark. Despite the most careful analysis, it is impossible to know for sure the effects or the costs of a policy. Policies often have completely unintended consequences. Uncertainty about outcomes also prevents risk-

Originally published in R. Morgenstern and P. Portney (eds.), *New Approaches on Energy and the Environment: Policy Advice for the President* (Washington, DC: Resources for the Future, 2004).

averse policymakers from giving many potentially useful policies a fair hearing.

In a more perfect world, perhaps, there would be a way for policymakers to implement a policy, observe its effects, and then rewind the clock, allowing a fine-tuned policy to be reimplemented at no cost or inconvenience to anyone. Unfortunately, such do-overs are possible only in children's games and marriage annulments. But the American political system potentially offers the next best thing: a federal structure, in which the central government is supreme but the individual states and local governments still retain significant powers in many policy areas. A federal system permits policies to be tailored to local preferences and conditions, moving government closer to the people, and it allows thoughtful policy experimentation by states or local governments. As Justice Louis Brandeis wrote in a famous Supreme Court opinion 72 years ago, "It is one of the happy incidents of the federal system that a single courageous state may, if its citizens choose, serve as a laboratory; and try novel social and economic experiments without risk to the rest of the country" (*New State Ice v. Liebman* 1932). Following are some examples of such experiments.

- *Worker safety.* In 1836, Massachusetts enacted the first child labor law in the United States. Several states followed this example, but it was not until 1938 that Congress enacted legislation that effectively ended child labor in manufacturing.
- *Human rights.* Massachusetts was also the first state to outlaw slavery, which the state Supreme Court ruled was incompatible with the state constitution in 1793. The territory of Wyoming was the first place in the country to extend the right to vote to women; when admitted to the Union in 1890, it became the first state to grant female suffrage.
- *Road finance.* Oregon imposed the first gasoline tax in 1919. By 1932, when the first federal gasoline tax was authorized, all states had imposed such taxes.
- *Environmental regulation.* California implemented the first motor vehicle emissions standards in 1959, anticipating the national gov-

ernment by more than a decade.
- *Land use.* New York City enacted the nation's first comprehensive zoning regulation in 1916. Today few local jurisdictions of any size do not have a zoning ordinance. (Houston, Texas, is the most notable example.)
- *Electricity restructuring.* In 1996, California became the first large state, after New Hampshire, to pass legislation opening retail electricity markets to competition and initiating a transition to full deregulation of the price consumers pay for electric energy. Several states, including Massachusetts and New York, followed suit. After a disastrous summer of high prices and rolling blackouts in 2000, California suspended retail competition in 2001, effectively halting all fed eral efforts to impose retail competition nationwide. Meanwhile, other states remain optimistic that their approach to retail competition will continue to avoid the problems California encountered. As this example shows, not all experimental policies are successful.

The Supply of Policy Experiments

Despite the numerous examples of interesting state-initiated policy experiments that we can cite, one thing is certain: there have not been enough of them. When a state undertakes a policy intervention of this sort, its citizens bear the risks of a failed experiment, as California did, yet they capture only some of the potential benefits. The policy produces new knowledge of what works and what does not, and this can be used by other states and the federal government to design further policies. The potential value of this information is large, yet few of the information benefits accrue to the state implementing the policy—and paying the price. Other states are likely to want to free-ride. This "market" failure leads to states underexperiment-ing with new policies.

Examples of Incentive-Based Policies

Economists have long advocated the use of pricing

mechanisms, either directly by taxes or indirectly by construction of artificial markets in permits that are limited in supply, to solve persistent social or economic problems. One particularly compelling example where this approach might work is the imposition of user fees on congested roadways.

Traffic congestion, an inevitable by-product of modern life, is much worse than it has to be. Urban road access is a scarce, valuable resource, yet it is freely accessible to all drivers. Not surprisingly, demand exceeds supply, inevitably leading to lengthy queues, meaning that motorists "pay" for road access even when it appears to be free. But payment in waiting time is an utter waste, whereas payment in road tolls most assuredly is not. Tolls are a revenue source that can be used to provide travel alternatives such as transit, build new roads, or even reduce other taxes.

A more complete explanation of the virtues of road pricing can be found in Parry and Safirova (Chapter 15). Despite these virtues, true congestion pricing—road tolls high enough to appreciably affect use—is uncommon, not only in the United States, but throughout the world. Most examples of successful road-pricing experiments are either "HOT" (high-occupancy/toll) lane experiments, which open up high-occupancy vehicle (HOV) lanes to single drivers willing to pay a toll, or levies on newly constructed lanes. HOT lane experiments are valuable and instructive. To make a real dent in the urban transportation problem, however, pricing reform is needed for the great majority of existing road capacity that is currently free.

Despite its theoretical promise, motorists, and therefore politicians, evidently remain very leery of road pricing. At least part of this leeriness is fear of the unknown. A few real-world experiments could go far toward either allaying these fears or validating them in the most direct possible way.

Another example is using an incentive policy such as a combined product tax and recycling subsidy to reduce solid waste, increase recycling, and promote product stewardship. Recently, managing household waste has become a greater challenge than in the past. After peaking in the early 1990s, recycling rates for many of the products tradition-

ally collected in curbside recycling programs—aluminum cans, PET plastic containers, and glass bottles—have declined. This decline has taken place at the same time that the percentage of the U.S. population with access to curbside recycling has grown. And while these traditional components of the household waste stream are recycled less, the quickening pace of technological change has contributed to a growing number of new products in the waste stream—obsolete computers, cell phones, and other electronic products. Used electronics pose a concern for disposal because of their sheer volume and the fact that they contain hazardous materials. Addressing waste and recycling problems has always fallen to states and localities in the United States; the federal role is minimal. In light of these issues, the time has come for states to be more creative in how they address solid waste and recycling concerns.

One particularly attractive approach for dealing with many waste problems is a combination of an up-front per-unit tax on products and a back-end subsidy, or refund, for recycling. The refund would provide consumers with an incentive to return products for recycling, and the tax should lead to source reduction. If the tax varies with the weight of the product or, perhaps, with the amount of toxic inputs, it would provide an incentive for manufacturers to decrease product weight or substitute away from toxic chemicals that pose a problem at the time of disposal. Research suggests that a tax-and-subsidy policy of this type is a more cost-effective way to reduce waste and promote recycling than many other alternatives, including a so-called advance recycling fee (ARF), used to pay for collection and recycling programs, such as the one considered in the National Electronics Product Stewardship Initiative. And the tax-and-subsidy approach avoids the illegal disposal problems that might occur with disposal fees and landfill bans.

Where combined tax-and-subsidy schemes have been used, they are typically very effective in promoting recycling. For example, states that have deposit-refund programs for beverage containers—so-called bottle bill states—typically achieve higher recycling rates than do states without such pro-

Update

One of the dubious pleasures of writing an article like this is finding out, after publication, that the program or policy you propose already exists in some form or another. In our case, that program is the EPA State Innovation Grant Program. Just as we recommended, the grant process is a competition among state governments to receive federal funds to implement an innovative program in environmental protection. In its first two years, the program awarded 15 grants totaling $2 million.

But while very similar concepts, the State Innovation Grant Program and our proposal differ substantially in size and scope. For one thing, the former is limited so far to new ways of writing environmental permits. In contrast, we envision a broader use that encompasses not only all environmental policies, such as farmland preservation or the protection of endangered species, but any other policy area, from criminal sentencing to highway safety. More importantly, whereas in the current EPA program the total grant monies are about a million dollars per year, we propose a much larger program, one that would possibly make individual awards in the billions, for we believe that states are likely to need large inducements indeed to implement large-reward/high-risk policies such as those we discuss in the article (e.g., congestion pricing on roads or tax/subsidy policies to reduce solid waste).

Nonetheless, there is surely much to learn from the State Innovation Grant Program; it may yield useful insights into the design of a larger and more comprehensive program.

motor oil and related products such as filters and containers that rely on an up-front fee charged at time of sale, combined with a return incentive paid to collectors of used oil and oil products. The programs in both locations have been very effective, with recycling rates in excess of 70 percent. Lead-acid car batteries and tires, both of which have deposits and refunds applied to them in many states, also have high recycling rates. On the other hand, product-specific recycling programs that fail to provide incentives have met with limited success. Rechargeable batteries are a good example. These are voluntarily recycled by industry, but no incentives are provided for consumers to return them for recycling. We estimate that recycling rates for rechargeable batteries are approximately 12 to 13 percent.

Despite the benefits of incentive-based programs, many states are reluctant to adopt them because of a lack of political will to increase taxes and potential practical difficulties and administrative costs. The waste problem created by many products—particularly new products such as electronics—is becoming widely recognized, but instead of incentive-based policies, states are instituting or proposing bans on particular products in landfills, ARFs that do not vary with product weight or material content and are not accompanied by a return incentive, and so-called producer responsibility schemes mandating that producers provide collection and recycling services for their products.

Implementing policies that combine up-front product fees with subsidies for recycling has some challenges. High transaction costs associated with administering refunds and sorting returned containers by brand have plagued bottle bill programs in several states. California avoids some of these pitfalls by having all containers returned to redemption centers and not requiring brand sorting. Transaction costs can be lowered even more when the tax-and-subsidy system bypasses final consumers and instead pays the refund to collectors, like the used-oil programs do. Collectors in a system such as this have the incentive to collect as much as possible and would be likely to provide

grams. The aluminum can recycling rate in 1999 was 80 percent in bottle bill states but only 46 percent in non-bottle bill states. California and several provinces in western Canada have systems for used

incentives to consumers to return products. Tax-and-subsidy policies also face challenges when the products being targeted are durable ones, such as computer monitors or cell phones, or when products include small mounts of hazardous materials. Policies that are set up and implemented differently in different states and applied to a variety of products could provide valuable information to other states thinking of putting such policies in place. They also could provide information about whether the federal government should play a role in policy coordination.

Encouraging State Experiments with Incentive-Based Policies

The very limited use of incentive policies to combat traffic congestion or to encourage recycling indicates that despite the potential payoff, the perceived risks of failure from implementation of these policies are often too great for any state or local politician to take up. If the federal government can persuade a few states or local governments, or even one, to implement experiments in congestion pricing or the use of combined tax-and-subsidy instruments to promote recycling, the knowledge gained will benefit all states.

However, for some policies, it will take a lot of federal persuasion. We know this is true for congestion pricing, because the federal government has tried to get states to adopt experimental road-pricing programs before, with little result. The Department of Transportation began the Congestion (later Value) Pricing Pilot Program in 1991. This program would provide a local Metropolitan Planning Organization (MPO) up to $25 million in planning grants if the plan led to the actual use of road pricing to reduce congestion. An MPO is the local government entity responsible for preparing transportation plans for the metropolitan area, usually consisting of representatives from each of the area's local governments. The program proved singularly ineffective as a promoter of the use of pricing to ration road access. Although a number of local planning studies led to concrete

road-pricing proposals, political opposition to these plans prevented almost all from being implemented.

The federal government has not, to our knowledge, offered financial incentives to states or localities to adopt particular waste and recycling policies. It has, however, put much effort into promoting one incentive-based policy: "pay-as-you-throw" (PAYT) pricing of household solid waste collection and disposal. EPA's Office of Solid Waste has had an active program since the early 1990s, which includes fact sheets, promotional materials for communities, brochures and videos with advice on how to set up PAYT programs and avoid pitfalls and problems such as illegal dumping, and case studies of communities that have implemented such programs. Although waste reductions have been well documented in PAYT communities, only about 4,000 communities nationwide use this form of pricing.

Recommendations

We recommend that the federal government use a policy auction to encourage states to develop and formulate incentive-based policies to alleviate urban traffic congestion and to promote recycling, and to stimulate other policy experiments as well. The policy auction should include the following key elements:

- *A competitive proposal process.* State and local government agencies will be invited to submit proposals for funds to implement actual policy demonstration projects. Because it is difficult to say how much enticement a state or local agency will require to submit a proposal, there should be no limit on the amount requested. Proposals will be judged on the basis of their likely contribution to practical knowledge, their credibility with respect to ultimate implementation, their generalizability to other states or localities, and their budgets.
- *An independent proposal review panel chosen by a committee of qualified experts outside the political process.* With the increase in funding suggested here, some safeguards would be needed to

ensure that the program does not degenerate into a pork barrel program. To ensure that the proposals are evaluated on a scientific and not a political basis, they should be evaluated by natural and social scientists knowledgeable about the particular subject area. For example, the Transportation Research Board, as a subsidiary of the National Research Council, is the one organization with the expertise, independence, and prestige to evaluate congestion pricing proposals.

- *No transfers of grant moneys (beyond initial planning grants) before the policy begins to take effect.* As the whole point is to gain actual experience with the incentive policy, and because decisionmakers in the state or local government agencies responsible for implementing a policy will likely face intense pressure to renege, their feet will need to be held to the fire.
- *Ex post facto evaluation by independent researchers.* To get the most out of the experiment and to maximize its credibility, scientific evaluation of outcomes is important. One possibility would be to have the National Academy of Sciences (NAS) convene a group of experts to perform an assessment. NAS has the independence and prestige that would permit a credible assessment; however, the evaluation committee should be unconnected to the committee choosing the applications.

Innovative state policies, such as the use of pricing instruments to reduce congestion or to reduce waste and increase recycling, can be a tough sell. The fact that we have little experience with these instruments in the United States to help inform future policy design is another drawback that contributes to policymakers' reluctance to try them in the future. The program proposed here will use federal money to break through this logjam.

Suggested Reading

Arha, Kaush, and Barton H. Thompson, Jr. (ed.). Forthcoming. *Endangered Species Act and Federalism: Effective Species Conservation Through Greater State Commitment.* Washington, DC: Resources for the Future.

Palmer, Karen, Hilary Sigman, and Margaret Walls. 1997. The Costs of Reducing Municipal Solid Waste. *Journal of Environmental Economics and Management* 33 (June): 128–150.

See also Chapter 15 in this volume: Ian W.H. Parry and Elena Safirova, "Pay as You Slow: Road Pricing to Reduce Traffic Congestion."

Part 6

Resource Management and Conservation

23

Catching Market Efficiencies
Quota-Based Fisheries Management

James Sanchirico and Richard Newell

The success in New Zealand of a comprehensive system of individual fishing quotas (analogous to cap-and-trade programs) points the way to a new approach to addressing the disastrous depletion of the world's fisheries.

Too many boats are chasing too few fish, it is said of U.S. fisheries. To address this problem, policymakers must stop treating the symptoms—by restricting gear, seasons, and areas—and focus instead on the incentives fishermen face. Individual fishing quota (IFQ) programs are a promising tool to cut economic and ecological waste in fisheries.

IFQ programs are analogous to other cap-and-trade programs, such as the sulfur dioxide allowance-trading program. They limit fishing operations by setting a total allowable catch, which is then allocated among fishing participants, typically based on historical catch. When fishermen have access to a guaranteed share of the catch, they have an incentive to stop competing to catch as much as possible and start improving the quality of their catch. When shares are transferable, inefficient vessels find it more profitable to sell their quotas than fish them. The result will be fewer and more-efficient vessels.

Worldwide, IFQs are used to manage more than 75 species, including 4 in the United States. Although assessments of these programs are generally positive, their future is unclear. In the United States, a six-year moratorium on implementing new IFQ systems expired in September 2002, but policymakers continue to debate program elements. Legislation introduced in 2001 by Senators Olympia Snowe (R-ME) and John McCain (R-AZ), for example, prohibits the selling and leasing of quota shares (S.637 §2.6.a). Many questions remain, in part because there have been limited opportunities to study the current programs.

The system in New Zealand—the world leader in imple-

Originally published in *Resources*, No. 150, Spring 2003.

menting IFQs—provides a standout opportunity for research. Richard Newell and James Sanchirico of RFF and Suzi Kerr of Motu Economic and Public Policy Research in Wellington, New Zealand, are documenting and measuring the changes in New Zealand fisheries. Here we describe New Zealand's IFQ system and discuss its effects—distributional changes, market efficiency and economic gains, biological health, and political and administrative changes. We then suggest some implications for U.S. fisheries policy.

New Zealand's Quota Management System

In 1986, New Zealand adopted IFQ programs for 26 marine species; the system now includes some 45 species. This system thus provides a wealth of information, covering more than 15 years and a large number of species that are diverse in both economic and ecological dimensions. For example, average life spans range from 1 year for squid to 125-plus years for orange roughy. Some species, such as abalone, occupy inshore tidal areas and are caught using dive gear; others are found offshore in depths over 1,000 meters and require specialized nets and large vessels.

Seafood is New Zealand's fourth-largest source of export income, and more than 90% of fishing industry revenue is derived from exports. As of the mid-1990s, the species managed under the IFQ system accounted for more than 85% of the total commercial catch taken from New Zealand's waters. We estimate that the quota markets have an estimated market capitalization of about NZ$3 billion, which is approximately equivalent to US$2 billion.

Under the IFQ system, the New Zealand exclusive economic zone was divided into ten quota management regions for each species based on the locations of major fish populations. Quotas for catching fish were set for each species in each region, creating a number of fishing quota markets. In 2000, the total number of fishing quota markets stood at 275, ranging from 1 for hoki to 11 for abalone. The quota rights can be split and sold in smaller quantities, and any amount can be leased and subleased. There are limits, however, on the number of quotas that any one company or individual can hold.

The New Zealand Ministry of Fisheries sets an annual total allowable catch for each fish stock. The goal is to have fish populations that can support the largest possible annual catch—the "maximum sustainable yield"—with adjustments to account for environmental, social, and economic factors. Compliance and enforcement are undertaken through detailed reporting procedures that track the flow of fish from a vessel to a licensed fish receiver to export records, along with satellite monitoring and an at-sea surveillance program that includes on-board observers.

Distributional Changes

Throughout the world, the debate about whether to implement market-based approaches in fisheries has concerned their distributional implications—the potential concentration and industrialization of the fishery. Critics of quota management systems argue that such systems will harm small-scale fishermen, a claim analogous to those made for preservation of the family farm. Proponents counter that current management practices are not sustainable, and the survival of any fishing industry—whatever the proportions of small versus large players—is better than nothing. Of course, an IFQ system could be designed to maintain a socially desirable composition of participants—in the Alaskan halibut fishery, for example, there are restrictions on who can trade with whom—but such constraints can reduce the potential efficiency gains.

In New Zealand, there has been a 37% decline in the number of quota owners, mostly in fisheries that were overfished and had overcapacity problems. Although some small fishing enterprises have exited, many remain. In fact, the typical quota owner holds the minimum required to participate in the fishery. And although about one-quarter of the fishing quota markets are concentrated (as defined in U.S. Department of Justice antitrust regulations), these fisheries were concentrated before the introduction of IFQs. In short, the industry started out with a few big players (vertically inte-

Update

Two significant events have occurred in New Zealand since the publication of this article. First, the NZ quota management system has expanded to include 93 species with plans to introduce 7 additional species in 2005. This brings the number of fishing quota markets to 550, up from 275 in 2000. Eventually, all living marine resources (including invertebrates and some seaweeds but not marine mammals) that are commercially valuable will be brought into the system. Second, the Hoki stock, which is the largest in terms of catch volume and value, is below its biological target as a result of several years of poor survival of young stocks. The total allowable catch has been reduced by 50 percent to allow the stock to rebuild itself.

In the United States, under the White House's Ocean Action Plan released in 2004, IFQs will have a role in U.S. fisheries management. Unlike NZ where a general mandate came from the government, adoption of programs in the United States will happen one fishery at a time. There are currently many regional fishery councils that are developing plans or are studying whether to adopt IFQs, such as the red snapper fishery in the Gulf of Mexico, the West Coast groundfish fishery, and the Gulf of Alaska groundfish fishery. The debate in the United States is focused on various distributional concerns; the result is a number of proposals for restrictions on trading to preserve the social fabric of fishing communities.

We concluded the article with the hope that by 2013, U.S. fisheries would double in value and fish stocks would be either rebuilt, rebuilding, or sustainably managed. While progress is taking place on all these fronts, the goal is unlikely to be met without a sea change in attitudes toward fishery management.

grated catch and food-processing companies) and many small fishing enterprises, and it looks much the same today. The size of holdings of the larger companies, however, has increased.

Market Efficiency and Economic Gains

Whether tradable permits are being applied to fish, pollution, or other resource problems, the ability of firms to buy and sell quotas in a well-functioning market is necessary for achieving efficiency gains.

We find that the New Zealand IFQ markets have been very active, with about 140,000 annual leases and 23,000 sales of quotas as of 2000—an annual average of about 9,300 leases and 1,500 sales. In the typical quota market, the percentage of the total allowable catch that is leased in any given year has risen considerably, from 9% in 1987 to 44% in 2000.

A majority of the transactions between small and medium-sized quota owners are handled through brokers; larger companies typically have quota managers on staff. Brokers advertise quota prices and quantities for sale or lease in trade magazines, newspapers, and on the Internet and charge a brokerage fee of 1% to 3%. Differences in prices paid for the same quota in a given month have fallen, indicating that market participants are learning and these markets are developing.

We also find evidence of economically sensible behavior in the relationship between quota lease and sale prices and fishing input and output prices, quota demand, ecological variability, and market rates of return. Moreover, after controlling for relevant factors, we estimate an increase in the value of quota prices, consistent with an increase in the profitability of the fisheries. Furthermore, as theory would predict, we find larger gains for fish stocks that were initially overcapitalized and overfished and that faced significant reductions in total allowable catches; in these cases, sale prices rose at an average annual rate of 9% and lease prices, 4%.

Overall, the results suggest that these markets are operating reasonably well, implying that market-based quota systems can be effective instruments for efficient fisheries management.

Health of Fish Populations

IFQ programs can be used to improve the biological health of populations if the cap on total catch is appropriately set. In New Zealand, evidence indicates that fish populations within the IFQ system are no worse off, in some cases show clear signs of recovery, and in other cases are apparently improving, given current catch levels. For more than half the fish stocks, however, sufficient data to measure changes in fish populations do not exist, and the program is too young for us to assess whether measures taken to improve fish stocks with very long life spans are succeeding.

The IFQ system has spawned some important behavioral changes in the fishing industry that might lead to an improvement in the health of fish populations. For example, New Zealand fishermen have reallocated fishing effort across locations and over the fishing year to minimize the incidental catch of nontarget fish (by-catch) and to avoid spawning aggregations. Because the system covers so many species, fishermen must reduce by-catch to avoid having to buy additional quotas in the market or pay fees to the government for the fish that are not in their quota portfolios.

Political and Administrative Changes

Proponents of IFQ programs see many political and administrative benefits from allocating shares of the catch to individuals. One way in which the political and administrative fisheries systems have changed in New Zealand is through the formation of quota-owner management companies. These companies invest in value-added research on marketing and processing and work with government scientists to improve data collection and fish stock assessments. As one government scientist told us, "Since the adoption of the IFQ system, things have not always gone smoothly with the industry, but the creation of the management companies has reduced the regulatory transaction costs in setting total allowable catch levels. Overall, there is less animosity between the government and the fishing industry than before the program."

How fishery management is funded has also changed. New Zealand fishermen now pay annual fees that cover many of the governmental management costs of the program. At the same time, the fishing industry has become more directly involved in these activities.

Implications for U.S. Policy

Our findings are relevant for the ongoing IFQ policy debates in the United States. We can infer from the behavior of the New Zealand fishermen that the flexibility of the system and the ability to transfer shares has high economic value. Furthermore, the option of short-term leases appears to have significant value, as revealed by the dramatic increase in leasing over time. In addition, the opportunity to operate in both sale and lease markets provides an additional dimension across which relevant economic and ecological information can be exchanged and rationalized.

We also find a significant decrease in the number of owners, even as the number of small players has remained high. For overcapitalized fisheries, a reduction in the number of vessels is beneficial. The United States is seeking to reduce capacity by offering fishermen money to retire their vessels, but often the takers are already on their way out, and very little reduction in fishing effort occurs. IFQ programs offer a market-based solution to this overcapacity problem that does not rely on direct payments from the government. The potential sociocultural costs associated with IFQ systems need to be weighed against the benefits of reducing fishing effort, but such concerns can be addressed within an IFQ framework.

The collaborative atmosphere among the stakeholders in New Zealand fisheries management contrasts with the approximately 100 lawsuits currently pending against the U.S. National Marine Fisheries Service (the agency responsible for fishery management). The ability to set fees that recover the cost of management is appealing, especially in this time of budget deficits. This additional revenue could be used to offset the costs of data collection, scientific research, onboard observer programs, and other enforcement programs.

Furthermore, the simultaneous adoption of so many species into the New Zealand system in 1986 provided the industry with incentives to avoid by-catch and improve its stewardship of the marine environment. U.S. policymakers attempting to design single-species IFQ programs that address by-catch might want to consider adopting IFQ systems for a much wider set of species rather than trying to engineer a solution one fishery at a time.

Finally, we estimate that the total value of New Zealand's IFQ fisheries, which account for more than 85% of the commercial catch taken in its waters, has more than doubled in real terms from 1990 to 2000, even as fish stocks are improving. Wouldn't it be great if, 10 years from now, the United States could cite similar statistics?

Suggested Reading

Council of Environmental Quality. Dec. 17, 2004. U.S. Action Ocean Plan. Available at: http://ocean.ceq.gov/actionplan.pdf.

Newell, R.J., J.N. Sanchirico, and S. Kerr. 2005. Fishing Quota Markets. *Journal of Environmental Economics and Management* 49: 437–462.

U.S. Commission on Ocean Policy. Sept. 20, 2004. An Ocean Blueprint for the 21st Century. Available at: http://oceancommission.gov/documents/prepub_report/welcome.html.

24

Marketing Water
The Obstacles and the Impetus

Kenneth D. Frederick

As water becomes more scarce, the need to allocate our limited water resources more efficiently is an increasingly pressing issue. The use of market incentives for managing the use of water has been slow to develop, but the potential benefits are great.

Water is becoming increasingly scarce in the United States. Demand is rising along with population, income, and an appreciation for the services and amenities that streams, lakes, and other aquatic ecosystems have to offer. In contrast, the options for increasing supplies are expensive relative to current water prices and often environmentally damaging. Furthermore, contamination and unsustainable rates of groundwater use threaten current supplies in some regions.

Ordinarily, Americans count on prices and markets to balance supply and demand and allocate scarce resources. When demand increases faster than supply, higher prices provide incentives to use less and produce more. And, as conditions change, markets enable resources to move from lower- to higher-value uses. Market forces, however, have been slow to develop as a means of adapting to water scarcity. Both the nature of the resource and the institutions established to control its use help explain why.

Market Obstacles

Efficient markets require that buyers and sellers bear the full costs and benefits of transfers. But interdependencies among the many users of a stream or aquifer make that difficult to do. Selling water rights, for example, is likely to alter the quantity of water in a stream or the location of a diversion or returnflow (water withdrawn from a stream or aquifer that is returned to a location where it can be used again). Third parties—people

Originally published in *Resources*, No. 132, Summer 1998.

benefiting from the water other than the buyer and seller—will be affected by the change. Third-party impacts might include a change in the recreational amenities provided by a free-flowing stream or the erosion of a rural community's tax base when a farmer sells water to a city.

Efficient markets also require well-defined, transferable property rights. But riparian rights, which are still the principal basis of water law in the Eastern United States, are poorly defined because water use is subject to regulatory or judicial interpretations as to what is reasonable or might unduly inconvenience others. Moreover, these rights are not directly marketable because they are attached— and their use is restricted—to the lands adjacent to a stream.

In the West, where streams are less common and flows are smaller and less reliable, "prior appropriation" quickly displaced riparian rights as the primary basis of water law. Appropriative rights are established by withdrawing water from its natural source and putting it to beneficial use. During drought, supplies are allocated according to the principle of "first in time, first in right." This principle provided a powerful incentive for the quick diversion of streamflows and allowed irrigators to acquire the highest priority rights to much of the water. While appropriative rights can be transferable, they are commonly attenuated in ways that limit how and where water can be used.

Water has traditionally been treated as a free resource to be harnessed to serve cities, factories, and farms. Anything less was seen as wasteful. Thus, subsidized water storage and distribution systems and irrigation projects contributed to a nine-fold rise in water withdrawals from 1900 to 1970. They also contributed to the loss of tens of thousands of miles of once free-flowing streams and, eventually, to a shift in national policy. To protect streamflows and recover forgone environmental and recreational values, Congress passed legislation such as the Wild and Scenic Rivers Act of 1968, the National Environmental Policy Act of 1969, and the Endangered Species Act of 1973.

In recent decades, these environmental laws have been used to block construction of many dams and in some cases to challenge previously established rights to divert water from streams and lakes. Domestic, industrial, and agricultural users continue to vie for water that is withdrawn from reservoirs and streams, and now all three groups must also vie with environmentalists and recreationists over how much water can be diverted. Conflicts also arise over the priority that dam managers should give to flood control, water supplies, hydropower production, fish habitat, and recreational opportunities. These conflicts are now generally played out in the courts or administrative proceedings rather than in the marketplace.

Overcoming the Obstacles

If water has been slow to be bought and sold like other commodities, the incentives to do so are strong. Most of the senior water rights in the arid and semiarid West are held by farmers and irrigation districts. They pay nothing for the water itself and generally only a modest amount to have it delivered to their farms. As a result, enormous amounts of water are applied liberally to relatively low-value crops and the marginal value of the water is likely to be well under $50 an acre-foot (af)—the quantity of water that will cover one acre to a depth of one foot. In some cases the value of the water could be increased simply by leaving more in the river to provide hydropower, fish and wildlife habitat, and recreation rather than diverting it for irrigation. In many other instances, the value of water would rise by selling some of it to urban areas that are spending more than ten times as much to augment supplies through recycling or other costly water projects.

Despite the obstacles, the impetus to move from lower to higher value use is driving some water transfers. Temporary transfers are becoming increasingly common to respond to short-term fluctuations in supply and demand. Precisely because they are temporary, short-term leases, options to purchase during dry periods, and one-time purchases through water banks blunt a principal third-party concern that a transfer will permanently undermine the economic and social viability of the water-exporting area.

Transfers among farmers within the same irrigation district are common and relatively easy to arrange because the third-party impacts are likely to be small and positive when the water stays within the community. But when farmers want to sell water to cities, irrigation districts resist, fearing the loss of agricultural jobs and income that accompanies rural water use.

A water bank provides a clearinghouse to facilitate the pooling of surplus water rights for temporary rental. If well-defined, its rules and procedures can reduce the costs and uncertainties associated with a transaction and increase the opportunities for both buyers and sellers.

California established emergency Drought Water Banks in 1991, 1992, and 1994 to reallocate water among willing buyers and sellers. Water purchased largely from farmers willing to idle land or pump groundwater rather than divert surface water for irrigation was sold to cities and farms or used to protect water quality in the state's delta region and meet instream fish needs. Any adverse third-party impacts on the water-exporting communities were probably insignificant compared with the overall benefits of moving water to higher-value uses. Sales exceeded $68 million in 1991; they averaged less than $11 million in the latter years when drought conditions subsided. Idaho and Texas have established permanent water banks and other states are now considering establishing them as well.

Transferring Permanent Water Rights

Temporary water transfers are particularly useful for adapting to short-run changes attributable to such things as climate variability. They are less effective in dealing with long-term imbalances between supply and demand resulting from changing demographic and economic factors, social preferences, or climate. At some point, the historical allocation of water becomes sufficiently out of line with current conditions to warrant a permanent transfer of rights.

The process of resolving the third-party issues associated with the transfer of a long-term shift in water use is often slow, costly, and contentious.

Update

The vast infrastructure of dams, reservoirs, canals, pumps, and levees that have been constructed to deal with the vicissitudes of the hydrologic cycle and to provide for rising water demands were designed and are operated on the assumption that future climate will be similar to the past. A greenhouse warming, however, will affect precipitation patterns, evaporation rates, the timing and frequency of runoff, and the frequency and intensity of storms as well as the demand for water. Even modest changes in climate can lead to changes in water availability outside the range of past hydrologic variability. But the magnitude and even the nature of these impacts on the supply and demand for water are highly uncertain.

The financial and environmental costs of building infrastructure in anticipation of uncertain climate and hydrologic changes would be high and the benefits uncertain. On the other hand, relaxing constraints on water transfers and reservoir operating rules and developing institutions to encourage exchanges of water through markets would create a system that is more efficient and better able to adapt to whatever the future may bring.

Proposed transfers face the hurdle of proving the negative, that a change will not harm others. This requirement stifles the development of markets in water rights. The Colorado–Big Thompson project (described in the sidebar), which has been able to avoid third-party issues, is the exception. The ongoing efforts (described below) of the coastal region of Southern California and the city of Las Vegas are more indicative of the obstacles to acquiring additional water.

Both of these geographic areas face the challenge of meeting growing demands for water at a time when their traditional sources are declining and environmental considerations restrict the devel-

Trading Water

The Bureau of Reclamation's Colorado–Big Thompson project brings an average of 230,000 acre-feet of water annually from the Colorado River Basin across the continental divide to northeastern Colorado. Rights to proportional shares of this water are traded actively within the Northern Colorado Water Conservancy District unencumbered by third-party concerns.

Under western water law, downstream users generally own the rights to the returnflows. But in this case the district is able to retain ownership of the returnflows because the water originates in another basin. As a result, rights to the water are traded within the district much like stocks in companies. This arrangement does not eliminate the third-party impacts associated with returnflows, only the need to consider them in transfer decisions. The benefits of being able to transfer water readily among agricultural, municipal, and industrial users exceed any likely third-party costs.

However, limiting sales to within the conservancy district precludes opportunities for even more profitable transactions. For example, an acre-foot of water in perpetuity has sold for $3,500 more in the neighboring Denver suburbs than in the conservancy district.

opment of new ones. Los Angeles has already been forced to reduce the amount of water it takes from the Mono Lake region and, to comply with a mandate to improve environmental conditions in Owens Valley, will have to further reduce the city's supplies. In addition, the Southern California Metropolitan Water District (MWD), a large water supplier servicing more than fifteen million consumers including the residents of Los Angeles, is losing access to surplus water (that is, unused entitlements of other states) from the Colorado River. Las Vegas, meanwhile, has been depleting its groundwater stocks, causing subsidence within the city.

Under a 1989 agreement, Southern California's MWD has invested more than $100 million in lining irrigation canals and other water conservation projects in the Imperial Irrigation District. In return, MWD received the right to use the conserved water, approximately 106,000af per year, for at least thirty-five years. Provisions were introduced to assure that neighboring irrigation districts in the United States did not lose their water rights as a consequence. But the impacts on irrigators across the border where groundwater recharge declined were ignored because the Mexicans lack a legal claim to the water.

San Diego receives about 90 percent of its water from the MWD and, as a junior claimant, is the first to be cut back in time of drought. To increase the quantity and reliability of its supplies, the San Diego Water Authority has agreed to fund additional conservation efforts in the Imperial Irrigation District in return for the conserved water. As originally proposed, 20,000af would be transferred in 1999, with the annual quantity increasing to 200,000af after ten years. Disputes with MWD over use of the Colorado River Aqueduct to transport the water, however, have delayed completion of the transaction.

Las Vegas, which is already using most of Nevada's legal entitlement to the Colorado River, is seeking to buy more shares of the river from states with unused entitlements. Legal issues have undermined earlier proposals for interstate and interbasin sales of Colorado River water and enabled Southern California's MWD to take unused entitlements for free. Rising water values, however, are creating new interest in such sales in Nevada, which lacks rights to surplus flows, and in states wanting to benefit from their unused shares.

In 1996, Arizona established a Water Banking Authority to purchase their own unused Colorado River water for storage in groundwater basins and possible sale to California and Nevada. Interstate sales, however, are tightly restricted; they are limited to 100,000 af/year and only when there is no use for the water in Arizona and there are no shortages on the Colorado River.

Las Vegas is also interested in buying water from Utah, which has not been using its full entitlement. However, a transfer between an upper basin state (Utah) and a lower basin state (Nevada) could require renegotiation of the 1922 Colorado River Compact dividing the river between the two basins.

The Federal Role

State institutions are primarily responsible for allocating waters within their borders. But the federal government—manager of much of the West's surface waters, supplier of water to about 25 percent of their irrigated lands, the source and enforcer of environmental legislation affecting water use, and trustee for Indian water rights—also has a critical role in breaking down the institutional obstacles to permanent water transfers. Some steps in this direction have been taken.

- In 1988, the Department of Interior adopted a policy of facilitating voluntary water transfers involving federal facilities as long as the transfers comply with federal and state law, have no adverse third-party impacts, and do not adversely affect facility operations.
- The Central Valley Project Improvement Act of 1992 authorized the transfer of federally-supplied water outside the project service area. Although no off-project transfers have been approved yet, the act is potentially significant because the project is the largest water storage and delivery system in California and most of the project water is allocated to agriculture under highly subsidized terms.
- A proposed federal rule from the Department of Interior (*Federal Register*, December 31, 1997) is designed to encourage and facilitate voluntary transactions among the three Lower Colorado River Basin states by establishing a framework for approving and administering interstate agreements.
- In addition, the federal government as well as some states have been acquiring water for environmental purposes, such as the preservation of endangered species. These purchases help establish markets as viable mechanisms for allocating water.

More steps are of course needed. Uncertainties surrounding large but unquantified Indian water claims, for example, hinder the assignment of clearly defined, transferable property rights in water. Providing the tribes with rights that could be sold for uses off the reservations would foster water marketing as well as tribal welfare.

Finally, water scarcity and the potential benefits of water marketing are not limited to the West. In the East, riparian rights are gradually being replaced by or supplemented with permits. The advantages of using markets to allocate these permits will grow as the resource becomes increasingly scarce. Indeed, auctioning and trading permits are innovative approaches that might facilitate a more efficient allocation of water. It is unlikely, however, that markets resembling the ones we use to allocate most goods and services will ever become commonplace to transfer water. Finding expeditious ways to deal with the third-party effects that plague nearly all water rights transfers is critical if traditional market forces are ever to thrive. In the meantime, the enormous potential benefits of water marketing still wait to be tapped.

Suggested Reading

Frederick, Kenneth (ed.). 2002. *Water Resources and Climate Change*. Northampton, MA: Edward Elgar.

Gleick, Peter et al. 2000. Water: The Potential Consequences of Climate Variability and Change for the Water Resources of the United States. The Report of the Water Sector Assessment Team of the National Assessment of the Potential Consequences of Climate Variability and Change for the U.S. Global Research Program. Oakland, CA: Pacific Institute for Studies in Development, Environment, and Security.

Wahl, Richard W. 1989. *Markets for Federal Water: Subsidies, Property Rights, and the Bureau of Reclamation*. Washington, DC: Resources for the Future.

Ecosystem Management
An Uncharted Path for Public Forests

Roger A. Sedjo

The U.S. Forest Service traditionally relied on multiple-use management approaches that balanced demands for timber, recreation, and wildlife habitat. But more recent pressures from the Endangered Species Act have led to "ecosystem management," which places the highest priority on the health of forests. This raises a number of difficult issues concerning potentially conflicting objectives in forest management.

Should public forests be managed to reduce all traces of modern human activities or to produce goods and services? Recently, the U.S. Forest Service seemed to answer that question by saying that it would like to restore the forests of the northern Rockies to presettlement conditions—that is, to the way the forests were at the start of the nineteenth century. This is indicative of the Forest Service's new philosophy of ecosystem management and reflects its shift away from multiple-use management, which has been the practice on public forestlands since the 1960s.

The impetus for both approaches is the desire to sustain forests. Concern about the rapid rate of logging on public lands following World War II led to congressional legislation that called for multiple-use management. This legislation explicitly recognized the worthiness of a range of goods or services provided by public forests—including market goods, such as timber, and nonmarket services, such as habitat for wildlife. Congress charged the Forest Service with managing forests to produce a mix of both within the context of sustainability.

In recent years, however, the leadership of the Forest Service has backed away from this goal as its attention has focused on forest ecology—the totality of relationships between forest organisms and their environment. This concern with forest ecology is embodied in the leadership's advocacy of ecosystem management. In accordance with this philosophy, the service has all but abandoned the notion of forests as primarily a vehicle for producing multiple goods (or "outputs") desired by society. Instead of practicing *multiple-use management,* which

Originally published in *Resources,* No. 121, Fall 1995.

emphasizes the sustainable production of myriad goods and services, the Forest Service has embraced *ecosystem management,* wherein the condition of forest ecosystems—the complex of forest organisms and their environment functioning as an ecological unit in nature—is considered to be the preeminent output.

Although an ecosystem-based approach has much to offer in the form of a broader, more integrated, and more comprehensive view of the forest—and thus contributes to the development of more effective management tools—its defect is its disregard for certain socially approved objectives. In essence, ecosystem management aims to restore forests to some biological condition that reflects fewer human impacts, but just *what* condition is a matter of arbitrary selection. Because ecosystem management has no real legislative mandate, decisions to seek any one of many possible conditions are being made by the Forest Service rather than by society at large, which makes its wishes known through the legislation of management objectives. More to the point from the perspective of taxpayers, these decisions are being driven almost exclusively by biological considerations, with little attention paid to economic and other concerns. In short, when identifying objectives, ecosystem management ignores the social consensus implicit in the congressionally legislated objective of producing multiple market and nonmarket forest outputs and, instead, attempts to achieve some arbitrary forest condition about which society has little say.

The comparison of ecosystem management and multiple-use management presented below highlights the pitfalls of the Forest Service's new philosophy. Despite these pitfalls, it would be unwise simply to dismiss ecosystem management. It has resulted in the development of some highly effective management tools and activities and reflects a concern for the health of ecosystems that traditional management may not sufficiently recognize. Management for multiple-use objectives should continue to be the practice on public lands, but perhaps with a view to incorporating some aspects of ecosystem-based management.

The Need for Clear Objectives

Management of public forestlands requires the identification of clear objectives and the development of a regime (procedures and tools) that will achieve the objectives without violating the constraints imposed by the availability of resources and the acceptability of actions and outcomes.

Forest management without objectives is meaningless. In the absence of stated goals, we cannot differentiate successful forestry activities from unsuccessful ones. And in the case of public forestlands, the ability to gauge the success of management efforts takes on added significance because these efforts are being financed by taxpayer dollars. Moreover, without specifying objectives, we cannot ensure that the preferences of society are being reflected in the way that our forests are managed. These preferences should inform goals as well as define the constraints within which a management regime will operate.

But where objectives dictate the management approach under multiple-use forestry, ends merge with means under ecosystem management. Indeed, in actual practice, the objective of ecosystem management is most often simply the application of an ecosystem, or ecosystem-based, approach that is concerned first and foremost with the state of the forest itself. Thus while the Forest Service has been embracing ecosystem management as its operating philosophy for several years, no clear vision of output goals, at least as traditionally understood, has emerged. What has emerged is a preoccupation with forest condition—that is, with biological attributes, such as a forest's structure (mixture of younger and older trees) and variety of tree species—rather than with the goods and services (particularly those consumed by humans) that forests provide.

Ecosystem Management versus Multiple-Use Management

Jack Ward Thomas, chief of the Forest Service, has said that ecosystem management means sustaining forest resources, from which will flow many goods and services. But our public forests have for decades been managed to sustain multiple uses. Is ecosystem management really different from multiple-use management?

The mandate for multiple-use forestry has been expressed by law since 1960, when Congress passed the Multiple-Use Sustained Yield Act. This act acknowledges that forests generate both market goods and nonmarket goods. The objective of multiple-use management is to produce the mix of these market and nonmarket goods that maximizes the value of forests to society.

If the objective of ecosystem management is simply the management of whole ecosystems for a variety of purposes, such management might be viewed as an expansion of the multiple-use approach. Under this expanded approach, the set of outputs under consideration would broaden to include the biological condition of the forest itself. In addition, the boundaries of the management unit would enlarge, because changes in forests affect the geographic area around forests. Finally, the potential uniqueness of each forest ecosystem would be recognized and new management techniques would be introduced. Conceptually, these considerations represent modest extensions of multiple-use management. The job of the public forest manager would continue to be producing the mix of outputs that would maximize the social value of the forest.

But proponents of ecosystem management are reluctant to treat such management as a mere extension of multiple-use forestry. Unlike multiple-use management, which focuses on distinct forest outputs, many of which are consumed directly by humans, ecosystem management focuses on forest condition as the dominant forest "output." In this context, timber, recreational opportunities, and other traditional forest goods are merely by-products of managing forests to achieve one of many possible forest conditions. Production of these other outputs is tolerable as long as it does not conflict with the primary objective of achieving one of these conditions. Thus, for example, timber harvests that improve the condition of a forest are acceptable. But while under multiple-use management such harvests could be decreased in order to

increase recreational opportunities, under ecosystem management such opportunities would not be augmented if they resulted in what was perceived as an undesirable change in forest condition. Under ecosystem management, forest condition—as the preeminent forest output—is not subject to trade-offs with other forest outputs, as it is under multiple-use management.

A clear statement of the objectives of ecosystem management appears in the Forest Service's proposed regulations dated April 13, 1995. In the proposed regulations, the management objective is stated as follows: "The principal goal of managing the National Forest System is *to maintain or restore the sustainability of ecosystems...*" (italics added). By this articulation, the goal of management is very similar to the constraints of other forest management sytems: sustainability. The proposed regulation goes on to suggest that the achievement of this goal will result in "...multiple benefits to present and future generations."

The Implications of Ecosystem Management

Given ecosystem management's focus on forest condition, the first question that arises is whether a given forest's current condition should be maintained or modified to some specified extent. Once such a decision is made, the vagueness of the management objective disappears. But, as I suggest below, the selection of desired or acceptable condition is essentially arbitrary. As a result, the objective chosen today may be sadly outdated in perhaps a few years.

Although not readily apparent, arbitrariness is reflected in the Forest Service's apparent preference for restoration, rather than maintenance, of forest condition. This restoration entails the return of forests to some state characterized by fewer human impacts—for example, the return of the forests of the northern Rockies to presettlement conditions. But why not aim for a forest condition that predates human activity?

On a philosophical level, such arbitrariness is perhaps easier to show if we compare the selection of desired condition for European forests with that for American forests. In the United States, landscape conditions before and after European settlement are readily distinguished, and the landscape conditions before European settlement often function as a model for desired forest condition. In Europe, however, the distinction between forests before and after human settlement is virtually impossible to make, and, as a result, determining desired forest condition is more difficult. Should forests there be returned to their pre-Celtic condition before about 1500 B.C., to their pre-Roman condition, to their condition in the Middle Ages, or what? This question inevitably raises more fundamental questions—namely, whether less human impact is always preferable to more human impact, and, if so, why. These questions do not have scientific answers.

At the same time, however, and despite assertions to the contrary, the perspective of ecosystem management is almost purely biological, with no serious attention given to social values and little real attempt made to relate forest outputs to human and social needs and desires. A critical question that is not being asked is whether achieving a particular forest condition is a sensible use of public funds. It is one thing to justify taxes to produce outputs, market or nonmarket, that are consumed directly by the public, but quite another for society to use its scarce tax dollars to achieve a biological objective that may or may not be valued by the majority of the taxpaying public.

Generating Benefits for Everyone

Public forests were established to generate benefits for all citizens, and in the past the objectives of forest management reflected a degree of political consensus. In recent decades, these objectives have been codified in congressional legislation: the Multiple-Use Sustained Yield Act of 1960, as well as the Resources Planning Act of 1974 and the National Forest Management Act of 1976. By contrast, forest management as practiced by the Forest Service in the mid-1990s has no clear political or social mandate. Indeed, ecosystem management marks a sharp shift away from legislatively supported multiple-use forestry—which recognizes

many biological, social, and economic values—focusing instead on an arbitrary forest-condition objective that, in essence, is defined by biological considerations only.

While the Forest Service's adoption of ecosystem management may be inconsistent with legislation mandating multiple-use management, it is not inconsistent with the Endangered Species Act (ESA). In fact, recent court rulings that earlier Forest Service actions were contrary to the ESA do provide a rationale for the service's shift to ecosystem management. These rulings do not, however, provide sufficient justification for jettisoning the multiple-use objectives called for in existing legislation, at least until such time as a national consensus on new forest management objectives is codified by Congress.

The practice of ecosystem management, however, has arisen partly as a result of the difficulties inherent in multiple-use forestry. Achieving the optimal social mix of outputs is, obviously, no easy task. The selection of outputs has been complicated further by court interpretations of the ESA that constrained management decisions. In this context, the current administration and the new Forest Service chief have promoted the shift to an ecosystem management approach.

Changes in the administration or the ESA are likely to alter the way that ecosystem management is practiced, however, perhaps making the forest conditions managed for today undesirable tomorrow. And changes are likely. Administrations come and go, after all, and with them the leadership of the Forest Service. Moreover, the ESA is expected to be amended. In the absence of any kind of legislative mandate, then, ecosystem management could go by the wayside or it could constantly alter the goods that forests provide and do so without reference to public opinion.

If ecosystem management is to be practiced on public lands, the application of democratic principles suggests that such management be made law. In the absence of new congressional directives, however, management for multiple forest outputs should continue on public lands. But ecosystem-based management should not be dismissed altogether. Its tools and activities could and probably should be used by the Forest Service to achieve the objectives of multiple-use forestry. And if there appears to be some public support for returning forests to a specified condition of fewer human impacts, this condition could be added to the list of existing management objectives, such as producing timber and providing recreational opportunities.

The advantage of multiple-use management is that it tries to accommodate additional objectives and make trade-offs among them in order to increase social values. Such an approach, although sometimes flawed, is much more likely to benefit all members of society than ecosystem management, which makes one objective dominant and essentially impervious to trade-offs. In retrospect, we can see that multiple-use management's chief strength lies in its flexibility and in its responsiveness to changing social desires. By comparison, ecosystem management is rigid in identifying objectives and essentially arbitrary.

Suggested Reading

Sedjo, Roger A. 1996. Toward an Operational Approach to Public Forest Management. *Journal of Forestry* 94(8) August.

Thomas, Jack Ward. Forthcoming. "Challenges to Achieving Sustainable Forests: Is NFMA Up to the Task," in K. Norman Johnson and Margaret A. Shannon, eds., *The National Forest Management Act in a Changing Society 1976–1996.*

26

Carving Out Some Space
A Guide to Land Preservation Strategies

James Boyd, Kathryn Caballero, and R. David Simpson

The legal instruments we use to conserve habitat are still evolving. Continuing experimentation is needed if we are to save more land at less cost. Meanwhile, the many conservation easements that have sprung up nationwide let us see how well some instruments are working.

In the years ahead, demographic and economic changes should fuel pressure for more land development in this country. In turn, the growing scarcity of natural habitats will increase the social value of preservation. Thus the need is growing for policies and institutions that can balance the requirements of economic development with the benefits of species, habitat, and open-space conservation.

If the challenge is balance, regulation does not look promising for the purpose. Many interested parties view the law as an "extreme" option that does a relatively poor job of weighing competing interests. Under the current version of the Endangered Species Act, for example, animal and plant species threatened by changes in land use are designated public trust resources, but most of the land on which they live is not. It is rather as if Solomon went ahead and cut the baby in two: neither wildlife conservationists nor private developers are satisfied with the result.

Alternatively, "market-based" incentives to motivate conservation are gaining favor. These incentives include full- and partial-interest land purchases, tax-based incentives, and tradable or bankable development rights. The key to their attraction lies in compensating private owners for putting restrictions on the use of their land for public benefit.

The Costs of Habitat Conservation

We can divide the costs of implementing any conservation pol-

Originally published in *Resources*, No. 136, Summer 1999.

icy between *transaction* and *opportunity* costs. The transaction cost is the amount of time, money, and effort needed to establish, monitor, and enforce such a policy. The opportunity cost is the difference between the value of land in its "highest and best" private use and its value when employed in ways compatible with conservation.

No policy can avoid the cost of foregone development. What may appear to be striking differences in the costs of alternative policies are actually differences in *who* bears the cost of conservation. While who deserves to pay will always be a subject of political debate, the cost itself cannot be avoided. Any policy that appears to provide for conservation without full compensation for the land's lost use for other gainful purposes is one whose costs are hidden, or one that will not be effective.

The fact that the opportunity cost of foregone development cannot be avoided does not mean that all conservation policies are equivalent, of course. From an implementation standpoint, and in terms of likely effectiveness, there are important differences. One way to organize an analysis of these differences is to focus on the institutions and actions necessary to implement the policies in the real world.

Such an analysis reveals differences in the information required by agencies, the types and difficulty of enforcement, and the structure—all of which add up to the transaction costs of a given conservation plan. The notions of opportunity and transaction costs can be used to characterize alternative approaches to habitat conservation, a number of which are described briefly below. Because of its relatively active use, we save our description of a conservation easement for last and devote the rest of this article to its discussion.

Purchase of full property interests. A "fee-simple" acquisition for conservation purposes requires a purchaser to pay the full value of the seller's use of the land. This arrangement can result in "overkill" if, for example, the price includes the value of agricultural or low-intensity activities that are actually compatible with conservation. On the other hand, purchase of full ownership obviates the need to specify future management practices and

engage in the expensive monitoring of other approaches.

Tax credits and penalties. Another way to keep land out of development is for the government to give owners tax credits or other subsidies for doing so. Tax-based incentives result in at least some of the opportunity costs of conservation being shared among other taxpayers, who must either make up the revenue shortfall resulting from conservation-related tax breaks or make do with fewer public services.

Private landowners have an incentive to over-represent the value of the lands they devote to conservation, to the extent that they receive tax breaks for doing so. However, the tax system typically "under-rewards" conservation donors. Because tax codes require payment of something less than the entire amount of income or value of property, relief from this tax payment incompletely compensates the donor for the claimed value of the donation. Tax-based conservation incentives also require monitoring in order to confirm that the taxpayer is maintaining the land.

Offering tax-based incentives is generally less effective than acquiring particular properties. The decision to make tax-deductible donations is a voluntary one, and it is generally impossible to predict exactly which landowners eligible to make such donations will choose to do so.

Tradable development rights. A tradable development rights (TDRs) program distributes "rights" to some fraction of the land in an area. Anyone who wishes to develop land in excess of the amount of TDRs he owns must purchase additional rights. The opportunity cost of these programs is minimized because the land set aside for conservation is also the land with the least value for alternative uses. Transaction costs may be low to the extent that private markets in TDRs work relatively efficiently, but the need to monitor and enforce preservation requirements on lands for which TDRs have not been issued remains. In addition, TDRs are similar to tax-based incentives in that one typically cannot know in advance which lands will be preserved.

Regulation. Regulation that prohibits development may appear to be costless at first glance; when

Anatomy of An Easement

While the conservation easement contracts we reviewed exhibited a fair degree of variability, most shared a basic set of characteristics.

• *A description of the subject property, its ecological conditions and known environmental hazards, and a broad "statement of purpose";*

• *An agreement by the owner to submit the land to an environmental assessment, identify and correct any encroachments, and identify and remove disamenities;*

• *A limitation on the owner's ability to develop the land or alter its existing uses, and a description of the land uses that are allowed;*

• *An agreement by the owner to meet certain standards in management of the property;*

• *A right granted to the conservator to enter the property to ensure through observation that the contract is being honored;*

• *A demonstration by the grantor that the property has no liens attached to it;*

• *Provisions for adjudication or arbitration in the event of an alleged breach of contract;*

• *Indemnification of the conservator against liabilities associated with the property;*

• *Application of the easement to all subsequent owners of the property (often, the conservator must be given right of first refusal if the property is sold); and*

• *A number of provisions that set out responsibilities, deadlines, and payments associated with the original easement sale itself.*

land use restrictions are imposed by regulation, no payments or subsidies are made to landowners. Nevertheless, regulation deprives a landowner of the opportunity to earn income from future development. It is for this reason that many consider such regulations to be "takings" of property. Like the other options, regulation entails monitoring and enforcement costs. Unlike TDRs and tax incentives, however, it has the virtue of being able to target specific habitat types. In fact, regulation may seem to be a particularly efficient way of approaching conservation, since it eliminates the need for intervening institutions such as markets or tax assessment and collection. The specificity of regulation can also be its greatest drawback, however. The involuntary and information constrained nature of regulation means that the properties whose opportunity costs are the lowest will not necessarily be selected.

Purchase of a conservation easement. In exchange for payment (or a tax deduction) a purchaser receives assurances that a landowner will not develop designated land any further. Since a con-

servation easement involves the purchase of a "partial interest" in the land, it is less expensive than acquiring fee-simple ownership. On the other hand, easements involve substantial transaction costs, both in writing a contract and in subsequently monitoring and enforcing it.

Conservation Easements

Considerable recent experimentation with conservation easements has afforded us an opportunity to see how such incentives work in practice. Thus, we have looked at a number of easement contracts in order to identify their common features, evaluate their effectiveness, and make suggestions for their improvement. Numerous conservation organizations and public agencies are currently engaged in easement acquisitions nationwide. More than thirty states have passed legislation specifically sanctioning conservation easements for conservation, scenic, or historic purposes.

Easements possess several advantages. First, partial interest in a piece of land is less costly to

What's An Easement Worth?

Unlike sales of full interests in property, easements are still relatively rare. Moreover, the particulars of each easement are unique. Thus, no typical "market price" exists on which to base tax deductions. To avoid fraud, tax authorities allow deductions only for donations of land made to bona fide conservation organizations. Regulations on appraisers and penalties for excessive appraisals also constrain abuses. Still, overappraisal can be difficult to prove. The Federal Tax Code penalizes excessive donations only if the appraisal is off by more than 100 percent.

Typically, easement valuations range from 20 percent of the land's estimated total value to upwards of 90 percent. The Florida easements purchased by the state water management districts range in value from 28 to 60 percent of the properties' total value.

acquire than full ownership. Second, compared with conservation tax incentives or tradable development rights, easements entail few new administrative burdens. Third, they necessitate few, if any, changes in environmental and property statutes. Finally, because they involve voluntary transactions, easements are more politically palatable than direct land use regulation.

Easements do present challenges, however. The money saved upfront in acquisition costs must be balanced against the higher, long-term costs associated with monitoring and enforcing the division of ownership rights between the primary landowner and the conservator (the owner of the easement). While these costs can, to an extent, be anticipated and reduced by drafting an enlightened initial contract, the process of contracting itself thus becomes more expensive.

We have assessed a number of easement contracts in the state of Florida. These agreements were signed between landowners and Florida Water Management Districts (WMDs) or the Nature Conservancy (TNC). While a couple of the TNC contracts were completed more than ten years ago, the rest are of more recent vintage and signify the

emergence of easements as a conservation tool in Florida. The properties concerned are dispersed throughout the state and are relatively large in size, in some cases encompassing over ten thousand acres.

Several aspects of these easement contracts are worth noting. First, they tend to be perpetual. Why? One reason is that bargaining for contract terms is costly. A short-term contract implies frequent bargaining (every time the contract expires) whereas a perpetual contract minimizes the activity. Balanced against a desire to avoid the costs of repetitive bargaining may be a desire to retain flexibility. Many contracts contain terms regarding their own termination.

Our review of the Florida easement contracts revealed many "optimal" characteristics. In the economic theory of contracting, the party that can best prevent or ensure against risks should be required to do so. In the cases that we looked at, that party was the landowner, who was responsible for two basic contingencies in the easement contracts. The first was degradation in a property's ecological condition, over which a landowner has the most direct control.

The second contingency consisted of pre-existing liabilities attached to the property, the most prominent examples being delinquent tax payments and environmental contamination. Once again, responsibility for these problems lay with the landowner, who was better positioned to anticipate and remedy them than was the conservator. The easement contracts acknowledged this ownership of responsibility by indemnifying conservators against such liabilities.

In addition, the review of the records showed that a property owner is typically required to conduct an environmental audit prior to transfer of the easement, to make a representation that the property is free of contamination sources such as leaking storage tanks, and to ensure that title to the property is free of any liens or encumbrances.

Of course contracts are never able to define every possible future contingency. The costs of identifying and allocating responsibility across a "complete" set of circumstances are prohibitive. For

How Should Properties Be Managed?

It is difficult to specify how easements to protect wildlife habitat should be managed. Easement contracts often call instead for standard Best Management Practices (BMPs) usually approved by federal or state organizations. Other contracts refer to a more general "duty of care." Unlike BMPs, such a duty has not been well defined. However, examples do exist, such as those pertaining to land management in the 1976 Federal Land Policy Management Act.

this reason, contracts often rely on underlying principles of law, precedent, or community custom to define what is acceptable. In the absence of explicit contract terms, it is left to the courts to decide whether or not a contract breach has occurred and to specify damages if one has.

The Need for Experimentation

Public support for, and increased government involvement in, land conservation initiatives call for an analysis of alternative preservation policies. Each of the alternatives raises a set of legal, institutional, and economic issues. Experimentation with these alternatives is essential if the greatest possible benefit is to be realized from scarce conservation dollars.

All land use policies are not alike, differing in the way in which they ensure conservation, the complexity of their execution, and their costs. Preferences for one policy over another must be rooted in the merits of implementation.

Conservation instruments are evolving toward accomplishing their objectives more efficiently, but continuing experimentation with innovative instruments will facilitate the goal of achieving more conservation at less cost.

Suggested Reading

For information on conservation easements, see Land Trust Alliance, 2004. National Land Trust Census. November 18. http://www.lta.org/census/index.shtml.

For further discussion of "transferable development rights" (TDRs), see Chapter 27 in this volume.

27 A Market Approach to Land Preservation

Elizabeth Kopits, Virginia McConnell, and Margaret Walls

As more and more agricultural land becomes developed, several areas in the United States are turning to a new tool to control sprawl and limit this development. This new instrument, transferable development rights (TDRs), effectively creates a market for a limited number of "development rights." A case study of Calvert County in Maryland provides some illuminating insights into how such a program works.

The rate of conversion of agricultural land and open space to development has accelerated over the past several decades. The combination of larger lot sizes, more affordable housing at distant locations from center cities, and increasing reliance on vehicles has encouraged the conversion to housing of nearly a million more acres per year, compared with 20 years ago.[1]

As development spreads, there is a growing concern about the lost farmland, open space, and environmentally valuable areas that may have public value beyond their private value to landowners. Many elected officials in state and local government have made controlling "sprawl" a priority, and the 2002 Farm Bill authorized nearly $600 million over six years for the federal Farmland Protection Program. In addition, private land trusts have purchased or placed easements on some 5 million acres in the United States since 1982.[2] An alternative policy to the purchase of development rights is a more market-based policy tool for preserving farmland and open space—transferable development rights (TDRs).

How TDRs Work

In TDR programs, the right to develop a parcel of land is severed from ownership of the land itself, and a market is created with buyers and sellers of development rights. Those who sell development rights permanently preserve their land in its current undeveloped state (e.g., as farmland); purchasers of development rights are typically developers who want to build

Originally published in *Resources*, No. 150, Spring 2003.

houses at a greater density than allowed by local zoning ordinances.

Often, TDR markets are used to try to channel development away from areas considered valuable for farming or other undeveloped uses (so-called sending areas), toward already-developed areas with infrastructure to handle additional development (receiving areas). In this way, TDR markets promote more efficient development patterns and compensate landowners for lost development potential.

TDRs were first used in the United States in the 1960s, but their use has grown in recent years. Pruetz (2003) counts about 140 programs, with such objectives as preserving farmland, safeguarding unique natural areas or historic landmarks, and protecting environmentally sensitive areas.

Difficulties in Implementation

Despite their potential, only a few programs have been successful in maintaining active and efficient markets for TDRs. One major difficulty in many programs is an imbalance between demand and supply. Often, large areas targeted for preservation are designated as sending areas and the zoning density is reduced, prompting a large number of landowners to sell development rights. Demand for these rights on the developers' side, however, may be insufficient if homebuyers are satisfied with the densities permitted in receiving areas under baseline zoning. This is common in many urban programs—Atlanta and Oakland are two examples—where TDRs have been used to direct growth to already-developed areas that have high-density zoning. An imbalance can also occur in low-density rural communities; programs in Chesterfield Township, Pennsylvania, and several Florida programs, for example, have had little demand for building beyond the relatively low baseline density limits.

Regulatory conflicts can impede TDR markets as well. Lack of infrastructure in receiving areas or binding environmental and development regulations often prevent densities from exceeding baseline zoning. Island County, Washington, and

Springfield Township, Pennsylvania, for example, reportedly lack the sewer service necessary to achieve the density bonuses allowed under TDR use. In Lee County, Florida, hurricane evacuation restrictions prevent building in some TDR receiving areas. In many suburban counties, development moratoria imposed under adequate public facilities ordinances can delay development and dampen TDR demand. Demand for TDRs is further decreased if communities allow density increases through other mechanisms, such as clustering, planned unit developments, or in-lieu fees.

Another common problem is the lack of good information about the market. Since TDRs are primarily a county-level planning tool, the potential size of a TDR market is limited. Transactions in most programs number only tens or at most a hundred a year, making it difficult to establish a record of transactions that provides critical information for potential entrants. In addition, there is often no central clearinghouse for TDR transactions where information about sales that do take place would be available. As a result, it is possible that either developers on the demand side or landowners on the supply side—can dominate and prevent the TDR market from operating efficiently.

In addition, complex and time-consuming TDR transfer processes can deter potential market participants. In Boulder County, Colorado, for example, densities can be tripled with the use of TDR, but because receiving areas are not predetermined, a lengthy and uncertain public hearing process deters developers. Even when receiving areas are predetermined, as in Montgomery County, Maryland, opposition from existing residents can make it difficult for developers to achieve the by-right TDR density bonuses. In the New Jersey Pinelands, the establishment of a credit bank and a strong marketing campaign helped to simplify the complexities of the transaction process, making TDRs more attractive to developers.

A Policy Alternative: PDRs

The difficulties of maintaining active TDR programs have caused some communities to rely more heav-

Update

Calvert County's TDR program continues to be one of the most active in the nation, and TDR prices in the county have risen since this article was originally published. In August 2005, prices are ranging between $6,000 and $9,000. At the same time, the county is rewriting its zoning code and some significant downzoning of both "sending" and "receiving" areas appears likely.

The state of New Jersey has been the most active in the TDR arena in recent years. In March 2004, the state passed Assembly Bill 2480 authorizing the use of TDR programs at the local level throughout the state and holding out the programs in Burlington County, New Jersey, as examples to be adopted by others. The bill also authorized the implementation of joint TDR programs among jurisdictions—an important innovation, in our view—and offered the assistance of the state's Smart Growth Office in setting up local TDR programs.

The township of Chesterfield in Burlington County has adopted some innovations in the use of TDRs, including the first ever TDR auction and the designation of a receiving area that is a master planned new community in a "greenfield" location. All developers building in the community – known as Old York Village – are required to use TDRs. This idea seems to have taken hold in other locations as well, including Montgomery County, Maryland, where developers in the new community of Clarksburg are using TDRs. Time will tell whether or not such planned, mixed-use communities are a success and whether they can work in conjunction with a market-based TDR program.

ily on alternative tools to preserve farmland and open space. In a purchase of development rights, or PDR, program, as in a TDR program, landowners voluntarily sell their rights to develop a parcel of land, but the development rights are not used to increase density elsewhere. Rather, the government or a private land trust buys the development rights and essentially retires them.

An obvious drawback of a government PDR program is that the government must raise the revenue to purchase the development rights. On the other hand, the vagaries of the housing market and baseline zoning in residential areas do not affect the government demand for development rights, as they do in private TDR sales. Moreover, the government may be able to target lands for preservation more easily in a PDR program.

RFF has examined one program that is primarily a TDR program but includes some elements of a government PDR program—a farm and forestland preservation effort in Calvert County, Maryland, established in 1978. This program, after 25 years of operation, appears to be working well: the market for development rights is active and stable, and TDR sales have permanently preserved over 13,000 acres of land.

The Calvert County Program

Calvert County is within commuting distance of Washington, D.C., Baltimore, and Annapolis, and has seen the fastest rate of housing growth of all Maryland counties in past decade. The county's preservation program has several unique features. First, any farming or forested property with productive soils is eligible to enter the program and sell TDRs; its location does not matter. Unlike many other TDR programs, there was no "downzoning"—reduced density—of targeted preservation areas.

Second, although some farming regions targeted as most valuable for preservation were designated as sending areas only, much of the rural land can become either a sending or a receiving area. In these regions property owners have the option of selling TDRs and preserving their farmland, developing their properties, or buying TDRs and developing their properties beyond the baseline zoning limits.

Figure 1. Cumulative Preserved Acreage, Calvert County TDR Program

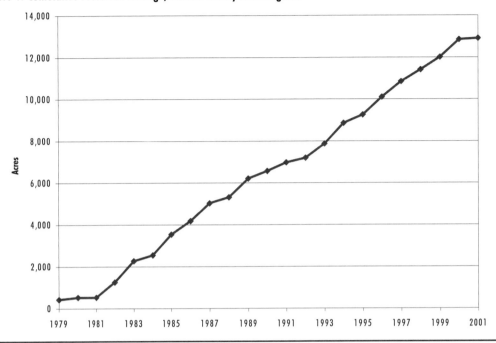

A third unusual feature of Calvert's program is that an entire parcel of land is preserved when a single TDR is sold off that parcel. In other words, a farmer may be allocated, based on her acreage, 50 TDRs, but once she sells the first TDR, her entire farm is in permanent easement status.[3] Hence, the timing of decisions can be critical. Moreover, landowners' trust in the viability and longevity of the program must be strong.

Finally, Calvert's current program has evolved as a combination TDR and PDR program. The county government became a direct participant in the TDR market in 1993 when it began buying and retiring TDRs. Purchases fluctuate from year to year, but in 2000, the county bought 252 TDRs in 21 transactions. Although private developers are more active in this market—in 2000, buying 989 TDRs in 43 transactions—the county is still a significant force in the marketplace. Figure 1 shows the combined acres preserved in the program, through both county and private market transactions, from 1978 through 2002.

Results of the RFF Analysis of the Calvert Program

The Calvert County program's minimal constraints on both sending and receiving areas have affected the spatial patterns of preservation and development. We found that TDR demand was highest in "Rural Communities" where baseline zoning permitted one housing unit per 5 acres, but the purchase of development rights would allow density to be increased to one unit per 2 acres. There was little use of development rights in residential or town center areas, where baseline density ranged from one to four houses per acre.[4] This suggests that the failure to create higher-density development through TDR programs may be partly due to low demand for such housing.

Our analysis also found that the TDR program is permanently preserving farmland in the regions identified as most important for preservation—the rural areas that were designated as sending areas only. Of the TDRs sold to date, about 80% were

Figure 2. TDR Price Distribution, Calvert County TDR Program

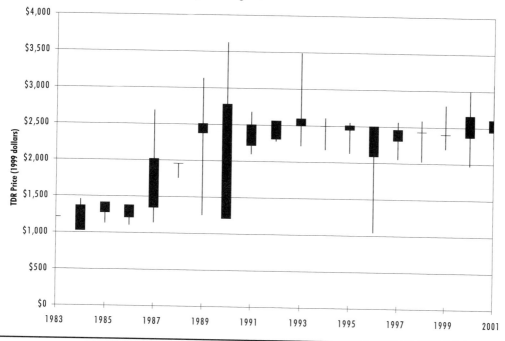

sold from these sending areas.

We found that most of the TDR supply —especially in the early years of the program—has come from farms in the central and southern part of the county—areas less valuable for development because they are farther from Washington, Baltimore, and Annapolis. TDRs are being used to increase development relatively more in the northern part of the county.

Political opposition to the program has been minimized in several ways. When the TDR program was implemented, the county zoning designations were not changed. Development is not completely prohibited in any part of the county: all rural land may still be developed at a rate of one dwelling unit per 5 acres. This flexibility reduces the potential for local opposition to the adoption of TDRs and the legal battles that have often followed in other TDR programs after a targeted sending area is downzoned.[5] The county downzoned in 1999, but did so everywhere as a way of reducing overall development.

The county's role as TDR purchaser has

enhanced the working of the market in several ways. It has provided information about TDR prices to potential market participants and helped establish a market with less price variability. The county government has announced its intention to buy TDRs each year at a price that is slightly above the average of the previous year's price. Figure 2 shows how prices have stabilized. The length of each line shows the range of prices for the year, while the bar on the line shows the range of prices in which 50% of transactions in that year occurred. In 1999, for example, the minimum and maximum prices were $2,200 and $2,800, respectively, and 50% of all transactions in that year occurred at prices between $2,400 and $2,600. In 1990, the range was much greater, and 50% of all transactions occurred at prices between $1,209 and $2,780 ($1999).

Government participation in the program has also increased overall market activity. Figure 3 shows the number of TDRs sold over time. Between 1980 and 1992, the average was 9 transactions and 417 TDRs per year. After 1992, when the county

Figure 3. TDRs Sold by Year, Calvert County TDR Program

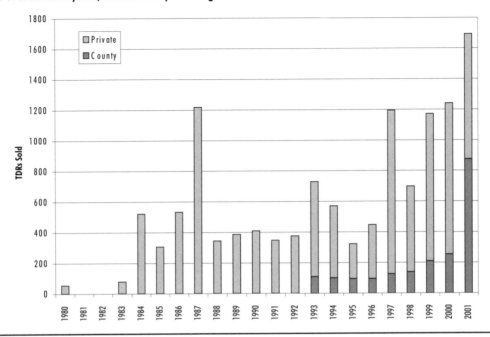

began purchasing, these numbers increased to 29 and 675, respectively. Thus, the county's role in the market appears to have increased private market confidence in the program and helped alleviate the problems of "thin" markets and lack of information.

Although Calvert's private TDR program is not as prescriptive about the location of sending and receiving areas as most programs, the PDR program does allow the county to target certain parcels for preservation. The cost to the county is reduced because an entire parcel is preserved once a single TDR is sold. To date, the county has spent a little over $2.7 million to bring new farms into the program, preserving more than 13,000 acres at an average cost of approximately $800 per acre preserved.

Conclusions

The Calvert County farmland preservation program shows that a transferable development rights program can be a cost-effective way of managing land

uses on the urban fringe. With a relatively straight-forward structure, good underlying market fundamentals, and the county taking a role in purchasing development rights in the TDR market and providing information to market participants, the Calvert TDR program appears to be achieving its goals.

Because the market seemed to stabilize when the county became a participant, combining PDR with TDR may work better than using either type of program alone. Having a market-based TDR program leverages government PDR funds for more acreage preservation, and at the same time allows private markets to channel development to locations where more density is demanded. More research is needed on the potential of these programs and other policy tools for managing land use.

Suggested Reading

American Farmland Trust. Heimlich, Ralph, and William Anderson. 2001. Development at the Urban Fringe and Beyond: Impacts on

Agriculture and Rural Land. U.S. Department of Agriculture ERS Agricultural Report 803. Available at http://www.ers.usda.gov/publications/aer803/.

Fulton, William, Jan Mazurek, Rick Pruetz, and Chris Williamson. 2004. TDRs and Other Market Based Land Mechanisms: How They Work and Their Role in Shaping Metropolitan Growth. Brookings Institution Discussion Paper, June. Available at http://www.brookings.edu/dybdocroot/urban/pubs/20040629_fulton.pdf.

McConnell, Virginia, Elizabeth Kopits, and Margaret Walls. 2003. How Well Can Markets for Development Rights Work? Evaluating a Farmland Preservation Program. RFF Discussion Paper 03-08.Washington, DC: Resources for the Future. Available at: www.rff.org/disc_papers/2003.htm.

Mills, David E. 1980. Transferable Development Rights Markets. *Journal of Urban Economics* 7: 63-74.

Pruetz, Rick. 2003. *Beyond Takings and Givings: Saving Natural Areas, Farmland, and Historic Landmarks with Transfer of Development Rights and Density Transfer Charges*. Marina Del Rey, CA: Arje Press.

Thorsnes, Paul, and Gerald P.W. Simon. 1999. Letting the Market Preserve Land: The Case for a Market-Driven Transfer of Development Rights Program. *Contemporary Economic Policy* 17(2): 256–266.

Notes

The views expressed in this paper are those of the authors and do not necessarily represent those of the U.S. Environmental Protection Agency. No official Agency endorsement should be inferred.

1. Economic Research Service, USDA. www.ers.usda.gov/Briefing/LandUse/Gallery/devland.htm.

2. Land Trust Alliance. 2000. www.lta.org/census.html.

3. In other TDR programs, an owner can sell an easement for part of a property but retain full development rights for the remainder.

4. Until a zoning change in 1999, one residential category allowed developers to build townhouses and apartment buildings at densities up to 14 units per acre.

5. Several lawsuits have been brought arguing that TDRs constitute a taking of property rights.

28 Preserving Biodiversity as a Resource

Roger A. Sedjo

Preserving the existence of the variety of global plant and animal life promises important benefits for developing new drugs and other products as well as sustaining life itself. But in the absence of property rights, many wild species are threatened with extinction. Certain kinds of contractual agreements have the potential to foster habitat protection.

The rationale for the preservation of the world's biodiversity runs from the highly spiritual to the pragmatic. On the spiritual side is the growing feeling among some groups that wholesale disturbances of natural systems are somehow unethical or immoral. On the pragmatic side, it is well recognized that the genetic constituents of plants and animals have substantial social and economic value from which all members of the global community may potentially benefit. Genetic information provides direct and indirect inputs for plant breeding programs, development of natural products (including pharmaceuticals and drugs), and increasingly sophisticated applications of biotechnology. The substantial increase in world agricultural output since the early 1970s has been due primarily to the ability of plant breeders to develop high-yielding varieties of the various food and feed grains by utilizing genes drawn from often overlooked plant species. More recently, recognition of the potential of wild genetic resources in development of drugs has led the National Cancer Institute to initiate a massive plant collection project that seeks to identify plants with chemical constituents effective against a variety of cancers. In recent years a number of widely used drugs have been developed from plants, including two important anti-cancer drugs derived from the now well-known rosy periwinkle found in tropical Madagascar.

Originally published in *Resources,* No. 106, Winter 1992.

Making Use of Wild Species

The benefits of using wild plants (or animals) as a resource may be obtained in three general ways. First, a species—or its phenotype, the individual plant or animal—can be consumed directly or it can be a direct source of natural chemicals and compounds used in the production of "natural" drugs and other natural products. Second, a species' natural chemicals can provide information and ideas—a blueprint—indicating unique ways to develop useful synthetic chemicals and compounds. For example, aspirin, an early synthesized drug, is a modification of the natural chemical salicylic acid (found in plants), which is too strong to be taken orally. And third, a wild species can be the source of a gene or set of genes with desired genetic traits that can be utilized in breeding or in newly developed biotechnological techniques. For example, germplasm from wild species is used to maintain the vitality of many important food crops. The latter two utilizations are essentially nonconsumptive, employing the genotype—the characteristics embodied in the genetic constituents of plant and animal species—as a source of information.

One recently publicized example of a useful natural chemical is taxol—a promising anti-cancer compound occurring naturally in the Pacific yew tree found in western North America. In 1985 taxol was found to shrink tumors in many ovarian cancer patients. In addition, its unique anti-tumor properties have been demonstrated in about 50 percent of advanced breast cancer patients treated with the drug. In two recent studies taxol has proved successful in treating tumors that had not responded to conventional treatments such as chemotherapy. It is the first and, to date, only member of a new class of anti-tumor compounds whose unique mechanism of action is distinct from the action of any currently used cytotoxic agent.

The current process for extracting taxol—peeling the bark of the yew—destroys the trees involved. It is anticipated that naturally occurring yews will provide most of the taxol through the mid-1990s, after which other sources will gradually be developed. These could include the conversion of compounds similar to taxol into taxol, the generation of taxol from plant tissue cultures, and biosynthesis. Synthetic production of taxol may also be possible, although this could be difficult due to the complexity of the compound.

With recent breakthroughs in biotechnology, the potential for development of useful products from wild plant and animal species would appear to be limitless. Species that have no current commercial application, contain no useful natural chemicals, or are as yet undiscovered, nevertheless may have substantial value as repositories of genetic information that may someday be discovered and exploited. The ability of modem biotechnology to transfer genes to unrelated natural organisms opens the possibility for the development of a wide variety of engineered plants and animals with hitherto unattainable sets of traits. As biotechnology develops, the scope for utilization of genetic information embodied in wild plants and animals will almost surely increase. Moreover, the ability to utilize the information from different organisms is likely to increase as genetic engineering expertise grows. The benefits of sustaining a rich and diverse biosystem are likely to be large since technology and natural genetic information may well complement each other in economic activity.

Loss of Genetic Resources

Despite the acknowledged social value of sustaining wild plants and animals, destruction of natural habitats in which they are found is widespread, posing a serious threat to genetic resources. Species with potentially useful characteristics for biotechnological innovations may be lost through tropical deforestation, for example. It has been estimated that 70 percent of the 3,000 plant species known to have anti-cancer properties are found in tropical forests. Considerable criticism has been directed at Third World countries with large areas of tropical forest for not protecting and properly appreciating the values of their native forests, particularly the values of biological diversity.

If preservation were without cost, then all genetic resources would be preserved. However, as the pressures on natural habitats rise due to alternative uses for the land, such as cropping or grazing, the costs of protection and preservation also rise. In earlier periods of human existence preservation of genetic resources was essentially costless. Recently, in situ and ex situ approaches have been used to protect the acknowledged values of genetic resources. The in situ approach involves protection of species in their natural habitats, whereas the ex situ approach involves protecting plants and animals in permanent collections such as zoos and botanical gardens, and preserving seeds and other genetic material in controlled environments such as germplasm banks. Although the ex situ approach has the advantage of lower costs, it is feasible for only a small fraction of species. This approach obviously cannot be used for species as yet unknown. Furthermore, the ex situ approach preserves selected species, not ecosystems, and thus risks the longer-term loss of species that are reliant upon the symbiotic relationships within ecosystems.

Although the destruction of a unique genetic resource base can occur from the consumptive use of a particular plant or animal itself, in practice a much more ominous threat comes from the process of land-use change. Land-use changes that destroy existing habitat and individual phenotypes can inadvertently drive to extinction potentially valuable genotypes, many as yet undiscovered, that are endemic to certain ecological niches.

Sustaining and Preserving Wild Genetic Resources

One way to view conceptually the problem of sustaining wild genetic resources is to think of these resources as a lottery containing a vast number of genetic "tickets," each with a different potential payoff. The timing and size of their economic returns vary greatly. Some of these tickets are currently generating payoffs. Others could or might generate future payoffs if the habitat is preserved long enough to allow their discovery and development. Still others would have to await further biotechnological developments before their potential returns could be realized. Although most of the lottery tickets will ultimately provide no payoff in terms of new chemicals, compounds, or transferable genes, a few will eventually result in substantial payoffs—jackpots—in the sense that these genetic resources will eventually generate large social benefits. However, it is difficult to differentiate in advance between those with significant potential future value and those with none.

Today, no ownership of the genetic lottery tickets exists. Individuals and countries, having no unique claim to the returns of the genetic information embodied in the wild plants or animals on the land they are developing, will tend to ignore the potential economic value of the existing habitat. The destruction of genetic resources thus becomes an unintended consequence, an external effect, of land-use changes that destroy natural habitats.

Although the costs of investing in habitat protection and preservation can become substantial, the industrial world has argued that such investment is needed because wild genetic resources are global resources from which the development of better lines of food grains, new medicinal products, and other advances generate global benefits that accrue to inhabitants of all countries. Nevertheless, a landowner—public or private—whose land provides the habitat for a unique genetic resource has no unique claim to its benefits.

The paradox is not hard to comprehend. Most public goods lend themselves readily to investments by the national state. The state perceives itself as readily capturing the returns to goods such as defense and lighthouses. However, it is much more difficult for the state to capture the returns to a global public good such as genetic resources. There are two reasons for this. First, international law does not recognize property rights to wild species or wild genetic resource genotypes, and hence any rents associated with valuable natural genetic resources typically cannot be captured simply through domestic management of the resource, even by a national authority. Second, the tradition that natural genetic resources are the common heritage of mankind and thus should be available without restriction provides an obstacle to the introduction of barriers to the unrestricted flow of wild genetic resources out of a country.

Protecting Public Goods

One result of the lack of private or national property rights to wild genetic resources is that, to date, most efforts to preserve and protect these resources have been altruistic. Most proposals for protecting them have involved actions by governments and the international community to preserve habitat. The usual approach is for environmental groups and the governments of industrial countries to try to persuade governments of developing countries to protect habitats rich in biodiversity, such as tropical rain forests. Some progress is being made—for example, in maintaining plant genetic resources used for breeding food and feed crops. An international system of germplasm preservation, commonly called seed banks or germplasm collections, has been developed. The collections are in both public and private ownership, with the private collections often being held by plant breeders who capture returns through the development of improved stocks to which some forms of exclusive rights exist. However, the system of collections is much less well developed for genetic resources that might have potential for drugs and pharmaceuticals than is the system for plant genetic resources used in crop breeding. In either case, collections can preserve only a small fraction of the total genetic resource base.

Progress in preserving this base is being made as individual countries, often in concert with international organizations, protect unique lands and habitats, including tropical forests, wetlands, and coral reefs. The world total of protected land doubled between 1970 and 1980 and increased another 50 percent in the first half of the 1980s. By the mid-1980s there were more than 400 million hectares of protected land (1 billion acres or 7 percent of land worldwide, excluding Antarctica), up from about 100 million hectares in 1960.

Altruism has motivated greater protection of unique lands and habitats in developed countries than in developing countries, many of which have been indifferent to seriously protecting habitat pre-

serves and have pursued protection haphazardly at best. This situation is beginning to change as the "common property" difficulty is recognized and various attempts are made to address it. For example, the Keystone International Dialogue Series on Plant Genetic Resources (talks among a high-level group of scientists and researchers from around the world) has identified as a "gap" the failure to develop an institutional framework for dealing with issues of plant genetic resource conservation related to ownership and intellectual property right (IPR) systems for plant genetic resources. In a June 1991 workshop on property rights, biotechnology, and genetic resources, held in Nairobi as part of the preparation for the United Nations Conference on Environment and Development (UNCED), the participants reached consensus on two key points. First, it was found that, as presently practiced, the treatment of biodiversity and genetic resources as a common heritage of humankind may have the unintended effect of ultimately undermining steps to conserve the resource. Second, it was agreed that any international negotiation on intellectual property rights should ensure that countries are free to decide whether or not to adopt IPR protection for genetic resources. Given this degree of interest, it is virtually certain that property rights for plant genetic resources will be an important item on the UNCED agenda.

A Coasian Solution

Perhaps the most exciting developments in the search for vehicles to facilitate protection of genetic resources and to ensure that some portion of the benefits accrue to developing countries are changes in legal arrangements, driven in part by market forces. It was first recognized by Ronald Coase, a Nobel laureate in economics, that external social benefits can often be "internalized" or captured through the simple legal instrument of the contract if transaction costs are small. In the last few years, contractual arrangements have begun to appear that allow developing countries to capture some of the rewards associated with the development of commercial drugs and other products that utilize genetic constituents of wild genetic resources found in their countries. These contractual arrangements require no new property rights. Rather, they utilize the ordinary legal instrument of a contract to, in effect, trade the right to collection in return for a guarantee of some portion of the revenues generated by the commercial development of a product that utilizes a genetic constituent from a unique wild genetic resource collected within the country. The judicious use of contract arrangements can allow for the capture of at least some benefits without de jure property rights to the individual natural genetic resources.

Organizations are also modifying their practices to allow them to enter into contractual arrangements with tropical countries to transfer the development rights to unique wild genetic resources to institutes in developed countries. For example, the National Cancer Institute in the United States is developing transfer agreements with tropical countries that have provisions for compensation, or revenue sharing, or both.

In addition, private collector firms are beginning to enter into contractual arrangements with tropical countries to offer royalties from revenues generated by future product developments in exchange for collection rights to wild plants. The most advanced activity of this type is occurring in Costa Rica, which recently created the National Biodiversity Institute to identify all of the wild plant species in the country, undertake preliminary screening of the various natural plants, and make agreements with pharmaceutical companies for further utilization of promising plants and natural chemicals. In 1991 the institute signed an agree-

ment with the Merck pharmaceutical firm, whereby Merck would provide $1 million over the next two years to help the institute build its plant collection operations. In return, Merck would acquire exclusive rights to screen the collection for useful plant chemicals and extracts. Indonesia is currently investigating the possibility of establishing a similar system that would allow for the capture of some portion of product benefits derived from its biological resources.

Whatever emerges from UNCED, those concerned with biodiversity will confront an extremely complex and rapidly evolving resource issue. In addition to the traditional approaches to protecting areas where biodiversity is high, innovative approaches are evolving that give promise of providing financial incentives for protecting habitat where biodiversity can be preserved and for returning some of the proceeds of the successful development of a natural-based product to the country that provided the genetic constituents. The challenge for UNCED will be to serve as a catalyst for facilitating further development of these innovations, while being careful not to advance procedures and controls that inhibit, rather than promote, such constructive processes.

Suggested Reading

Barton, John H. 1991. Patenting Life. *Scientific American* 264: 40–46.

Sedjo, Roger A. 1992. Property Rights, Genetic Resources, and Biotechnological Change. *Journal of Law and Economics* XXXV, April.

Cost-Effective Conservation
A Review of What Works to Preserve Biodiversity

Paul J. Ferraro and R. David Simpson

The preservation of biodiversity is a pressing issue, but the costs are high and the debate is heated. It is crucial that we learn what kinds of incentives work best to protect threatened species.

Humanity has never had a greater impact on the world's land use than we do at the present. As a result, some natural scientists predict that a third or more of the species on earth could become extinct in this century. Such losses are encountered in the geological record only at times of astronomical cataclysm. Half of all terrestrial species can be found in the 6% of the world's land area covered by tropical forests, and these species face the gravest risk. In developing tropical countries, the social agenda is dominated by the pressing needs of poor and growing populations.

Despite the difficulties inherent in influencing behavior in other countries, international efforts to preserve biodiversity have been under way for many years. Aggregate statistics are difficult to come by, but some numbers are indicative of the commitment. The World Bank has dedicated well over a billion dollars toward biodiversity conservation. A number of donors have allocated the same amount toward retiring developing country debt under debt-for-nature swaps. A recent study of conservation spending in Latin America reported approximately $3.3 billion in expenditures. Private foundations have contributed more than $10 million per year to conservation in developing countries.

Over the past two decades, conservation funding has shifted away from the "parks and fences" approach toward one attempting to integrate conservation and development projects. This new approach has been harshly criticized. "Integrated conservation and development projects," as they are called, have

Originally published in *Resources*, No. 143, Spring 2001.

Box 1. A Taxonomy of Habitat Conservation Policy Options

Direct approaches pay for land to be protected. Examples include:
- *Purchase or lease—Land is acquired for parks or reserves.*
- *Easement—Owners agree to restrict land use in exchange for a payment.*
- *Concessions—Conservation organizations bid against timber companies or developers for the right to use government-owned land.*

Indirect approaches support economic activities that yield habitat protection as a by-product. Examples include:
- *Payments to encourage land use activities that protect habitat and supply biodiversity as joint products. These payments can take several forms:*
  - *Subsidies to ecofriendly commercial ventures: Subsidies assist ecotourism, bioprospecting, and nontimber forest product entrepreneurs with facility construction, staff training, or marketing and distribution.*
  - *Payments for other ecosystem services: Payments for carbon sequestration, flood and erosion protection, or water purification provide incentives to maintain the habitats that both provide these services and shelter biodiversity.*
- *Payments to encourage economic activities that direct human resources away from activities that degrade habitats. This "conservation by distraction" approach provides assistance for activities such as intensive agriculture or off-farm employment. These activities may not be eco-friendly, but their expansion can reduce local incentives to exploit native ecosystems.*

been labeled as little more than wishful, and generally ineffectual, thinking in works such as John Terborgh's *Requiem for Nature* (Island Press, 1999). Calls to return to a parks and fences approach have sparked another backlash from critics who regard it as little better than stealing indigenous peoples' land at gunpoint. While these debates are raging, other groups are cataloguing, extolling, or sometimes lambasting a variety of innovative approaches to conservation finance.

The conservation need is urgent, the stakes are high, and the debate is heated. There has never been a greater need for both a clear understanding of the principles involved and a careful investigation of the facts.

Direct vs. Indirect Approaches

Biodiversity conservation is largely a matter of preserving the habitats sheltering imperiled species. Effective conservation requires that people who would destroy such habitats be provided with incentives to preserve them. Equitable conservation requires that we identify the people who have a rightful claim to such habitats and compensate them. People who do *not* have rightful claims must be prevented from destroying imperiled habitats.

People will generally do what is in their own interest. If they can receive more benefits from protecting an area of habitat than they could from clearing it for other uses, they will preserve it.

Box 1 identifies a number of conservation policy options. We've grouped them into *direct* and *indirect* approaches. Direct approaches are straightforward. The conservation organization pays for conservation. Payments may be in the form of outright purchases or purchases of "partial interests" such as easements or concessions, but the basic idea is to pay for actual conservation.

Indirect approaches are more complicated. Subsidies are provided to activities that are felt to be conducive to conservation. A conservation organization might, for example, assist a local entrepreneur in constructing a hotel for ecotourists, or

Box 2. Profitable Ecofriendly Enterprises

Landowners in many parts of the world are "doing well by doing good."
- *Some ranchers in Zimbabwe and other African nations earn more money managing native species than they would from cattle.*
- *Scores of landowners in Costa Rica choose to maintain their land as private nature reserves.*
- *Earth Sanctuaries Limited, a private firm operating game reserves in Australia, became the first conservation-related enterprise to be publicly traded when it was listed on the Australian Stock Exchange.*

These developments are to be applauded. The question remains, though, "What should we do when local landowners do not perceive biodiversity conservation to be in their own interest?"

training people to evaluate native organisms for their pharmaceutical potential. Indirect approaches raise two questions:
- If the activities local people undertake are profitable, why is assistance from conservation organizations necessary?
- If the activities are not profitable, might direct approaches be more effective in motivating conservation?

Ecofriendly enterprises have proved profitable in many parts of the world (see Box 2), so subsidies are not always required. Many millions, if not billions, of dollars have been devoted to assisting ecofriendly enterprises, however. The wisdom of these subsidies is suspect for a number of reasons.

First, such subsidies are generally an inefficient way of accomplishing a *conservation* objective. Consider two options facing an organization that wishes to preserve a certain area of land. First, it could pay for land conservation. If an ecofriendly enterprise can profitably be operated on the land, the conservation organization could sell a concession to operate the enterprise. The *net* cost of conservation under this option would be the cost of buying the land less the income received from the concession.

Under the second option, the conservation donor would subsidize the ecoentrepreneur by, for example, investing in hotel facilities to be used by tourists. The ecoentrepreneur would then acquire land for the ecotourism facility. The conservation

donor may be able to motivate the protection of more land by providing a higher subsidy. The conservation organization's net cost of conservation under this option would be the value of the subsidy it offers.

The second approach is more expensive. The basic principle at work is that "you get what you pay for," and the cheapest way to get something you want is to pay for *it*, rather than things indirectly related to it. While it is extremely difficult to estimate reliably the earnings of ecofriendly projects, we have been able to construct a number of examples that demonstrate dramatic differences in costs under the alternative approaches. The cost of the direct approach can be no greater than the forgone earnings that would have arisen from land conversion. If any earnings can be generated from ecofriendly activities, they can be subtracted from the cost of protection in computing the net cost of conservation. The cost of the indirect approach can, on the other hand, be several times higher than the cost of outright purchase or lease.

A number of other considerations also weigh against indirect approaches:
- There is no guarantee that subsidizing ecofriendly activities will motivate more conservation. Organizations offering such subsidies often assume that their effects will be positive, but if, for example, nicer hotel facilities induce would-be ecotourists to spend more time in their rooms than outdoors, the invest-

ments would prove counterproductive.

- Activities intended to be ecofriendly can have unintended consequences. Careless tourists may damage the sites they visit. Projects to commercialize local collection of forest products may induce overharvesting, or encourage local people to cultivate particular plants at the expense of their region's broader biodiversity.
- Integrated conservation and development projects may fail to achieve development objectives. Many developing nations would be better served by broader investments. Spending on public health or primary education is likely to pay greater dividends than training specialists in taxonomy or hotel management.

What Works in Practice?

Theory and practice can, of course, be very different things. It's one thing to advise conservation organizations to pay to preserve imperiled habitats, but it can be quite another thing for them to implement such a policy. One of the problems often observed in implementing conservation policy in developing countries is that the legal institutions for establishing and defending property rights are absent. Nevertheless, there is evidence that direct approaches are working at least as well as the alternatives:

- A recent paper in the respected journal *Science* by a group of researchers from Conservation International and the University of British Columbia demonstrates that many areas derided as "paper parks" are, in fact, effective in protecting imperiled habitat.
- Organizations in several tropical countries have initiated apparently successful programs to provide direct payments for habitat protection.
- There is no reason to suppose that indirect approaches will be any more effective than direct ones when property rights cannot be enforced. Whether it is an ecoentrepreneur or a park ranger, *someone* needs to guard against incursion.
- Payments for habitat conservation can create incentives for institutional change. When local

Update

National and international conservation donors have continued to experiment with incentives for conservation, and direct payment programs are now in place on all of the inhabited continents. One of the authors (Ferraro) has assembled information on, and examples of, direct payment programs on a website. For this, see http://epp.gsu.edu/pferraro/special/special.htm.

Policymakers are recognizing the potential of such payment systems. For example, a "Main Finding" in the Policy Responses volume of the recently completed Millennium Ecosystem Assessment is that "direct incentives for biodiversity conservation usually work better than indirect incentives." However, the Ferraro-Simpson question of "what works to preserve biodiversity" can only be answered empirically. While evidence is beginning to accumulate from well-designed evaluations of the effectiveness of alternative approaches, conservation success is necessarily a long-term proposition. The question of how best to advance the linked goals of conservation and development will remain controversial at least until more experience and a wider body of data can be assembled and analyzed.

people stand to gain from instituting clear property rights, they are likely to respond by doing so.

Conservation Finance

Just as there are a number of approaches to spending money for conservation, there are also a number of ways to raise money to spend. It is important to think clearly about each. While innovative approaches are to be applauded, one must also maintain realistic expectations because "if it sounds too good to be true, it probably is." A number of options have been suggested (see Box 3). Some

Box 3. Financial Instruments for Habitat Conservation

Financial Instruments may be used to fund either direct or indirect approaches. Examples include:
- *Debt-for-nature swaps—A conservation organization purchases and retires the loan of an indebted nation in exchange for the country's promise to conserve more biodiversity.*
- *Environmental funds—Public or private investors provide debt or equity financing for conservation projects.*
- *Securitization—Debt or equity issued to support conservation-related activities is bought and sold in organized financial markets.*

financing approaches that have received considerable recent attention may be no more effective than existing options, or could even perpetuate inefficiencies.

- A debt-for-nature swap may be no more effective than simply allocating money for conservation directly. Exactly the same outcome would be achieved if the conservation organization paid the indebted government to preserve habitat. The government could then, if it chose, use the money to retire its debt.
- Ecofriendly enterprises might "securitize" their financial obligations by combining them in negotiable stocks or bonds. In order to do so, they must meet the standards of the organized financial exchanges on which they hope to list them.
- A number of investment companies already offer their clients socially responsible options. When conservation donors subsidize funds for eco-friendly investment, it raises the questions regarding the efficacy of indirect approaches that we addressed above.

Risky Bargains

Conservation donors are intrigued by programs that would afford them leverage: small investments with big payoffs. There is, however, an irreducible cost of conservation. If people are to preserve the habitats under their control, they must receive benefits as least as large as they would have from con-

verting them to other uses. Some conservation donors find these costs daunting, although we have found that they are often surprisingly affordable.

The costs of conservation would only be lower if local people misunderstand the benefits conservation would afford them or cannot organize to realize them. These possibilities hold out a glimmer of hope to those who would achieve conservation on the cheap. There might be "demonstration effects." For example, one landowner might devote holdings to tourism rather than farming after observing that another has done so successfully. Or there might be "spillovers," if, for example, one landowner's property is a more attractive tourist destination if a neighbor chooses to keep his or her land in its natural state as well.

Is wagering the success of conservation policy on demonstration effects and spillovers wise? Perhaps it is, if one truly believes that only a spectacular reduction in conservation costs will suffice to assure the meaningful preservation of biodiversity. If one is not quite so pessimistic, though, three considerations argue against seeking such risky bargains.

- The simplest explanation of a phenomenon is not always right, but it should be the first considered. The simplest explanation for why local peoples do not maintain biodiversity is that they find destructive options are more attractive.
- The track record is not good. A number of programs have failed to achieve exactly these

demonstration and spillover effects.

- Conservation is often not as expensive as it seems. Over vast areas of the developing world, people can be dissuaded from converting natural habitats for a pittance.

The world's biodiversity is at risk and we ignore this fact at our own peril. Desperate times may, however, call for *thoughtful* measures. Different strategies may work in different circumstances, and there are exceptions to every rule. Mounting evidence suggests, however, that direct conservation measures are generally most effective.

Suggested Reading

Paul J. Ferraro. 2001. Global Habitat Protection: Limitations of Development Interventions and a Role for Conservation Performance Payments. *Conservation Biology*.

Paul J. Ferraro and R. David Simpson. 2000. The Cost-Effectiveness of Conservation Payments. Discussion Paper 00-31. Washington, DC: Resources for the Future.

Part 7

Energy Policy for the Twenty-First Century

30 Setting Energy Policy in the Modern Era
Tough Challenges Lie Ahead

William A. Pizer

National energy policy must address a range of challenging issues: improved reliability and efficiency, reduced dependence on foreign sources for reasons of national security, the containment of carbon emissions to limit global climate change, and the improvement of air and water quality. Energy markets (and electricity markets in particular) are complicated markets subject to extensive regulation. There is no single, easy answer to our energy problems; they will require market reforms, effective intervention, and public support of research in new energy technologies.

More than 50 years ago, when RFF was founded, there was widespread concern about potential shortages of crucial energy and natural resources that might jeopardize economic wellbeing in the United States. RFF scholars, among others, helped to disprove that myth, showing that free markets, free trade, and technological innovation would alleviate pressure on resource constraints, an idea that seems almost clairvoyant today. The United States has experienced remarkable economic growth since then, with the real gross domestic product increasing by more than 400 percent. Our domestic reserves of natural gas and petroleum are virtually unchanged over the same period, and global reserves have roughly doubled in the past 25 years alone.

Today, the United States finds itself facing a very different set of issues over energy supplies, focusing mainly on security and the environment. We currently spend more than half a billion dollars *a day* on imported oil, overwhelmingly from the Middle East, even as we fight a war on terrorism centered in that region. We are increasingly concerned about the reliability and resiliency of the electricity grid to both unintentional and intentional disruptions. We are also the largest emitter of greenhouse gases, primarily from the burning of coal, oil, and natural gas, which are believed to cause changes in the earth's climate.

The perceived problem 50 years ago, resource scarcity, is one best solved by letting free markets work out how to efficiently extract and allocate limited supplies, simultaneously signaling both conservation and innovation, and the development

Originally published in *Resources*, No. 156, Winter 2005.

of new technologies. But the new problems of energy security and environmental challenges result from a fundamental failure of energy markets to address issues that fall outside the market framework. This time, government clearly must intervene to correct these problems.

The government's role should be to intercede in ways that allow the private sector the most flexibility to trade off equally effective actions in the face of incentives that promote security and environmental protection. Such interventions could include an emissions trading program for greenhouse gases, a petroleum tax to address concerns about oil use, or clear rules for cost recovery associated with new electricity transmission infrastructure.

Energy "Problem" or Functioning Marketplace?

Popular discussions of energy problems today tend to focus either on increases in consumer energy prices or on high-profile news events, such as the Northeast blackout in 2003 and the California energy crisis in 2000. Natural gas prices, which stayed consistently in the range of $2 per million British thermal units (MBtu) for virtually all of the 1980s and 1990s, have been above $4 since January 2003. Crude oil, which similarly hovered in the $20 per barrel range from the mid-1980s until 2002, has been above $40 since July 2004. Adjusting for inflation, crude oil prices are still lower than the levels experienced during the early 1980s, but both the suddenness of the runup and the gut-level reaction to gasoline prices above $2 per gallon have propelled concern over energy policy to higher levels.

But what kind of energy policy do we need? The reliability and performance of electricity markets (as well as related demand for natural gas) are clearly something that needs to be addressed cooperatively by both federal and state agencies. Higher prices, on the other hand, may be part of a new balancing of supply and demand and something that energy policy can do little to relieve. In 2000, the

Energy Information Administration (EIA) forecasted prices of $2.50–3.00 per cubic foot (pcf) of natural gas and $20 per barrel of crude oil by 2020. These estimates have clearly been exceeded, and the higher prices are expected to continue. Government has little room to intervene here.

Although many people express concern about national security and environmental issues, few see the connection to national energy policy and especially to their own patterns of energy use. The historical trend in new vehicle sales toward less fuel-efficient pickup trucks, minivans, and SUVs and away from more fuel-efficient cars has continued unabated despite the events of 9/11 and hoopla surrounding the Kyoto Protocol. (One promising trend is the 89 percent annual growth in hybrid sales since 2000, though they are still a tiny fraction of the new-vehicle market.)

Security

Our ongoing national debate over energy security has so far focused on the steady growth in oil use in the transportation sector, the consequent rise in imports of oil from the Middle East, and the threat of economic calamity should our oil supplies be disrupted. But there are emerging concerns that deserve equal attention, namely the resilience of the domestic energy infrastructure— oil and gas terminals and pipelines, nuclear power plants, and the electricity grid—to terrorist attacks and, in the future, the same problems for natural gas imports as there are for oil. The former requires somewhat conventional security policies—building stockpiles, fortifying installations and control networks, and creating redundant back-up systems. The latter requires thinking about how various policies will affect natural gas supply and demand in the future.

In this vein, electricity generation accounts for about half the forecast growth in natural gas use over the next 20 years, with about two-thirds of that supply coming in the form of imported liquefied natural gas. Policies that emphasize coal, renewables, and nuclear power generation—three energy sources with abundant, secure domestic supplies—will reduce pressure on natural gas imports. Similarly, efforts to encourage and diversify natural gas supplies can diminish the kinds of security concerns that are associated with oil imports.

Our large and increasing dependence on oil—supplied in growing part from the Middle East—to fuel the transportation sector nonetheless remains the 900-pound gorilla seated at the policy table. As economists struggle to put a dollar value on the risks posed by oil imports from the Middle East, two broad categories of consequences often emerge: economic dislocation from actual or threatened supply disruptions, and the diplomatic and military costs associated with safeguarding access to Middle East oil supplies. With the ongoing war on terrorism, another concern has arisen: some of the oil revenue flowing into the Middle East makes its way into the hands of the very terrorists we are fighting.

The global nature of oil markets makes it impossible for the United States to discriminate against oil from particular sources. The idea of completely isolating ourselves from these markets is also unappealing: despite costly fluctuations, international markets still provide us with much cheaper oil supplies than we could ever access domestically. The solution, then, is for the government to encourage broad-based reductions in petroleum use, reducing our exposure to supply disruptions, our need to intervene diplomatically or militarily, and the flow of funds into the Middle East. A particularly simple (but politically unlikely) approach is to set a petroleum tax at a level that reflects the estimated consequences—risk and cost of a oil shock, diplomatic and military expense to maintain global market access, and indirect support of terrorism—associated with additional oil consumption.

A broad tax has the advantage of both encouraging less fuel use and encouraging the development of energy-saving technologies, which are now more valuable. A second-best alternative might be to focus solely on energy-saving technologies through a broad, market-based performance standard for all vehicles or other incentives. In this scenario, the new-vehicle fleet is forced to meet a

miles-per-gallon standard on average but can offset production of less-efficient vehicles with credits gained from producing more-efficient vehicles. Under such a standard, the new vehicle fleet is forced to meet a miles-per-gallon standard on average, but production of more fuel-efficient vehicles generates credits that can be used to offset production of less fuel-efficient vehicles by any manufacturer. This approach focuses on the "technology" margin of reducing fuel use per vehicle mile traveled, rather than the "behavioral" margin of encouraging people to drive fewer miles.

Climate Change

Global awareness and acceptance of the problems associated with carbon dioxide emissions are growing, but considerable disagreement remains over what to do about it. Many nations have embraced the idea of national caps for greenhouse gas emissions embodied in the Kyoto Protocol, and most notably, Europe has implemented an emissions trading scheme for carbon dioxide. Other countries, including the United States, have instead focused on voluntary programs and federal spending on technology—even as emissions trading proposals sporadically appear in Congress and some states attempt to implement regional programs.

U.S. technology programs center on nuclear, renewables, coal with carbon capture and sequestration, and hydrogen as a future energy carrier. Meaningful government efforts to push these technologies will go only so far, however; government also needs to provide incentives to encourage private-sector investment in them. A flexible emissions trading program or emissions tax sends a clear signal to the market about the value of emissions reductions both now and in the future. In a competitive environment, firms cannot invest significantly in emissions-reducing activities or R&D designed to lower the cost of these activities in the future if their competitors do not; that reality will confound effective voluntary programs. Most analysis also suggests that technology policy alone is unlikely to displace entrenched carbon-emitting technologies.

Markets and Innovation

Maintaining and expanding the efficiency of underlying energy markets poses a different set of challenges. Electricity markets in particular exist somewhere between regulation and competition with a great deal of uncertainty about their future. Because electricity generation constitutes a large source of natural gas demand, gas markets are also affected by this uncertainty. Federal and state governments need to work out a clearer roadmap for the future of these and other energy markets.

Government support for technological innovation is just as important now as it was 50 years ago. Investment in research and development tends to be undervalued because many of the economic benefits of new discoveries are not captured by those who discover them, but instead accrue to firms that imitate successful innovations. In the case of research into oilsaving and greenhouse gas–reducing technologies, it is likely that these innovations are further undervalued because policies to directly address those problems (such as petroleum taxes and emission caps) may be weaker than security and environmental concerns justify.

Moving Forward

Part of the guidance we need to tackle today's energy problems lies in the suggestions put forward by RFF researchers decades ago. Then, as now, concern over scarcity and price will be best addressed through well-functioning energy markets and government support for technological innovation. However, concern over newer issues, where the market fails to incorporate broader societal concerns over security and the environment, requires government intervention, ideally through flexible, market-based approaches. But the devil is often in the details. Energy markets and particular fuel choices are complicated by a variety of features. Market-based approaches, because they raise prices, often face political resistance.

There is no magic bullet for our energy problems, no single way to address our security and environmental concerns. Effective intervention and market reform requires attention both to the peculiar features of energy markets and fuel choices, as well as to broad incentives that promote society's security and environmental goals.

31 Petroleum: Energy Independence is Unrealistic

Ian W.H. Parry and J.W. Anderson

> Like it or not, Americans must confront the reality that oil prices are set by worldwide markets that respond to many economic and political factors beyond the control of the U.S. government.

With heightened concerns about energy security and global warming, governments and businesses around the world are beginning to think seriously about a transition away from oil as the crucial fuel for transportation. The central policy question is how to push this transition forward without slowing or destabilizing the growth of economies to which rapid and convenient transportation has become essential.

Crude oil prices have tripled over the past three years and this has sparked predictions of ever-rising prices in coming decades and exhaustion of the world's oil reserves. We heard these predictions before, during the energy crises of the 1970s; however, the subsequent two decades in fact saw falling prices, increasing world oil production, and expanding reserves as the market responded to higher prices. On the demand side, energy conservation and fuel-switching measures reduced the amount of oil used per unit of gross domestic product (GDP), in the U.S. case, by half over the last three decades (see Figure 1). And on the supply side, higher prices encouraged oil exploration and development of known but previously uneconomic fields, through improved technologies for locating and extracting reserves.

It is possible that we will see some reversal of recent price rises as these types of economic forces come into play again; for example, breakthroughs in converting oil shales and tar sands could significantly add to global supply. However, other factors appearing on the horizon suggest that things may turn out rather different, this time around. Most importantly, is large

Originally published in *Resources*, No. 156, Winter 2005.

Figure 1. Trends in oil import share and oil intensity of GDP

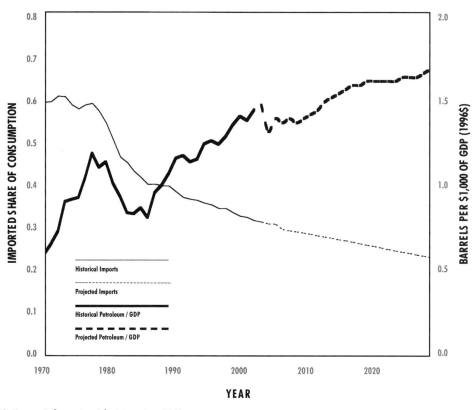

Source: U.S. Energy Information Administration, 2002.

developing countries, particularly China but also India, are beginning to embark on the path of wider automobile ownership that nearly all countries experience as they get richer. With four times the population of the United States, China currently has eight vehicles per thousand people compared with the U.S.'s 780 vehicles per thousand—a ratio suggesting relentless pressure on oil markets in coming decades.

And now that U.S. electricity generators have more or less dispensed with oil, further reductions in oil intensity are harder to come by. Two-thirds of American oil consumption is now in transportation—highways, air routes, and long-distance railroads—and there is no other fuel that is currently viable.

Growing Dependence on Foreign Oil

Concerns about the economy's dependence on oil are compounded by three trends. First is the steadily rising volume of imports, which currently account for well over half of the 20 million barrels a day that we consume (see Figure 1). Domestic American production is falling as long-worked fields are exhausted, but demand keeps rising relentlessly.

The problem of dependence on foreign oil is frequently misunderstood and overstated: for a given amount of domestic oil consumption, the increase in cost of producing goods in the economy following an oil price shock depends on the amount of oil used per dollar of production, and

Ethanol at every pump? Not Quite Yet...

Consumption of ethanol has been rising fast in recent years, pushed by subsidies—currently set at 52 cents a gallon from the federal government, plus additional help in several states—and federal requirements for oxygenation of gasoline. Ethanol's only competitor as an oxygenating agent is MTBE (methyl tertiary butyl ether), which has been banned in some states and may soon be banned nationally as a carcinogen that finds its way into drinking water supplies.

Ethanol represents only a tiny fraction, 2.4 percent, of the total automotive fuel used last year. However, it enjoys powerful political support from farm lobbies because nearly all of it is made from corn, using one-tenth of the corn crop and lifting corn prices. Developing technologies may soon make it possible to make ethanol more cheaply from cellulosic biomass, such as cornstalks, sawdust, and waste paper. One question for the future of ethanol is whether its political support will continue if the industry moves away from grain toward less expensive raw materials.

not the share of oil consumption that is imported. And given that the world oil market is fully integrated, the price we pay for imports is the same whether imports come from the Persian Gulf or from reliable sources such as Canada and Mexico. But import dependence does exacerbate the macroeconomic disruptions caused by oil price shocks: because extra dollar payments for imports go out of the economy to OPEC and other foreign suppliers, rather than being recycled within the economy to domestic oil companies, oil dependence leads to a further reduction in aggregate demand for U.S. goods.

The second trend is that production is expected to become increasingly concentrated in the Persian Gulf region, where an estimated two-thirds of global reserves are located; by contrast, estimated reserves for the United States are only about 2 percent of the global total. Intense concentration of supply in any one region would be cause for concern, but that concern is intensified by the history of political upheaval and violence in the Middle East.

A third trend is the growing U.S. trade deficit, which many economists consider unsustainable. For 2004 it will come to about $650 billion dollars, well over 5 percent of GDP. At the current price—at this writing, over $40 a barrel— U.S. oil imports total about $175 billion a year, more than one-fourth of the total trade deficit. Pressure on the exchange rate of the U.S. dollar is rapidly becoming another prominent reason for Americans to look for

ways to reduce oil use.

Calls for energy independence are unrealistic, to put it mildly, for the foreseeable future; cutting oil consumption to current domestic production would severely derail an economy in which cheap and rapid transportation is taken for granted. Like it or not, Americans must confront the reality that oil prices are set by worldwide markets that respond to many economic and political factors beyond the U.S. government's control. Of all the reasons that make policymakers uneasy about dependence on oil, the most immediate is the impact of sudden and prolonged price swings on economic growth, particularly given evidence that price increases harm the U.S. economy more than price reductions benefit it.

Environmental Concerns and Technological Challenges

Many of the traditional environmental concerns associated with oil use have been alleviated through a combination of regulation and technological improvements; for example, new passenger vehicles are more than 90 percent cleaner than 20 years ago and will become cleaner still with more stringent emissions standards.

The big environmental issue is the carbon dioxide emissions that form during fuel combustion and remain in the globe's atmosphere as a heat-trapping gas for hundreds of years. As consensus has solidified among most scientists that human-

Oil Prices and Foreign Policy

When oil prices rise or fall, there's an impact on both world politics as well as on country economies.

In 2002, Russia was working to attract foreign investment and appeared to be concerned about its reputation in the West. Net direct foreign investment in Russia that year was a little over $3 billion. But when oil prices rose by 50 percent in 2004, increasing Russian oil export earnings at a rate of perhaps $30 billion a year, one consequence was to diminish Russian need for foreign capital. Last year, President Vladimir Putin renationalized the major part of the country's largest oil company, tightened the government's grip on the press, and accelerated the trend toward centralized authority in Russia.

Several European governments have been urgently negotiating with Iran to dissuade it from pursuing nuclear weapons. One of the most attractive incentives they could offer has been investment capital. But Iran suddenly needs it much less: since early 2005, Iran's oil export earnings are now projected to run about $15 billion a year higher than the year before.

Throughout the Middle East, this wave of unexpected new revenue has flooded into countries not all of which are well equipped to keep it from reaching violent political factions and terrorists. The routes and magnitudes are unknown. But it is evident that the insurgency against the U.S. presence in Iraq is not suffering from inadequate financial support.

To the extent that Americans have contributed to the rise in oil prices through their steadily rising demand for oil, they appear to have undercut their own foreign policy goals and their own national security interests.

---

induced climate change is now occurring, calls for policies to slow down and eventually stop further increases in atmospheric greenhouse gas concentrations have intensified. But the challenges are immense: reductions in global emissions of 70 percent or more below current levels would ultimately be required if atmospheric greenhouse gas concentrations were ever to be stabilized before a doubling of atmospheric concentrations over preindustrial levels is reached.

No sector faces a tougher challenge for drastically reducing carbon dioxide emissions than the transportation sector, where, unlike power generation, the possibility of substituting existing low-carbon fuels or capturing exhaust gases for underground or deep-ocean storage is simply infeasible.

As documented in a 2002 National Academy of Sciences report, there appears to be a wide range of emerging technological possibilities for raising the fuel economy of new passenger vehicles through improvements in engine efficiency, reduced rolling resistance, and so on. Better fuel economy for conventional gasoline engines would be a significant, though not dramatic, help in reducing oil con-

sumption and carbon emissions. So would a shift toward diesel passenger vehicles that use less fuel per mile driven. Although widespread in European countries, diesels have been held back in the United States for two reasons: they would complicate auto manufacturers' compliance with stringent federal emissions standards for their vehicle fleets, and unlike in most European countries, diesel fuel is not taxed at a lower rate than gasoline.

Hybrid vehicles, which supplement a conventional gasoline engine with an electric drive train, promise a significant increase in gasoline mileage, particularly when the vehicle is used in urban stop-and-start driving. Toyota led the way with its hybrid Prius, and within the next few months, several major manufacturers will have hybrids in their showrooms. However, whether hybrids will achieve substantial market penetration in the foreseeable future is unclear; unless gasoline prices reach unusually high levels, their cost, including fuel costs over vehicle lifetime, is likely to exceed that of equivalent all-gasoline models.

General Motors has put several experimental automobiles powered by fuel cells on the streets of

Washington, D.C. However before the fuel cell becomes a standard source of power on the highway, chemists and engineers will have to resolve a number of formidable technological challenges.

Policy Responses

In an ideal world, the instrument of choice to reduce the country's use of oil would be a modest tax on all oil uses, perhaps $5 a barrel to begin with and increasing gradually thereafter, to accelerate energy conservation measures and remind people of the full costs to society from oil use. The potential costs of global warming are not currently reflected in U.S. oil prices. Neither are the full costs of oil dependence: although businesses may try to account for risks to themselves from oil price volatility in their investment and inventory strategies, they may not consider other risks, such as the cost of temporarily idled labor and capital following energy price shocks. An oil tax would have implications for productivity in an economy that depends on rapid and flexible transportation. Still, society would benefit overall, so long as the tax were appropriately scaled and revenues were used productively—for example, in other tax reductions or deficit reduction.

The Bush administration wants to expand domestic oil production, particularly through legislation to open the Arctic National Wildlife Refuge to drilling. The oil reserves there could produce, at peak, some 1 million to 1.3 million barrels a day, according to administration data. That would be a substantial contribution to correcting this country's foreign trade deficit. But since that new production would represent slightly over 1 percent of world oil consumption and 5 percent of American consumption, it would be unlikely to have a significant effect on oil prices.

Furthermore, unlike an oil tax, increased domestic production does nothing to reduce the overall oil intensity of GDP, and hence our exposure to oil price shocks. A broad oil tax would also be much more effective at reducing oil use than a hike in the federal gasoline tax or higher fuel economy standards for new passenger vehicles, since the broader tax would encourage energy conservation measures and innovation throughout the economy, rather than just in motor vehicles. Although the chance of new energy taxes in the next few years appears very remote, it is conceivable that this situation may change down the road, given continuing pressures to "do something" about U.S. greenhouse gas emissions and our looming deficit problems. There is strong opposition to raising any tax in the United States; however, unlike for income taxes, there is at least some support for in favor of higher energy taxes because of the security and environmental benefits.

But over the long haul, the problem is really a technological one: developing transportation vehicles with low or zero conventional fuel requirements that can be manufactured for prices consumers are willing to pay. Although the market is engaging in some early R&D efforts on its own, in response to higher oil prices and future anticipations of a carbon-constrained world, a case can be made for strengthening and expanding grants and tax breaks for the development and adoption of clean vehicles. Without such incentives, manufacturers are likely to underinvest in innovative efforts, as they are not fully compensated for the environmental and energy security benefits from cleaner vehicles, and the benefits to other firms in the United States that may adapt their innovations, let alone potential benefits to vehicle manufacturers in China and other parts of the world.

Suggested Reading

Bohi, Douglas R., and Michael A. Toman. 1996. *The Economics of Energy Security.* Boston: Kluwer Academic Publishers.

National Commission on Energy Policy. 2004. *Ending the Energy Stalemate: A Bipartisan Strategy to Meet America's Energy Challenges.* Washington, DC.

Parry, Ian W.H., and Joel Darmstadter. 2004. Slaking Our Thirst for Oil. In *New Approaches on Energy and the Environment*, edited by R. Morgenstern and P. Portney. Washington, DC: Resources for the Future, 23–27.

32 Coal: Dirty Cheap Energy

J.W. Anderson

Coal as a major energy source poses a basic challenge. It is the source of enormous amounts of air pollution, and the carbon emissions it generates appear to be a major contributor to global climate change. But it is a relatively cheap source of energy, and projections suggest that the use of coal will increase in coming decades with continuing economic growth. New technologies may provide ways to alleviate some of the adverse environmental effects of the use of coal, but this will require forceful public action to support the development and use of these technologies.

Despite the pollution that it causes, coal will probably continue to meet nearly one-fourth of the world's steadily rising demands for energy in the coming decades. World consumption of coal, 5.3 billion tons in 2002, will go up to about 8.2 billion tons by 2025, the U.S. Energy Information Administration recently projected.

Almost all of that increase will come from three countries—in order, China, the United States, and India. All three have large and easily accessible deposits of coal, a major consideration for governments concerned about the instability of oil prices and the insecurity of oil imports. This increase is contrasted with Western Europe and other regions, where coal use is expected to decline, partly because of a greater availability of natural gas. See Table 1 for projected coal use worldwide.

The policy challenges of reconciling rapid economic growth with clean air and reduced risks of climate change will be met—or evaded—with the deepest consequences for the planet's richest country and the two biggest of its poor countries.

In the United States, where it is used almost exclusively to generate electricity, coal has been competing recently with a cleaner fuel, natural gas. Partly for environmental reasons, the electric power sector swung to increased use of gas in the 1990s. One result was a rapid rise in gas prices, which have more than doubled since 1999. And that, in turn, is currently causing the power companies to swing back toward greater reliance on coal.

Originally published in *Resources*, No. 156, Winter 2005.

Table 1. Coal consumption, 2002 to 2025 (in millions of tons)

Region	2002	2025	Percent Change
United States	1,066	1,505	41.2%
Western Europe	573	459	-19.9
Japan	179	177	-1.1
Former Soviet Union	397	480	20.9
China	1,422	3,242	128.0
India	421	736	74.8
Rest of the world	1,204	1,627	35.1
Total world	5,262	8,226	56.3

Source: U.S. Energy Information Administration, 2004.

This shift is not without consequences to human health and the environment. Coal smoke contains fine particulates—soot, ash, and gases such as oxides of sulfur and nitrogen—that threaten the health of those who breathe them. Coal is also a prolific source of carbon dioxide, which, of all the greenhouse gases generated by human activity, is the one that contributes by far the most to global warming.

Some writers have speculated that shortages of fossil fuels might soon push the world toward cleaner sources of energy. In the case of coal, that is highly unlikely. Current production amounts to 0.5 percent a year of the world's proven and economically recoverable coal reserves, and the United States is in no danger of running low in the future.

A shift to renewable energy or other cleaner sources would require strenuous pushing by governments. The necessary political will and financial support will emerge only when societies decide that the negative effects of coal smoke on health, human welfare, and the environment outweigh the benefits of power at the lowest possible price.

China is beginning to consider action against the air pollution that coal causes. It has chosen Taiyuan, a city notorious for its bad air, as the site of an experiment in cutting emissions of sulfur dioxide with a cap-and-trade program based on the highly successful American model. With support from the Asian Development Bank, the Chinese government proposes to cut emissions in Taiyuan

by half, allowing the sources of these emissions to trade permits among themselves to hold the cost down. One question is whether this American concept can be transferred to a country with a very different economic and political system. The 50 percent goal is ambitious and the proposal is complex, but the fact that Taiyuan is thinking seriously about reducing emissions that dramatically is itself evidence of changing attitudes.

Here in the United States, energy policies sometimes work at cross-purposes with one another. Deregulation of electricity, for example, promises lower prices to consumers. But that leads to less use of natural gas, which is cleaner but more expensive, and more use of coal, which is cheaper but dirtier.

When utilities were regulated, state authorities were able to encourage electric companies to reduce pollution by guaranteeing them a return on their outlays. But under deregulation, the competitive pressure to push down prices is relentless. It is possible to combine deregulation with policies to curb emissions, possibly through a cap-and-trade program or a federal carbon tax. But in either case, one effect would be to raise the price of electricity.

One popular response to the rising emissions of carbon dioxide is the renewable portfolio standard, which typically requires a certain level or percentage of electricity to be produced from renewable sources. In the United States, since the mid-1990s, about 20 states have imposed such standards on electricity producers or retailers. But policies to promote clean technologies such as renewables may not have a large effect on coal consumption. A renewable portfolio standard will decrease usage of natural gas more than coal, in part because of the price differential. For that reason, encouraging renewables will not have a large effect on coal use or carbon dioxide emissions from the electricity sector in the absence of other policy measures, such as a tax on carbon.

The cost of pollution reduction will be heavily influenced over the coming decades by technological developments. One promising avenue is the integrated gasification combined-cycle (IGCC) process, which chemically turns coal into a synthetic gas that can then be burned in a turbine. This method permits the segregation and capture of most of the pollutants, including carbon. In the form of carbon dioxide, it can be injected underground for permanent storage in geological formations that are common throughout most of the United States, without harming the environment.

But the IGCC technology has yet to be shown to work reliably at the scale of a large utility power plant. In a deregulated market, investors appear unwilling to risk the cost of a big plant based on an uncertain process. Experience so far indicates that substantial public subsidies will be required to put this concept into actual practice.

To demonstrate how this would all work, the Department of Energy is currently pursuing a project it calls FutureGen, a partnership between the federal government and industry to design and build an industrial-scale electric power plant with carbon emissions pushed close to zero. It is to run on gasified coal, with the carbon dioxide to be injected into permanent underground reservoirs. When the project was announced in early 2003, the department estimated that the investment in public and private funds would come to about $1 billion over a decade.

According to one careful estimate, carbon capture and storage would become profitable at a price of roughly $200 to $250 a ton of carbon—that is, the point at which public policy, through regulatory limits or taxation, pushed the cost of emitting a ton

of carbon into that range. That is approximately the price that would result from public action in this country to comply with the Kyoto treaty on climate change, which would have required the United States to cut its emissions of carbon dioxide by about 30 percent from the amount that it would otherwise reach in 2010. The United States has dropped out of the Kyoto treaty on grounds, among others, of the cost. But Kyoto continues to set a marker, in general terms, of the cost of a serious effort to protect the global climate from accumulating greenhouse gases.

At present there are no nationwide restrictions on carbon dioxide emissions in the United States, although most of the state governments have begun to move toward controlling them. To raise the cost to $200 per ton of carbon would require a very substantial change in national policy. But most studies indicate that the cost of carbon capture and storage is likely to come down significantly with technological improvements. The Energy Department announced in late 2004 that it would provide up to $100 million in federal subsidies over the next four years for field-testing promising carbon sequestration technologies.

Of all the fuels, coal poses the basic policy questions in their simplest form. The first choice is between dirty and cheap or clean and less cheap—possibly a good deal less cheap. Conservation is always highly desirable, but in a society in which the demand for electricity is growing steadily, voluntary conservation alone does not offer a way out of the hard choices. A serious effort to combat regional air pollution and global climate change will require the development of new technologies, probably with public financial support. It will also require forceful public action, through regulation, to ensure that power producers, if they burn coal, adopt these new technologies.

Suggested Reading

Anderson, Soren, and Richard Newell. 2003. Prospects for Carbon Capture and Storage Technologies. RFF Discussion Paper 02-68. Washington, DC: Resources for the Future.

Chow, J., R.J. Kopp, and P.R. Portney. 2003. Energy Resources and Global Development. *Science* 302 (Nov.): 1528.

Morgenstern, Richard D., et al. 2004. Emissions Trading to Improve Air Quality in an Industrial City in the Peoples' Republic of China. RFF Discussion Paper 04-16. Washington, DC: Resources for the Future.

Palmer, Karen, and Dallas Burtraw. 2005. Cost-Effectiveness of Renewable Electricity Policies. RFF Discussion Paper 05-01. Washington, DC: Resources for the Future.

U.S. Energy Information Administration. International Energy Outlook 2005. Available at: www.eia.doe.gov/oiaf/ieo/index.html.

33 Nuclear Power: Clean, Costly, and Controversial

Paul R. Portney

Nuclear power, which presently provides about 20 percent of the electricity in the United States, is a carbon-free source of energy that reduces our dependence on foreign oil. But concerns about safety and the disposition of spent fuels raise serious concerns. If we push ahead on nuclear power, the major issue is whether the government should take steps to facilitate the construction of at least some plants, or whether this decision should be left to the privately owned companies that build virtually all of the nation's electricity-generating capacity.

For quite some time, nuclear power was the United States' most controversial energy source. While it is still not without its share of problems, some of them formidable, nuclear power has enjoyed a clear resurgence of late. For the first time in many, many years there is at least talk about beginning construction on a new nuclear power plant in the United States, something that hasn't happened here since the 1970s. Even some environmental advocates, once among the most implacable opponents of nuclear power, are casting a less jaundiced eye its way.

Nuclear power—harnessing the energy that results from the splitting of atoms—enters the energy mix in the United States in the form of electricity generation. Currently, slightly more than 100 operating nuclear power plants together provide about one-fifth of the electricity we use to power our factories, office buildings, homes, schools, and shopping malls. This makes nuclear power the second-largest source of electricity generation in the country; coal accounts for more than half of electricity generation, and natural gas (the fastest-growing source) for about one-sixth. Among all the developed countries in the world, nuclear power accounts for almost a quarter of electricity generation, a slightly larger share than in the United States.

The Case for Nuclear Power

What accounts for the second look that nuclear power is getting from energy experts and even some environmental advocates?

Originally published in *Resources*, No. 156, Winter 2005.

First, it is free from some of the serious air pollution problems that can accompany coal-fired and, to a lesser extent, natural gas-fired electricity generation. This includes both conventional pollutants, such as sulfur dioxide, hydrocarbons, and nitrogen oxides—and mercury, cadmium, and other heavy metals present in coal. While emissions of all these pollutants have been reduced significantly since the 1970 Clean Air Act took effect, electricity generation is still a major source of them all. Much more importantly, nuclear power is carbon-free. That is, unlike coal, natural gas, and petroleum, it does not release carbon dioxide into the atmosphere in the process of generating electricity. At a time when there is growing concern about the link between carbon dioxide and other greenhouses gases on the one hand and the warming of our planet on the other, this advantage of nuclear power has begun to loom larger.

A second advantage of nuclear power has to do with energy security. Concerns have existed since the early 1970s about the extent to which the United States is dependent upon petroleum imports to fuel our transportation sector, particularly from countries in the Middle East. For the first time, a concern about possible import dependence has begun to extend to the electricity generation sector. This is not because petroleum is used for electricity generation—its role there has almost disappeared. Rather, the concern is that natural gas production in the United States cannot keep pace with demand growth and that an ever-greater share of the natural gas we use for home heating and industrial production, as well as for electricity generation, will have to come from abroad (including from some of the same countries whose share of world oil reserves makes us nervous). Because of the likely adequacy of North American uranium reserves, this is not a concern for nuclear power (nor is it for coal, for which domestic reserves are ample).

There is a third attraction to nuclear power, though it currently pertains only to those plants that are already in operation— once built and paid for (a big qualification, as we'll see below), these plants are extremely inexpensive to operate. Indeed, the incremental cost of generating electric-ity from an existing nuclear plant is on the order of 1.5 cents for each kilowatt-hour (kWh) of electricity generated. This compares with about 2 cents/kWh for a conventional coal plant and, at current natural gas prices, about 3.5 cents/kWh for a natural gas plant. With the average retail cost of electricity in the United States currently standing at 7.5 cents/kWh, the 100 or so nuclear units in the country are quite profitable.

The Case Against

Despite those advantages, the last nuclear plant to commence operation in the U.S. began generating electricity in 1996, and no new plant has been started since 1973. Four liabilities have accounted for this disappointing record.

First and perhaps foremost, although nuclear plants are cheap to operate once they are up and running, they are by far the most expensive to build. Based on recent construction costs in Japan and Korea and on estimates from the vendors who would likely build plants in the U.S., a new 1,000-megawatt (MW) nuclear power plant would cost on the order of $2 billion and take five years to build. By contrast, a new 1,000-MW pulverized coal plant would cost $1.2 billion and take three to four years to build, and a new clean coal plant (one in which the coal is first converted to cleaner-burning natural gas) would cost about $1.4 billion and take four years. Illustrating why natural gas has been the fuel of choice for most of the recent growth in electricity generation, a new 1,000-MW combined-cycle gas turbine can be built in less than two years at a cost of $500 million.

The longer construction time and higher capi-tal cost of a new nuclear plant currently more than offset its operating cost advantage. According to a recent report by experts at the Massachusetts Institute of Technology, the "all-in" costs (capital plus operating) of electricity from a new nuclear plant operating for 40 years at 85 percent capacity would be 6.7 cents/kWh. This compares with 4.2 cents/kWh for a coal plant and 4–5.6 cents/kWh for a new gas turbine, depending on the assumed price for natural gas. Even if it faced no other obstacles, then, nuclear power would have a formidable economic challenge to overcome.

Other obstacles exist, however. For one thing, there has long been concern in the United States about the safety of commercial nuclear reactors, concern that predated the accident at Three Mile Island in 1979, where the core of one of the reactors was damaged. The operating record of the U.S. nuclear industry has improved significantly over the past 20 years, with safety and other related downtime having been reduced to as little at 10 percent at many plants. However, problems still arise from time to time, such as those at the Davis-Besse reactor in Ohio, where shoddy maintenance could have led to a serious accident had it not been caught in an inspection. Unless plant safety continues to improve, not merely stay the same, nuclear power faces an uphill climb, economics aside.

Opponents of nuclear power also point to the risk that the spent fuel from nuclear plants could be stolen and diverted to the production of so-called dirty bombs or even thermonuclear weapons. While this is a risk that must be taken extraordinarily seriously everywhere, it is a much larger concern outside the United States—especially in countries that have no obvious need for nuclear power. One example is Iran, which has vast natural gas reserves that could be used for electricity generation, but which has elected not only to build nuclear power plants, but also to do so using a fuel cycle providing easier access to the plutonium required for nuclear weapons.

The final challenge associated with nuclear power has also to do with spent wastes—namely, where in the world they will be stored. Currently, almost all the wastes that result from nuclear-powered electricity generation are being stored on the grounds of the power plants. No one believes this is the best place for these wastes, and at some plants storage capacity has been or soon will be exceeded. For this reason, the federal government committed long ago to build and open a high-level nuclear waste repository. Yucca Mountain, Nevada, was chosen as its site, and the repository has now been completed, at an eventual cost to the public of $50 billion, perhaps more. There's just one problem.

Update

The Energy Tax Incentives Act of 2005 contains provisions that will make the construction of new nuclear power plants considerably more attractive to electric utilities. For instance, the Act establishes a tax credit for electricity produced by a new advanced nuclear plant for the first eight years it operates. It also authorizes the Secretary of Energy to provide loan guarantees for up to 80 percent of the cost of building new plants, which will reduce the interest rates that must be paid by those undertaking new construction. In addition, the Act provides for the federal government to reimburse plant owners/operators for some of the costs associated with delays resulting from litigation and/or deliberations by the Nuclear Regulatory Commission. The Act also provides subsidies of nearly $1.2 billion related to the costs of construction of new advanced nuclear plants. Finally, the Act extends to 2025 the liability limits for nuclear plant operators established under the Price-Anderson Act.

Nevadans have no interest in being home to these wastes and have been successful in preventing the first shipments to Yucca Mountain, aided by a recent U.S. Court of Appeals ruling that the Environmental Protection Agency erred in establishing a safety standard Congress had directed it to set. Until and unless this stalemate is resolved, the future of new nuclear plants—not to mention the continued viability of the existing ones—is uncertain.

Policy Issues

Yucca Mountain notwithstanding, the real policy question facing the United States is this: should the government take steps to facilitate the construction of at least some new nuclear plants in the United States, or should it leave this decision solely to the privately owned companies that build virtually all of the nation's electricity generating capacity? Some type of federal assistance would enable the companies building the first handful of plants (likely in consortia) to overcome the "first-of-a-kind" costs that can make them much more expensive than subsequent units. If these latter units then became as cheap as some vendors suggest, their upfront costs would be quite competitive with new clean coal and even pulverized coal units and perhaps even competitive with natural gas plants on an all-in basis if gas prices remain high. Not surprisingly, the industry is seeking such government assistance in the form of a contribution toward the cost of building the first new plants, like the production tax credit afforded to wind power and other emission-free sources, as well as possible loan guarantees and other protections.

Government subsidies are not the only way to ensure that nuclear power gets to compete as a clean and secure source of electricity generation, of course. In the same way that the conventional air pollution problems associated with coal-fired generation have been substantially internalized through federal emissions-control requirements, so too could the comparable externalities associated with climate change. This could be done through a carbon tax or through a mandatory cap-and-trade program that forced both coal- and gas-fired plants to reduce their carbon dioxide emissions. Similarly, the energy security costs associated with an increasingly international market for natural gas could be internalized through an appropriate tax. Once these external costs had been internalized, along with those associated with nuclear power and other sources of electricity generation, of course, the government could step aside and let nuclear battle with coal, natural gas, wind, biomass, solar, and any other means of power production one could think up.

Far from the "dead duck" nuclear power was once proclaimed to be, it has arisen phoenix-like from the ashes. Whether this revival will extend to a new fleet of commercial nuclear reactors in the United States depends in large part upon how the inherent problems are resolved and how a nuclear program would be implemented.

Suggested Reading

National Commission on Energy Policy. December 2004. *Ending the Energy Stalemate: A Bipartisan Strategy to Meet America's Energy Challenges*. Washington, DC. Section IV.C ("Nuclear Energy") provides a brief, but useful, overview of how nuclear power may contribute to expanded supplies of energy. Available at: www.energycommission.org.

Smil, Vaclav 2003. *Energy at the Crossroads*. Cambridge, MA: MIT Press. Smil provides a useful treatment of the political burdens and risks of nuclear power.

The Future of Nuclear Power: An Interdisciplinary MIT Study. 2003. Cambridge, MA: Massachusetts Institute of Technology.

See also Allen Kneese's treatment of these risks in Chapter 8 of this volume.

34 Renewable Sources of Electricity
Safe Bet or Tilting at Windmills?

Joel Darmstadter and Karen Palmer

Renewable energy sources are not on a level playing field with fossil energy. Competing coal and gas-fired power generators do not pay for the pollution and other external costs they impose on society. Such external costs are far lower for solar and wind power generation.

Excluding hydropower, renewable energy makes up a tiny portion of the nation's overall electricity supply—its roughly 2.2 percent share is dwarfed by fossil energy, nuclear power, and hydroelectric dams (see Table 1). But given all the environmental and safety caveats associated with more traditional energy sources, a lot of people are paying closer attention to how renewables can play a larger role in the domestic energy mix.

Hydropower continues to overwhelm all other renewable resources in magnitude, but even existing dams, much less newly built ones, are widely seen as unpopular because of their effect on commercial and recreational fishing and on ecosystems as a whole. Virtually no one expects any meaningful addition to the nation's current hydropower capacity.

In the current marketplace, the dominant renewables are wind power, wood products (used mainly as a fuel source in manufacturing), municipal solid waste, and geothermal resources. Wind power has taken the lead in this race, with an 11 percent rate of growth since 1990, pushing it from 4 percent of total renewables generation (excluding hydropower) in 1990 to 13 percent in 2003. This trend of relatively strong growth for wind power is likely to continue.

Virtues

A useful way to appreciate the virtues of renewable energy resources is to look at some of the disadvantages associated with their conventional counterparts—fossil fuels and nuclear

Originally published in *Resources*, No. 156, Winter 2005.

power—that dominate today's world energy scene. Coal is cheap and abundant but its combustion produces pollutants only partially controlled by prevailing regulations, while control of carbon dioxide emissions awaits sequestration technologies not yet at hand. As trade in liquefied natural gas increases, natural gas may in time present challenges similar to those currently associated with large dependence on oil imports. And even if the nuclear option could be revived on technological and economic grounds, the public remains divided about such problems as disposal of nuclear waste.

In all of these respects, renewables are attractive. Typically, though not invariably, their use produces far less environmental damage than conventional energy. Burning of biomass such as fuelwood does result in some air pollution. At some locations, wind farm opponents cite danger to wildlife and aesthetic affronts. But these environmental concerns pale in comparison with those associated with conventional fossil fuels. Even with current pollution controls, coal-generated power still causes more pollution than that produced by wind turbines, and that's before considering greenhouse gas emissions from coal combustion. Also, renewables are largely insulated from the rising costs that, in time, may hit depletable resources like natural gas. Finally, in moderating the demand for fuels imported from unstable parts of the world, renewables indirectly provide an energy security benefit.

To be sure, these contrasting features are a bit overdrawn: some renewable energy sources might be limited while some nonrenewables are effectively inexhaustible. For instance, among renewables, geothermal resources are, strictly speaking, exhaustible because a given site may lose heat after a number of years of extraction. Among alternatives to renewables, one could cite nuclear power, whose underlying resource requirements are effectively limitless.

Drawbacks

Although renewable energy might be beneficial overall, it is no panacea. Renewables clearly do have some drawbacks. The most important is their high cost. To compare costs, we use each technology's "levelized" cost—that is, the real cost of generation, including capital cost, over the estimated life of the plant. For some technologies, such as solar photovoltaics, the levelized cost of producing a kilowatt-hour (kWh) can be three or four times as high as that for a new natural gas-fired combined-cycle plant, even at today's high natural gas prices. For other technologies, such as wind power, the cost differential from natural gas is smaller.

Except for biomass, renewable technologies have no fuel costs, and their other operating costs are typically low as well. So the bulk of their expense consists of the fixed costs of equipment and land. The translation of that fixed cost into a levelized cents per kilowatt-hour (kWh) of generation depends on how much electricity the facility generates. Because wind generation depends on when and how fast the wind blows, wind power installations typically operate at capacity factors of 10 to 35 percent, contributing to higher costs per kWh.

Renewables aren't on a level playing field with fossil energy. Competing coal and gas-fired generators escape having to pay the full cost of the air pollution and other consequences they impose on society. Such externalities, though they certainly exist, are surely minimal for wind and solar generation.

The potential for renewable sources of energy to gain a larger share of the electricity market is further constrained by physical and political limits. Geothermal energy sources are limited primarily to southwestern states. Biomass generation requires conversion of large quantities of land from other uses. A relatively strong, steady wind and land that is amenable to construction of wind towers and reasonably accessible to the transmission grid are prerequisites for the development of commercial-scale wind projects. Many offshore sites can meet these criteria, but efforts to develop offshore windpower in the United States have met with substantial resistance from those—like residents of Cape Cod—whose coastal views would be affected. Meeting the demands of opponents could add to the cost of developing windpower resources in particular sites.

All that being said, the costs of renewables have come down over time. Research at RFF shows that over the course of the 1980s and 1990s, the *cost* of electricity from renewable sources typically fell faster than had been predicted. But despite achieving lower costs, renewables failed to meet prior expectations regarding trends in the *volume* of future generation. This is because a simultaneous decline in the cost of competing fossil fuels and generation continued to place renewables at a relative cost disadvantage. How renewables fare in the years ahead will, similarly, be at least partly determined by concurrent trends in fossil or nuclear generation costs.

Policy Options

A totally hands-off governmental stance—subjecting renewables exclusively to market discipline—has abstract appeal. But fairness would dictate that other fuels would then no longer qualify for subsidies and also be subject to their full environmental costs, something that is not likely at the present time.

A number of different policies have been implemented in the United States to promote greater use of renewables in the electricity sector. One approach that gained currency with the introduction of competition in electricity markets in numerous states is "green power" marketing. Under this voluntary approach, renewable generators seek to appeal to households willing to pay a premium—ranging, in 2003, from 0.6 to 17.6 cents per kWh—to purchase power wholly or largely generated by renewables. Roughly 400,000 households participated in such green power programs by the end of 2003—an inconsequential 0.3 percent of nationwide residential power customers.

Another, far more significant trend is the evolution of renewable portfolio standards (RPS). As of early 2005, nearly 20 states had introduced requirements that a minimum amount of electricity distributed (as well as, in some cases, produced) in the state be generated using qualified renewable systems. These rules typically impose increasingly stringent requirements over the next several

Update

Among the issues sidestepped in the Energy Policy Act of 2005 are ones relating to automotive fuel efficiency, restrictions on emissions of greenhouse gases, and—with respect to this article—a Senate-mandated adoption of a nationwide renewable portfolio standard (RPS) along the lines discussed in our piece. Ironically, the very circumstances that prompted us to point out the net benefits of a *federal* RPS initiative in preference to the existing patchwork of varying state programs became the very basis for a counter-argument by critics: Why step in with a federal requirement when state initiatives are already in place? An additional reason for the opposition stemmed from the fact that a nationwide RPS would disproportionately enrich regions well-endowed with renewable potentials while penalizing those not so fortunate. Nonetheless, in a modest concession to the RPS idea, the 2005 Act does require that by 2013, renewables account for 7.5 percent of the *federal government's* electricity purchases. Additional incentives for generating electricity with renewables include a two-year extension of the renewable energy production tax credit, which applies to new electric-generating projects powered by wind, biomass, geothermal, solar, or landfill gas facilities. The incentive provisions of the Act also cover incremental generation resulting from efficiency improvements at existing hydroelectric plants.

decades. An important feature of RPS programs in numerous states is a tradable-credit system. Such credits, created whenever renewables-based power is generated, allow for the fact that electricity distributors vary in the ease with which they can incorporate a renewables component in their sales. In a system somewhat analogous to tradable emissions rights in sulfur dioxide and nitrogen oxides to meet stipulated clean air targets, utilities overcom-

Table 1. U.S. electricity capacity and generation, 2003

Fuel type	Capacity		Generation	
	1,000 megawatts	Percentage	Billion kwh	Percentage
Fossil/nuclear	823	89.4	3,493	90.7
All renewables	98	10.6	359	9.3
Hydropower	79	8.6	275	7.1
Non-hydropower	19	2.0	84	2.2
Wind	7	0.7	11	0.3
Geothermal	2	0.2	13	0.3
Solar	0.5	0.1	1	0.0
Wood/MSW	9	1.0	59	1.5
Total	920	100	3,852	100

Notes: Capacity refers to summer availability. Because of rounding, numbers may not add to totals. MSW = municipal solid waste. *Source*: U.S. Energy Information Administration, 2005.

plying in their renewables requirement can sell excess credits to those utilities failing to meet their quotas. Some state programs have an important provision that effectively sets an upper limit to the cost of such credits.

Undoubtedly, a nationwide renewable credit trading system would be more efficient than one limited to intra- or multi-state transactions. Indeed, several proposals to adopt a federal RPS have been introduced in Congress. For example, Congress considered comprehensive energy legislation in 2002 in the form of the Energy Policy Act, which included a requirement that renewables account for 2.5 percent of electricity generation in 2005, increasing 0.5 percent per year until reaching a 10 percent target in 2020. This legislation, similar to some state programs, included a cap of 3 cents per kWh on the price of renewable credits, but recent analysis suggests that the credit price for a 10 percent federal RPS would fall well below this threshold. But thus far, such congressional initiatives have failed—in part because of strong opposition from traditional electricity providers.

A production tax credit (or subsidy) has so far been the main policy instrument to promote renewable generation at the federal level. In 1992, Congress authorized a renewable energy production credit (REPC) now amounting to 1.9 cents per kWh of electricity produced from wind and dedi-cated closed-loop biomass generators, later expanded to include electricity from geothermal, solar, and landfill gas resources. The REPC could be claimed by new generators for the first 10 years of their operation. The federal government also grants a tax credit equal to 10 percent of investment costs for new geothermal and solar generation facilities; this policy has no expiration date. Both credits make renewables more competitive than they would otherwise be.

Net metering is another renewables-promoting program at the state level. Under net metering, customers who generate electricity on site—for example, through fuel cell or solar photovoltaic systems—can sell any excess electricity back to the supplier at the retail price, essentially running the meter backward. Many states have special programs to fund R&D on renewables, and the U.S. Department of Energy has been devoting $300 million annually to R&D into renewable energy sources, though that level may be hard to sustain under the administration's latest budget. A variety of technological improvements, in part benefiting from such support, might bring about a substantially expanded volume of wind-generated power. Possibilities include continuing advances in the aerodynamics of wind turbines, electricity storage systems to provide power availability during poor wind-speed periods, and cost-reducing breakthroughs to reduce line losses in long-distance electricity transmission.

What's in Store for Renewables?

Although technological advances and unexpected steep price increases for conventional energy can contribute to expanded use of renewables, their prospective growth—as projected in the just-released *Annual Energy Outlook 2005* by DOE's Energy Information Administration (EIA)—remains modest. Absent major energy policy departures, non-hydropower renewables are projected to increase their share of total electricity generation only to about 3.2 percent by 2025, compared to 2.2 percent in 2003. This is in spite of the fact that the *absolute volume* of renewables, mostly in the form of wind, will likely grow substantially over that period.

A strategy for a stronger renewables role would have to be embedded in a broader national energy policy. Dealing with the problem of greenhouse gas emissions from fossil-fuel combustion is one component of such a policy. Well-designed R&D funding for both renewable and other innovative energy systems is another. To stimulate renewables more directly, one of the most cost-effective approaches would be a national RPS, providing for an efficient nationwide trading system and a mechanism to protect against high costs. It will be very challenging to integrate a federal renewables policy with the variety of existing state-level RPS policies. Nevertheless, the RPS provision of the Energy Policy Act of 2002 provides a feasible approach. Complementary to other elements of a more rational energy policy than either political party or its leaders have given us in recent years, the RPS would represent one creative step forward.

Suggested Reading

National Commission on Energy Policy. 2004. Ending the Energy Stalemate: A Bipartisan Strategy to Meet America's Energy Challenges. Washington, DC. See especially Section IV.D on "Renewable Electric Technologies."

Palmer, Karen, and Dallas Burtraw. 2005. Cost Effectiveness of Renewable Electric Policies, Discussion Paper 05-01. Washington, DC: Resources for the Future.

Two useful online sources are:

http://www.nrel.gov/ [This site of the National Renewable Energy Laboratory provides a wealth of information, especially on technological problems and potentials.]

http://www.dsireusa.org/ [This "Database of State Incentives for Renewable Energy" provides detailed and up-to-date information on a state-by-state, as well as national, renewable policies.]

35

The Effectiveness and Cost of Energy Efficiency Programs

Kenneth Gillingham, Richard Newell, and Karen Palmer

The balance of evidence suggests that existing programs to increase energy efficiency have positive net benefits and can be a relatively inexpensive part of a more general effort to curb global climate change.

Energy efficiency plays a critical role in energy policy debates because meeting our future needs really boils down to only two options: increasing supply or decreasing demand.

However, in light of a range of energy issues—such as climate change, air pollution, and energy security—focusing exclusively on increasing supply is probably not the best way to go. Currently the United States emits approximately 1.58 billion metric tons of carbon equivalent (MMtCE) a year, and this number is rising steadily, presenting a daunting challenge to policymakers. Increasing energy efficiency holds the promise of providing a relatively inexpensive response to this challenge and other environmental effects of energy use, while continuing to meet demand.

The effectiveness and cost of government energy efficiency programs have, however, been the subject of a longstanding debate. To move beyond this point, two key questions need to be addressed. First, what types of energy efficiency programs have been implemented, and how much energy has been saved as a result? And, second, how much have these programs cost the public and private sectors, and how cost-effective have they been?

To look for answers, we evaluated the literature on a broad range of U.S. energy efficiency programs, with a focus on the adoption of energy efficient equipment and building practices (as opposed to transportation energy efficiency). Applicable programs and policies tended to fall into four general categories: appliance standards; utility-driven financial incentives—also referred to as energy demand-side management, or DSM; infor-

Originally published in *Resources*, No. 155, Fall 2004.

mation and voluntary programs; and management of energy use by the federal government, the nation's largest energy consumer.

Measuring the effectiveness or total energy savings from a conservation initiative or program can be difficult for a number of reasons and can lead to overly optimistic (or pessimistic) estimates. One problem is defining the baseline energy efficiency improvement that would occur in the absence of any program and avoiding double counting of the same energy savings attributed to multiple government programs. Another is accounting for "free riders," people who receive rebates for energy efficient equipment that they would have purchased anyway. There is also the rebound effect, where people increase their utilization of equipment (for example, leaving their fluorescent lights on) because it costs less to operate. Another consideration is whether all of the salient costs (costs to the government, business, and consumers and losses due to quality changes) and the benefits of the programs (including otherwise unaccounted-for spillovers to energy savings in other areas) are being accounted for.

Our review reveals a lack of detailed independent *ex post* analyses of conservation programs, with almost all available quantitative estimates coming from institutions either administering or advocating the programs themselves. Independent analyses are key to understanding the robustness of the effectiveness and cost-effectiveness estimates reported here. Detailed analysis is particularly important for classes of programs, such as appliance standards or utility DSM, that policymakers may use more widely in the future.

Despite these caveats, the balance of evidence suggests that these programs are delivering positive net benefits and are likely to be a relatively inexpensive part of the overall solution to climate change mitigation.

Appliance Standards

Minimum energy efficiency standards for appliances in the United States first appeared in response to the energy crises of the 1970s and early 1980s. Many states, particularly California and New York,

implemented appliance standards to cut the growth in energy demand. Leading manufacturers responded by putting pressure on the federal government to develop national standards that would supersede those of the states. Since 1987, the federal government has enacted a series of laws and regulations mandating minimum appliance energy efficiencies.

National standards have been established for an array of household appliances, including refrigerators, kitchen ranges and ovens, dishwashers, washers, dryers, and air conditioners. Standards have also been established for lighting fixtures and residential and commercial heating and cooling equipment. Cumulative federal government expenditures for the appliance efficiency program totaled $61 million in 2002 dollars in the period 1979 to 1993. The effectiveness and overall benefits and costs of standards are discussed below.

Demand-side Management

Utility-based programs cover a variety of energy conservation and load management policies that allow utilities to better match demand with their generating capacity. Federal regulators and state public service commissions began implementing policies that led to the creation of utility DSM programs after the energy crises of the 1970s. Initially most were information-and-loan programs, designed to educate consumers and businesses about the cost-effectiveness of energy efficiency measures and to provide low-cost subsidized financing for investments in those measures.

Utilities gradually learned that education alone produced limited energy savings. In addition, most consumers were not interested in subsidized loans. As a result, utilities moved toward programs with stronger financial incentives to convince consumers to make energy saving choices, typically rebates for purchasing designated energy efficient equipment, such as fluorescent light bulbs. Load management programs, another consistent element of utility DSM, aim to limit peak electricity loads, shift them to off-peak hours, or encourage consumers to change demand in response to changes in utilities' cost of providing power at different times of the day.

In the 1990s, utilities turned to market transformation strategies, whereby an attempt is made to change the market for particular types of equipment or energy services so that more efficient practices become the norm. This process usually consists of a coordinated series of demonstrations, training, or other information and financial incentives, with the hope that once a market is completely transformed, there will be substantially greater energy savings as the participation or market penetration rate approaches 100 percent.

Utility DSM evolved into standard operating procedure for a large number of power companies. For example, in 1990 over 14 million residential, 125,000 commercial, and 37,500 industrial customers nationwide were involved in DSM programs run by over a thousand utilities, large and small. While DSM policies matured in the mid-1990s, many state governments began to deregulate utilities. Diminished funds resulted in energy companies' suspending or curtailing these programs, although in recent years spending on them has leveled off.

Voluntary and Information Programs

The environmental protection agency (EPA) and the Department of Energy (DOE) jointly run the voluntary labeling program, Energy Star, which provides information on the relative energy efficiency of products. It was designed to reward manufacturers of the most energy efficient products with positive publicity, thereby encouraging consumers to buy those products and other manufacturers to improve the energy efficiency of their own products. The program now covers a wide array of products, including major appliances, computers and monitors, office equipment, home electronics, and even new residential, commercial, and industrial buildings. In addition to the labeling program, Energy Star also encompasses a range of public-private partnerships (for example, Green Lights), many of which began as separate programs and were moved under the auspices of Energy Star in the late 1990s. EPA spends around $50 million annually on administering all Energy Star programs.

DOE also runs two voluntary programs to report and reduce greenhouse gas emissions. The 1992 Energy Policy Act mandated the establishment of a national inventory of greenhouse gases and a national database of voluntary reductions in greenhouse gas emissions (commonly referred to as the Section 1605b program). Companies are required to report measures to reduce emissions on a yearly basis. Reductions could come from any of a variety of methods, including fuel switching, forest management practices, use of renewable energy, manufacture or use of low-emissions vehicles, greater appliance efficiency, and even nonvoluntary measures such as facility closings and governmental regulations.

In 2001 alone, 228 different companies or government agencies voluntarily reported reductions in greenhouse gas emissions for 1,705 projects. These reductions totaled 6.1 million metric tons of carbon equivalent from energy efficiency conservation projects not associated with other voluntary or DSM programs. The government administrative costs of the Section 1605b database and inventory system are currently less than $500,000 annually.

One factor that needs to be accounted for is that most entities reporting tended to be affiliated with one or more of the other government programs, and some percentage of their registered emissions reductions would have occurred anyway, without the Section 1605b program.

DOE also runs a complementary, voluntary program for utilities. The Climate Challenge program is designed to facilitate voluntary emissions reductions that make sense on their own merits. To take part, a utility must report to DOE annually on its progress, be willing to confer with the agency on possible strategies, and agree to one or more of six specified reduction commitments.

What Is the Effectiveness and Cost of These Programs?

Taken together, estimates indicate that the conservation programs we reviewed save up to 4 quadrillion Btus (quads) of energy per year and reduce annual carbon emissions by as much as 63 million metric tons of carbon equivalent. This represents about a 3.5 percent reduction in annual carbon emissions relative to what they would have been in the absence of these programs. These estimates typically reflect the cumulative effect of programs (that is, all appliance efficiency standards currently in effect) on annual energy consumption. These total energy savings—4 quads—represent at most 6 percent of annual nontransportation energy consumption, which has hovered around 70 quads in recent years.

Most of these energy savings come from reduced energy use associated with residential and commercial buildings (as opposed to more efficient industrial processes), so another relevant basis of comparison is total energy use in buildings, which accounts for 54 percent of the 70 quads of nontransportation consumption. Consequently, the 4 quads of energy saved represent approximately 12 percent of all building-related energy use and about a 3.5 percent reduction in current annual carbon emissions.

The table opposite summarizes energy savings, costs, and carbon emissions savings for the largest-scale conservation programs. The programs are listed in an order roughly reflecting our degree of confidence in the reliability of the estimates. Existing estimates suggest that minimum efficiency standards and DSM programs have provided some of the largest energy savings—about 1.2 and 0.6 quads, respectively, in 2000. Energy savings associated with the Energy Star and 1605b registry programs are also sizable (0.9 and 0.4 quads, respectively, in 2000), but it is less clear what portion of these savings would have occurred in the absence of these programs. Energy savings from other programs are relatively small or unavailable. We emphasize the use of quads for comparison among programs because many of the programs cover non-electricity reductions, which have a different heat rate than electricity.

Bringing the energy savings and cost estimates together provides our measure of cost-effectiveness, defined as the annual cost of each conservation program divided by the physical energy savings it achieves. Estimates of overall costeffectiveness are available only for efficiency standards for residential appliances ($3.3 billion/quad saved in 2000) and DSM ($2.9 billion/quad, including only utility costs for the energy efficiency portion of DSM). Note that higher dollars-per-quad cost-effectiveness estimates imply the program is *less* cost-effective (that is, it costs more per quad saved). If all energy savings were in the form of electricity, these estimates would translate to 3.8 cents/kilowatt-hour and 3.4 cents/ kWh for appliance standards and utility DSM respectively.

The price of the energy that is saved by these programs can be used as a measure of benefits to which one can compare the cost-effectiveness estimates. While this price varies over time, as a benchmark the average price of electricity in 2000 was $6.3 billion/quad of primary energy (or 7.4 cents/kWh of end-use consumption). As these energy savings are greater than the cost estimates cited above, this suggests that, as a group, efficiency standards are likely to have had positive net benefits (before environmental benefits are included). The cost-effectiveness of DSM is similar, but includes only utility costs. The average price we use for comparison is only a rough measure of benefits, however, and a more accurate measure would account for differences between this price and the *marginal* cost of the energy conserved.

The environmental benefits resulting from energy efficiency programs—from lower emissions of carbon dioxide (CO_2), nitrogen oxides (NO_X), sulfur dioxide (SO_2), and particulate matter (PM-10)—add value to these programs on top of the value of the energy they save. Based on national average emissions rates and available estimates of the dollar value of reducing air pollutants, we find that the environmental benefits of reduced energy consumption may add approximately 10 percent to the value of the energy savings relative to basing that value on the price of energy alone. That is, for every dollar energy efficiency programs save in reduced energy costs, they save about another 10 cents in reduced environmental harm. The majority (7 percent) of these benefits come from CO_2 reductions, with fewer benefits from NO_X (2 per-

Table 1. Summary of estimates of energy savings from largest conservation programs in 2000

Program name	Date	Energy savings (quads)	Costs (billion $2002)	Cost-effectiveness (billion $2002 per quad[d])	Carbon Emissions savings (MMtCE)
Appliance standards	2000	1.20	$2.51[a]	$3.28[a]	17.75
Utility DSM	2000	0.62	1.78[b]	2.89[b] (high 19.64)	10.02
Energy Star	2001	less than 0.93	0.05[c]	—	less than 13.80
1605b registry	2000	less than 0.41	0.0004[c]	—	less than 6.08
DOE climate challenge	2000	less than 0.81	—	—	less than 12.04

a. Indicates that total costs and cost-effectiveness estimates are for residential appliance standards only while the energy savings and carbon emissions are for commercial and residential standards combined.
b. Indicates only utility costs are included.
c. Indicates that only direct government administrative costs are included.
d. Billion dollars per quad of primary energy can be roughly converted to cents/kWh of end use consumption by multiplying by 1.166, which assumes all of the savings come from electricity using the average mix of generating facilities.

cent), and SO_2 and PM-10 (0.5 percent each). Including environmental benefits therefore strengthens the case for energy efficiency programs but does not dramatically change their value based simply on energy savings. Viewed as a means for addressing climate change, however, energy efficiency policies appear likely to be a relatively inexpensive option, as the energy savings alone can cover the cost.

The continued use of energy efficiency policies over more than two decades and the prospect of expanded and new policies on the horizon suggest that this approach to achieving energy and carbon reductions will have a lasting presence. This is particularly true if conservation programs have positive net benefits in their own right and therefore yield emissions reductions at zero or negative net cost. But even if these estimates are overly optimistic,

energy efficiency programs can be an important part of a low-cost, moderate climate policy, given that the effect of existing efficiency programs is of similar magnitude to what rough estimates suggest might come from a moderate carbon tax.

Suggested Reading

Gillingham, Kenneth, Richard Newell, and Karen Palmer. Forthcoming. Retrospective Review of Energy Efficiency Programs. In *2006 Annual Review of Environment and Resources*. [This is the complete study on which this chapter is based.]

Newell, Richard, Adam B. Jaffe, and Robert N. Stavins. 2004. The Economics of Energy Efficiency. In *Encyclopedia of Energy, Volume 2*, edited by Cutler Cleveland. Elsevier, 79–90.

Part 8

Global Climate Change

36 Climate Change and Climate Policy

J. W. Anderson

Global climate change is the most complex environmental challenge that governments have ever tried to address. Not only is it fraught with uncertainties of many kinds, but it requires global cooperation in a highly costly enterprise. In spite of the failure to reach a global policy consensus, many programs are underway in various parts of the world that may point the way to a coordinated effort to control the world's emissions of greenhouse gases.

In the summer of 1988, the possibility of worldwide climate change suddenly burst onto the agenda of international politics. It happened quite abruptly. The occasions were a U.S. Senate hearing in Washington and a world conference in Toronto. Among specialists, concern had been rising for some time as they gathered data on rising temperatures and the changing chemical composition of the Earth's atmosphere. Then came the moment in which the policymakers and the general public realized that something might, conceivably, be going dangerously wrong.

In the years since then every country on Earth has been drawn into the struggle over what to do about it. The politics of global warming, as it has developed over the past 17 years, provides the setting for the hard decisions that may lie ahead.

Climate change is, by a wide margin, the most complex environmental challenge that governments have ever attempted to address. This change is being driven by the emission into the atmosphere of gases that cause it to retain heat—and the most important of those greenhouse gases is carbon dioxide, generated chiefly by burning the fossil fuels that provide five-sixths of the world's energy. Reducing those emissions would require profound changes in the way the world produces and uses energy. The central challenge for policy is to accomplish it without unacceptable disruption of the world economy and reductions in standards of living. One key question is how much society should pay now to reduce a risk of severe and damaging climate change in the future—a risk that may be great, but that no one can quantify with any certainty.

A related question is the degree of urgency. Is it essential to begin reducing greenhouse emissions immediately, as most European governments believe? Or is it enough to rely on new technologies to phase down fossil fuel use over the next half-century, without resorting to unpopular mandatory limits, as the U.S. government believes?

Climate change is an unusually difficult issue for the people who make the decisions in democratic governments. The science contains important uncertainties, but governments have to make firm decisions—if only the decision to do nothing—long before these uncertainties can be resolved. Any serious attempt to cut emissions will have clear and immediate costs, but the benefits may not appear for a long time. To the extent that the benefits may be disasters that don't happen, they may never be obvious. But the costs will be. As the debate develops, much of it is being cast in terms of the restraint that the present generation owes to future generations.

Another dimension is the obligations that rich countries have to the poor, and that all countries have to the world. To achieve effective worldwide cooperation, policy has to find a fair way to allocate the burdens of emissions control between the rich countries and the poor. The United States, in 2001, generated 20 tons of carbon dioxide per capita. China generated 2.4 tons per capita, and India 0.9 tons per capita. China and India have argued that, in view of this wide disparity, the United States has an obligation to make deep reductions before it can ask the developing countries to assume any burdens. But the United States has said that it won't join any international agreement that doesn't put restrictions on developing countries' emissions.

Scientists demonstrated in the nineteenth century that more carbon dioxide in the atmosphere causes an increase in heat retention. But despite the gathering momentum of industrial development they gave little thought to the climate because most assumed that the carbon dioxide was being absorbed in the oceans. That assumption was not tested until the late 1950s, when it rapidly became clear that the concentration of carbon dioxide in the atmosphere was rising steadily. In the centuries before the Industrial Revolution began, the concentration of carbon dioxide in the atmosphere had hovered around 280 parts per million. By 1958, when scientists began to pay attention to it, the concentration was 316 parts per million. At present, in 2005, it is approaching 380 parts per million.

The average global temperature has also been rising, although less regularly. A period of steady warming began around 1910 and ended in the early 1940s, followed by a pattern of stable or slightly declining temperatures through the mid-1970s. From then through the present, temperatures have been rising steadily and rapidly. In a record that goes back to the mid-nineteenth century, all 10 of the warmest years have occurred since 1990. By the end of the twentieth century, the global average temperature was about 0.5 degrees Celsius higher than a century and a half earlier. In recent years the rise has clearly been accelerating.

These data have set off a series of furious debates among scientists, industrial interests, environmental advocacy organizations, and politicians. The first question was whether the rise in temperatures is real, or whether it might be the product of the way that the figures are collected. The publication of many studies over the years with increasingly broad data bases seems to have settled that issue: both the rise and the recent acceleration are real.

Another question was the break, from the 1940s to the 1970s, in the pattern of warming. If growing atmospheric concentrations of carbon dioxide mean a warmer climate, why didn't global temperatures rise in those decades? The answer seems to involve aerosols, the tiny particles of solids and liquids blown into the air along with carbon dioxide in the exhausts of furnaces and engines. Aerosols in the atmosphere tend to reflect heat from the sun, just as carbon dioxide tends to retain it. But with the legislation in Europe and the United States in the 1960s and 1970s to clean up the air, the load of aerosols dropped and no longer offset the effects of carbon dioxide.

A more difficult question is the basic relationship between carbon dioxide and warming.

Skeptics point out that the Earth's climate has swung from hot to cold and back many times in the geological record. How do we know that the current warming is not caused mainly by purely natural changes that would be little affected by any reduction, however drastic, in greenhouse gases? One example of such a natural cause is the multidecadal oscillations in ocean currents, which clearly affect climate.

The case for identifying carbon dioxide as the main cause of global warming is in large part statistical. Research organizations have constructed mathematical models of the global climate system, and run them backwards to see how closely they mimic the actual historical record. No one has yet come up with a reasonably accurate model that does not give high importance to carbon dioxide, generated by human activity, as a cause of what's happening. The present consensus among most scientists holds that the increasing carbon dioxide in the atmosphere is certainly a cause of global warming, and highly probably the major cause—but it may not be the only cause.

So far, so good. The next question is where we go from here. That one is much harder to answer, and research has not yet produced a reliable consensus. The amounts of carbon dioxide generated over the coming century will depend on rates of population increase, the speed of economic growth, and the nature of technological development—all factors impossible to forecast over a period as long as a hundred years. The scientists running climate models have sidestepped these variables by adopting a baseline scenario in which the concentration of carbon dioxide rises to 550 parts per million— roughly twice the pre-industrial level—by the year 2100. They have then tried to calculate the temperatures that would result from that rise. (This exercise, incidentally, was included by the journal *Science* in July 2005 in a list of the great questions now driving basic scientific research.)

A major international study published in 1995 found that a concentration of 550 parts per million of carbon dioxide in the atmosphere would increase the world's average temperature by 1 to 3.5 degrees Celsius. Since then results have shifted to a some-

what higher range, 1.5 to 4.5 degrees or even higher. This range is too wide to be helpful to politicians who have to make decisions. At 1 degree over a century, the change would affect some species but, to homo sapiens, it would hardly be noticeable. At 4.5 degrees, the change might well approach the catastrophic for much of humanity.

While scientists use temperature as the metric of change, the greatest impacts on human life might well come from the secondary effects. Higher temperatures would probably mean a shifting pattern of precipitation, with more storms and droughts, and perhaps less rain here and more there—with enormous implications for the world's food supply. Higher temperatures also mean a rising sea level, because water expands with heat and because the polar glaciers are already melting.

Because these projections are generated by computer models, they generally show temperatures rising along a smooth and predictable curve. But the actual record, over geological time, is very different, an erratic sawtooth pattern. Over the millennia the world has swung, sometimes with astounding speed, between ice ages and warm periods, suggesting the existence of powerful feedback mechanisms not yet known to climate science. A committee of the National Research Council recently published a study of abrupt climate change in which it concluded that "the only thing we can be sure of is that there will be climatic surprises." In the very distant past, the global average temperature has sometimes changed by much more than 4.5 degrees within a matter of decades, not centuries. In the past century, the world's climate has already moved beyond the range of recent historical experience, this study warned, and no one really knows how it will react to further warming. Feedback mechanisms might hold temperatures in the present range despite rising carbon dioxide concentrations. Or they could send temperatures shooting up well beyond anything in human experience.

Confronted with these risks and uncertainties, political leaders and governments have had great difficulty deciding what to do about them.

In June 1988 Senator Timothy Wirth (D-

Colo.), deeply exasperated by his inability to draw public attention to the subject, waited for a day that was forecast to be spectacularly hot. He then called a hearing at which several experts testified. With the temperature at 98 degrees Fahrenheit and anxiety rising about a drought gripping the Midwest and South, one of the witnesses, James E. Hansen, told the senators that the world was warmer than at any time in the century and he was 99 percent certain that the cause was human-made gases and not natural variations. Hansen's testimony had unusual force because he was the director of the National Aeronautics and Space Administration's Goddard Institute for Space Studies and the first scientist of his stature to declare flatly that rising temperatures were related to burning fuel.

"It is time to stop waffling so much and say the evidence is pretty strong that the greenhouse effect is here," he told a reporter for the *New York Times*, which put the story at the top of its front page.

Four days after that Senate hearing a conference opened in Toronto, attended by several hundred government officials and politicians as well as scientists. It started the push for action by calling for a 20 percent reduction in CO_2 emissions by the year 2005—a goal, incidentally, that was not set on the basis of any economic or scientific analysis, but rather because it was a number that seemed to the conferees to indicate a serious purpose. In December 1988 the U.N. General Assembly approved the establishment of an Intergovernmental Panel on Climate Change (IPCC) to provide expert reports on the science. The following year, at their annual summit meeting, the heads of state of the seven big industrial democracies called for a treaty—a "framework convention," as it became known—to limit the world's production of CO_2. Negotiations shortly got under way.

But strains between the United States and most of the western European countries soon became visible. President George H. W. Bush and his administration said that they were uneasy about the scientific basis for policy and wanted more time for research, while most Europeans thought the evidence was more than clear. Part of the difference arose from the relatively rapid rates of population

and economic growth in North America, which meant that limitations on fuel use would bind more tightly there. It is also true that the energy and automobile industries in the United States were more influential than in Europe, and the American unions more hostile to limits on their products. But basically it was a disparity in cultural attitudes. Second thoughts about the desirability of high economic growth are much more prevalent in Europe. In the United States, where people are accustomed to high consumption and inexpensive transportation, the idea of constraints on fuel use struck many people as an attack on their way of life and was widely unpopular.

In 1990 the first of the IPCC reports appeared, reflecting contributions of hundreds of scientists from dozens of countries. It showed a broad consensus among them that the possibility of global warming at least had to be taken seriously. If warming had not yet started, the IPCC said, rising emissions of carbon dioxide would certainly lead to it sooner or later.

At the huge and colorful 1992 United Nations Conference on Environment and Development, held in Rio de Janeiro, nearly every government in the world joined in agreement on a treaty that they titled the Framework Convention on Climate Change (FCCC). In it they set the goal of stabilizing the concentrations of greenhouse gases at a level that would prevent "dangerous" interference with the world's climate system. Most governments, including the United States, promptly ratified the treaty and put it in force.

But no one could say exactly at what point greenhouse concentrations in the atmosphere become dangerous. And the FCCC left any remedy entirely to voluntary action. No country was actually committed to make cuts in its greenhouse emissions. The FCCC was written with the intention simply of stating a goal in the expectation that, if necessary, further treaties would supplement it.

The 1992 elections brought to Washington a president more sympathetic to action on environmental issues than his predecessor. In April 1993, to celebrate Earth Day, Bill Clinton announced that he would reverse the government's previous position and work to stabilize greenhouse gas emissions at the 1990 level by 2000, as the Framework Convention urged. He called for a broad tax on all energy consumption to force conservation. But Clinton's own Democratic Party was split on this issue. A vigorous but relatively small part of his constituency was strongly in favor of action to protect the environment. A larger part was wary, and acutely concerned about the possible costs in terms of jobs lost.

Although the Democrats controlled both houses of Congress, the idea of the broad energy tax met a wall of hostility. Nothing emerged but a rise of 4.3 cents a gallon in the gasoline tax, a concession that had much less to do with reducing greenhouse emissions than with a need for more money for highway construction. When the president's Climate Change Action Plan appeared, it recommended nothing beyond voluntary cooperation. Many environmental advocacy organizations denounced it as futile, and they turned out to be correct. Emissions of carbon dioxide kept rising unimpeded.

But the earlier Bush administration turned out to have been correct when it suspected the Europeans of making promises that they could not keep. By the mid-90s it was clear that, of the world's major industrialized countries, only three would have lower emissions in 2000 than in 1990 —and in none of those three cases would the cause be deliberate environmental policy. Russia's emissions were lower because the Soviet economy had collapsed. Germany's were lower because it had been closing down the grossly inefficient plants it had inherited from the former East German regime. And the United Kingdom would succeed because, again for reasons of economic efficiency, it was shifting from coal to natural gas from the North Sea.

At this point, in 1996, the IPCC published its second survey of the science of global warming. Markedly more decisive than in the first report five years earlier, it concluded, in a line that became widely quoted, that the statistical evidence "now points towards a discernible human influence on global climate." But it followed that judgment with an admonition about the limitations on knowledge

and the need for further research. Like all of the IPPC assessments, it did not represent unanimity among researchers. The skeptics and dissenters included highly reputable scientists. But the report was, in effect, a textbook reflecting mainstream opinion among the specialists in climate science and the many other sciences on which it draws. It was not as sharply conclusive as many politicians would have liked. Its emphasis on the many questions still unanswered made it an uncertain foundation on which to build public policy. But the publication of this report marked the stage at which most of the scientists involved had decided that, to one degree or another, human activity was playing a part in global warming.

Because voluntary cooperation to diminish emissions was having no visible effect, the governments that had ratified the Framework Convention—formally, the parties to the treaty— began meeting to consider what might be done next. At the second conference of the parties, COP-2, in Geneva in the summer of 1996, the Clinton administration moved toward a firmer position. Timothy Wirth, the former senator, now Undersecretary of State for Global Affairs, announced that the United States would support legally binding limits on greenhouse gas emissions if other countries also did so. With that, negotiators went to work on a treaty—technically, a protocol to the Framework Convention—that could be signed at COP-3, scheduled for Kyoto, Japan, in December 1997.

President Clinton knew that the political base in the United States was inadequate for broad action, and he began a campaign to strengthen public understanding of the subject. He went to New York in June 1997 to address a special session of the United Nations. "The science is clear and compelling," he said. "We humans are changing the global climate." He spoke of new technologies and economic strategies such as emissions trading that would, he argued, reduce greenhouse emissions without damaging economic growth. In late July he returned to the issue, holding a White House conference in which he and Vice President Gore, the administration's ranking environmentalist, pursued their campaign for greater public awareness. "We see the train coming," the president said, "but most Americans in their daily lives can't hear the whistle blowing."

The following day, as if in response, the Senate passed, 95 to 0, a resolution warning the president not to agree at Kyoto to any treaty that would hurt the American economy or fail to commit the large developing countries to similar action. The chief sponsors were Robert Byrd, Democrat of West Virginia, a coal-mining state, and Chuck Hagel, Republican of Nebraska. While lobbyists from the energy industries had vigorously pressed this resolution, it also reflected wider concerns in the Senate about rising commercial competition from Asia and, specifically, from China.

President Clinton refined his plan in a speech in October at the National Geographic Society in Washington—the proper setting, he thought, for the topic. He said that the United States would support at Kyoto "the binding and realistic target of returning to emissions of 1990 levels between 2008 and 2012." The United States wanted the target to be a five-year average, not a single year's emissions, to avoid distortions arising from the ups and downs of the economic cycle. On one key point he agreed with the Senate. "The United States will not assume binding obligations," he said, "unless key developing nations meaningfully participate in this effort."

But Clinton avoided any reference to an issue that hung heavily over the whole discussion—the cost of emissions control. That meant the cost not only in dollars but, more important, the cost in jobs lost and in economic growth foregone. Oddly, there was very little serious economic analysis on emissions reduction before the middle 1990s, too late to have any influence on the drafting of the Kyoto text. The goals set by the negotiators continued to reflect political judgments as to what was desirable.

Although there was some discussion in the drafting process of policies and measures to achieve the cuts, it is fair to say that very few governments had any clear idea how they were actually to achieve the goals to which they were committing themselves. Most of the means involved profound changes in technology or in society's habits in the

use of energy. To make those changes the authors of Kyoto gave themselves only 11 years to the beginning of the commitment period in which the limits became binding. Beneath all the other debates over the treaty ran a current of doubt whether the goals were even possible in so short a time.

The Kyoto Conference identified six key greenhouse gases and plunged into the task of writing language to measure and reduce emissions of them. The conference was a huge affair, with some 10,000 officials from nearly every government on Earth as well as a large following of lobbyists and observers from a great variety of nongovernmental organizations representing both environmental and industrial interests. The work was immensely complex and, as generally happens, at the eleventh hour one relatively simple issue became the crucial symbol. The United States was willing to return its emissions only to the 1990 level, as President Clinton had proposed, while the Europeans demanded a much deeper cut.

President Clinton sent Vice President Gore on an emergency mission to Japan to save the day. Gore worked out a compromise under which the United States would reduce its emissions 7 percent below the 1990 level in the first commitment period, 2008 through 2012. With the increases that would result from expected economic growth, that meant a reduction of more than 30 percent below business-as-usual emissions in 2008-2012. Europe was to go to 8 percent below 1990. But the text put no obligations on the developing countries to cut emissions—a violation of the condition that Clinton himself had laid down in his October speech. The Kyoto text applied limits to the emissions of only the 38 industrialized countries. The developing countries had argued vehemently that the threat of climate change had been created by those countries as they had grown rich over two centuries of industrial growth, and that it was altogether unfair to impose the burdens of a solution on countries that were only beginning the same ascent.

Nevertheless, the Kyoto Conference declared the negotiations a success and opened the treaty to signature by the world's governments. A number of issues that had proved insoluble at Kyoto, not all of them minor, were quietly set aside for further discussions.

The United States had played a large role in shaping the Kyoto document. In particular it had pushed for flexibility as the world ventured into this altogether new realm of governance. Flexibility meant, among other things, introducing methods for trading emissions permits, modeled on the trading provisions in the U.S.'s Clean Air Act. This kind of trading was much less familiar in the rest of the world, and the rules were unfinished when the conference ended.

In the United States a spirited debate soon developed over the costs of meeting the Kyoto requirements. The administration produced figures showing that they would be low. The energy industries countered with studies showing that the costs would be devastatingly high. Both camps were using the same models and the same widely accepted methods of calculation. The differences lay in the assumptions that the analysts made about the future trading systems. Those differences could not be resolved because, of course, the rules for the trading systems had not yet been negotiated.

Because of the vast uncertainty over costs and the exclusion of the developing countries from any emissions limits, the chances that the U.S. Senate would ratify the Kyoto Protocol appeared close to nil. The Clinton administration signed it, a gesture of support with little legal significance. But in the administration's remaining three years in power, it never submitted the treaty to the Senate for ratification.

Around the world, supporters of Kyoto diligently put themselves to the task of working out the detailed regulations that would make it possible for the treaty to operate effectively. After nearly three years of planning and preparation they set the sixth conference of the parties to the Framework Convention—COP-6—to be held in The Hague, in late 2000, as the time and the place for the final decisions and agreements. But COP-6 was an ill-timed and dispirited affair. It convened a week after the election in which George W. Bush had defeated Kyoto's champion, Vice President Gore, for the American presidency. The American delegation, which had hoped to play a major role, was suddenly adrift and the hope of any serious agreements receded. The managers of the conference called a second session to be held in the spring.

American policy was changing. As a candidate, Mr. Bush had once said that he favored controlling CO_2 emissions. But as president, he moved in the opposite direction. He opposed the Kyoto Protocol because, he said, echoing the Byrd-Hagel resolution four years earlier, it would damage the American economy and would exempt the big developing countries. But, he emphasized, his administration "takes the issue of global climate change seriously." He closed his statement by declaring, "I am very optimistic that, with the proper focus and working with our friends and allies, we will be able to develop technologies, market incentives, and other creative ways to address global climate change."

President Bush's repudiation of Kyoto had a harsh impact on other governments, particularly in Europe. They knew that the United States would never ratify the current treaty. But they expected the U.S. to take an active part in further negotiations over the period after 2012 (when the first commitment period under Kyoto expires), and wrestle the regime around into something that it could support. Instead, the flat rejection meant that the world's most powerful economy, and biggest source of emissions, planned to do nothing in future talks but drag its feet.

Among some of Kyoto's supporters, especially in Europe, one effect was to increase their determination to put Kyoto in force and demonstrate that the United States could not, alone, control international action. They drew strength from the Third Assessment Report in 2001 of the IPCC, the international consortium of climate scientists. Once again the new report used sharper and less conditional language than the previous version six years earlier: "There is new and stronger evidence that most of the warming observed over the last 50 years is attributable to human activities." While many sources of uncertainty remained, the report said, accumulating evidence supported the new conclusion.

COP-7 met in the fall of 2001 in Marrakesh,

Morocco, under the leadership of people who badly wanted to finish, at last, the intricate job of writing the rules that would transform Kyoto from a statement of principles into a regime that could operate in the real world. It soon appeared that there was tension between the Europeans, who wanted tight enforcement of the emissions limits, and several other countries—most notably Japan, Canada and Australia—which felt that they needed greater latitude in making the substantial adjustments that Kyoto required.

Much of the dickering took place over the highly technical provisions regarding sinks, the natural reservoirs such as the soil or living plants like trees, into which carbon is absorbed from the atmosphere. A great deal depended on the precise wording of the provisions that gave governments credit for policies in agriculture and forestry that affected the amounts of carbon absorbed.

Another point of contention was the Kyoto text's omission of any means of enforcement. The negotiators finally agreed that any country emitting more than its limit in 2008-2012 would have to make up the excess, with a 30 per cent penalty added, in a subsequent commitment period. But since the treaty made no mention of a subsequent commitment period or the limits that might apply then, the attitude toward sinners seemed comfortably forgiving.

In the end Japan and Canada got most of what they felt they needed (although Australia decided to join the United States in dropping out the Kyoto treaty altogether). But the effect of all the fixes and flexes, together with the absence of the United States, was that the projected reduction in worldwide emissions would be very small.

The original targets were significantly loosened because the ratification process gave great leverage to countries that were having trouble making up their minds. The Kyoto text provided that it would go into force when it was ratified by 55 countries, including countries that represented 55 percent of the industrial countries' CO_2 emissions in 1990. Since the United States alone had contributed 36.1 percent of those emissions, ratification by nearly every other large industrial country was required to

carry the treaty into effect. In particular Russia, with 17.4 percent of the emissions, was crucial. After Marrakesh three years of hard bargaining ensued, in which the Russians withheld their ratification to see what advantages they could extract in exchange for it.

In the United States, the criticism of the Kyoto Protocol was by no means limited to the Bush administration. Many American economists pointed out the inefficiency of a regime that began by requiring very deep emissions cuts in the short term, but said nothing about the longer future. Scholars argued that it was unwise to impose dramatic goals with no idea how they might be achieved. Regarding the trading system, they contended, international law was not strong enough to protect the property rights represented by the emissions permits. These objections were widely influential in the United States, if not in Europe, but they left the critics with a question: If you don't like Kyoto, what do you like?

President Bush, unlike some of his supporters, was not prepared to say that the warnings of climate change were spurious and that nothing needed to be done. In February 2002, nearly a year after he had pulled the United States out of the Kyoto process, the president returned to the subject with a speech and a series of global climate change initiatives. Here the administration's reliance on long-term technical solutions, rather than immediate regulatory and coercive ones, became more explicit. But the president also showed an awareness that the issue of global warming was getting more important, and was not one on which he would want to have an empty record.

The key concept in Bush's new initiative was emissions intensity, the ratio of emissions to the total output of the American economy. His plan, the president said, would reduce intensity by 18 per cent over the next decade. Since that number was very close to the amount by which the normal processes of modernization and improvement had reduced intensity over the previous decade, neither Bush's friends nor his adversaries took the plan very seriously. The instruments were to be tax incentives, federal funds for industrial research and development, and voluntary cooperation. Analysts pointed out that, even should the plan succeed, normal economic growth would push the total volume of emissions to a level substantially higher at the end of the decade than in 2002. But the Bush administration was sticking with its calculation that the world had time to wait for the development of technologies—such as alternative fuels, and methods of capturing and sequestering carbon dioxide—that would make mandatory emissions cuts unnecessary.

The diplomacy of Kyoto often seemed to suggest a simple struggle between Europe and the United States, with the developing countries sitting off to one side. In truth, none of those blocs was as monolithic as it sometimes seemed.

Among the developing countries, views and interests were very mixed. The two big countries, China and India, were adamant that nothing interfere with their economic growth. Yet both were increasingly anxious about the effects of air pollution on public health, and were well aware that carbon dioxide is usually associated with the emissions of other gases and particles that carry a significant death rate. Most of the oil exporting countries vigorously fought Kyoto and any other scheme that might threaten their market. On the other side the small island nations formed a bloc of their own, plaintively observing that global warming, and the consequent rise in sea level, threatened their very existence. Some of the tropical countries, notably Brazil, were uneasy about Kyoto because of its possible implications for international control of land use. Clearing forest and jungle land sends huge amounts of carbon into the atmosphere.

In the United States, in the absence of federal leadership, state and municipal governments began to take action to cut greenhouse gases. At the state level the first reactions to Kyoto were negative. A number of states passed resolutions or legislation attacking it. But very soon, around the turn of the century, states started to move in the opposite direction. That was unexpected since the contribution of any one state alone has to be regarded, arithmetically, as negligible. But some state governments wanted to be seen to be at work protecting eco-

nomic interests, such as agricultural yields or a tourist industry's interest in scenic values. Some wanted to start a movement to build national policy from the bottom up. Some, although certainly not all, intended their programs to be a reproach to the Bush administration in Washington. As is usual in environmental politics, the alignments had more to do with geography than with political party allegiance.

In 2001–2002 the six New England states and five eastern Canadian provinces agreed on targets for reduction of greenhouse emissions. In 2002 California enacted legislation to control greenhouse emissions of cars (the implementing regulations are currently being challenged in litigation over the right of states to make climate policy). In 2003 George Pataki, the Republican governor of New York, took the initiative in forming the Regional Greenhouse Gas Initiative, a consortium of nine New England and Mid-Atlantic states that is currently developing a cap-and-trade program—that is, a program that will impose caps on major sources' emissions of greenhouse gases, and permit them to meet these requirements by trading emissions permits among themselves. In 2004 eight states sued five power companies to try to force them to reduce their carbon emissions.

By 2005 some 20 states had enacted legislation requiring various industries, usually the electric utilities, to obtain certain percentages of their power from renewable sources. And in June 2005 the Republican governor of California, Arnold Schwarzenegger, signed an executive order calling for sharp reductions in the state's greenhouse emissions over the next half century.

"As of today," Schwarzenegger declared, "California is going to be the leader in the fight against global warming. I say the debate is over. We know the science, we see the threat, and the time for action is now."

This activity had not gone unnoticed by Congress. In 2003, two leading senators, John McCain (R-N.M.) and Joseph I Lieberman (D-Conn.), introduced a bill with a national cap-and-trade program to impose controls on carbon dioxide. The bill got 43 votes—not enough to pass the

Senate, but enough to signal a substantial prospect for passage in the future. In 2005 the Senate passed a resolution declaring that the time had come for Congress to enact national limits on greenhouse gas emissions to slow, then stop and reverse their rise. The vote was 53 to 44. Compared with the 95 to 0 vote on the Byrd-Hagel resolution six years earlier condemning the principles on which the Kyoto Protocol was built, these more recent votes showed that climate change was no longer an unfamiliar subject for American politicians and that concerns about warming were growing stronger. The comparison also suggested that, in the judgment of these senators, the first step in a response to the threat of climate change was a national program using tools already familiar in federal anti-pollution legislation, rather than a highly complex international treaty and all the unknown and untried instruments that it required.

The European Union had become increasingly irritated by the delays and defections in the process of moving the Kyoto treaty forward, and had agreed on a cap-and-trade program of its own that would be legally separate from it and not dependent on it. The European Trading System (ETS) went into effect on Jan. 1, 2005, placing emissions limits on carbon dioxide. The principle was simple, largely modeled on the American program to control sulfur dioxide emissions under the Clean Air Act. But the actual operation promised to be far more complicated. The ETS covered perhaps four times as many sources of emissions as the American model, in a much wider range of industries. Many of the key policy decisions had been handed down to the 25 national governments with the prospect that rules would be far from uniform. But despite all the possible pitfalls ahead, the ETS was a serious and determined effort to reduce greenhouse emissions, conducted by the governments that felt most deeply about the need to do it.

In February 2005, the Kyoto Protocol finally went into force—more than seven years after it had been drafted and opened for signature. Over the previous year the Russians abandoned their hopes of making large sums of money by selling emissions permits. In the end they struck a bargain with the Europeans in which, in return for the crucial Russian ratification of Kyoto, the Europeans would support the Russian application for membership in the World Trade Organization.

In its immediate impact on emissions, Kyoto was likely to prove a disappointment to its authors. With the United States out of the regime, and with the concessions to other countries to bring them in, Kyoto's effect on the rising curve of worldwide emissions promised to be mild at best. But it was equally true that, under mandatory restrictions, the world would begin to get real experience in managing an international control system. That would include reliable data on what emissions reductions actually cost. All of that might well prove monumentally important if, in the future, the world decided that it needed a more comprehensive system to get faster cuts.

While the Bush administration continued to cite scientific uncertainty as the reason for opposing carbon emissions controls, the scientific consensus was moving in the opposite direction. In June 2005 the heads of 11 countries' academies of science, include Bruce Alberts, the president of the U.S. National Academy of Sciences, published a statement addressed to a forthcoming meeting of the leaders of the major industrial countries in which they said: "The scientific understanding of climate change is now sufficiently clear to justify nations taking prompt action. It is vital that all nations identify cost-effective steps that they can take now, to contribute to substantial and long-term reduction in net global greenhouse gas emissions." Other signers included the heads of the Chinese, Indian and Brazilian academies, as well as those in the highly industrialized countries.

At this time of writing, in mid-2005, nearly half a century has passed since researchers found that concentrations of carbon dioxide in the Earth's atmosphere were rising. It has been 13 years since nearly every government on Earth agreed, in the FCCC, on the goal of "stabilization of greenhouse gas concentrations in the atmosphere at a level that would prevent dangerous anthropogenic interference with the climate system." In that period public policy has yet to show any substantial impact on the steadily rising curve of carbon dioxide in the atmosphere, or on the more irregular but similarly accelerating rise in the planet's average surface temperature.

But a great deal of preparatory work has in fact been accomplished, through the hard work of many people around the world and the expenditure of a good deal of money. The results are visible in three areas.

First, in the science of the Earth's climate, much more is known than in 1992, at the time when the FCCC was signed. There are still skeptics, but many more scientists are prepared to say that accumulating evidence strongly indicates that recent warming has been caused mainly by human activities—that is, by burning fossil fuels. Much of the present uncertainty concerns the future consequences of a continued rise in carbon dioxide concentrations.

Second, two grand experiments are now under way to control carbon dioxide emissions by regulation, and a third is being organized. The Kyoto Treaty binds 36 industrialized countries to clearly quantified limits in the period 2008 through 2012. The European Trading system commits the 25 states of the European Union to similar emissions limits, and is already in force. In the United States nine northeastern states are working toward an agreement on carbon emissions. Some other states and even municipalities have set their own emissions goals. None of these efforts is likely to change the volumes of emissions sharply in the near term, but they will all produce valuable demonstrations of what works and what doesn't—and at what cost. In retrospect the year 2005 may well turn out to be the point at which the debates over the economics of emissions control ceased to be based on hypotheses and projections, and began to have a foundation in actual experience in many different societies.

Third, around the world but especially in the United States research is well under way on radically new technologies to keep carbon out of the atmosphere. Two well advertised examples are the development of hydrogen as a highway fuel, and an industrial-scale electric power plant in which all

pollutants including carbon dioxide are captured and sequestered. Neither will have a rapid pay-off. By most estimates it will be a matter of some decades before any significant number of Americans are driving on hydrogen, and the emissions-free generating plant will take at least a decade to put in operation. But these initiatives, if they prove successful, would have the capacity to alter fundamentally the carbon economy.

For public policy the question is how much of an effort to make, as a form of insurance against the risk of global warming. Disrupting the growth of economies that run mainly on fossil fuels would have a high cost. But allowing carbon dioxide concentrations to rise unimpeded might, at some unknown level, lead to disaster.

As the world works its way toward answers, it will be strongly influenced by actual events, changes in climate, and the scientific understanding of them, that make the risk seem greater or smaller. Perhaps the most important development in the past couple of decades, in political terms, is that millions of people now understand that climate can change, and science offers no guarantee that change will be either gradual or benign.

Suggested Reading

Alley, Richard B., et al, comprising the Committee on Abrupt Climate Change of the National Research Council. 2002. *Abrupt Climate Change: Inevitable Surprises.* Washington, DC: National Academy Press.

Grubb, Michael, with Christiaan Vrolijk and Duncan Brack.1999. *The Kyoto Protocol: a Guide and Assessment.* London: Royal Institute of International Affairs.

Intergovernmental Panel on Climate Change (IPCC). 2001. *The Third Assessment Report: Climate Change 2001.* 4 vol. Cambridge: Cambridge University Press. Available at the IPCC Web site, www.ipcc.ch/pub/online.htm. A fourth assessment is due in 2007.

Kruger, Joseph A., and William A. Pizer. October 2004. Greenhouse Gas Trading in Europe. *Environment.*

Mintzer, Irving M., and J. A. Leonard (eds.). 1994. *Negotiating Climate Change: The Inside Story of the Rio Convention.* Cambridge: Cambridge University Press.

Victor, David G. 2001. *The Collapse of the Kyoto Protocol and the Struggle to Slow Global Warming.* Princeton and Oxford: Princeton University Press.

37

How Much Climate Change Is Too Much?
An Economics Perspective

Jason F. Shogren and Michael A. Toman

Evaluating the benefits and costs of the mitigation of global climate change is a challenging enterprise, especially because of the high degree of scientific uncertainty and the long time span over which the process is expected to occur. But a systematic consideration of these projected benefits and costs provides some illuminating insights if not conclusive answers.

Having risen from relative obscurity as few as 10 years ago, climate change now looms large among environmental policy issues. Its scope is global; the potential environmental and economic impacts are ubiquitous; the potential restrictions on human choices touch the most basic goals of people in all nations; and the sheer scope of the potential response—a significant shift away from using fossil fuels as the primary energy source in the modern economy—is daunting. The magnitude of these changes has motivated experts the world over to study the natural and socioeconomic effects of climate change as well as policy options for slowing climate change and reducing its risks. The various options serve as fodder for often testy negotiations within and among nations about how and when to mitigate climate change, who should take action, and who should bear the costs.

Lurking behind these policy activities is a deceptively simple question: How much climate change is acceptable, and how much is "too much"? (The other key question is, Who is going to pay for mitigating the risks?) The lack of consensus on this issue reflects the uncertainties that surround it and differences in value judgments regarding the risks and costs.

In this chapter, we review the economic approach to the question of how much climate change is too much. The economic perspective emphasizes the evaluation of benefits and costs broadly defined while addressing uncertainties and impor-

Originally published in Michael Toman (ed.), *Climate Change Economics and Policy: An RFF Anthology* (Washington, DC: Resources for the Future, 2001).

tant considerations such as equity. We also consider some important criticisms of the benefit–cost approach. Then, we discuss the key factors that influence the benefits and costs of mitigating climate change risks. This discussion leads to a review of findings from the many quantitative "integrated assessment" models of climate change risks and response costs. This review does not lead to a simple answer to our overarching question about how much climate change is too much. But we do identify several good reasons for taking a deliberate but gradual approach to the mitigation of climate change risks.

The issues we cover are both diverse—ranging from the economics and philosophy of long-term cost-benefit analysis, to modeling strategies for representing climate change risks and greenhouse gas abatement costs—and, at times, somewhat complex. We have tried to be fairly comprehensive while seeking to make the discussion as accessible as possible.

Overview of the Risks and Response Costs

Life on Earth is possible partly because some gases such as carbon dioxide (CO_2) and water vapor, which naturally occur in Earth's atmosphere, trap heat—like a greenhouse. CO_2 released from use of fossil fuels (coal, oil, and natural gas) is the most plentiful human-created greenhouse gas (GHG). Other gases—which include methane (CH_4), chlorofluorocarbons (CFCs; now banned) and their substitutes currently in use, and nitrous oxides associated with fertilizer use—are emitted in lower volumes than CO_2 but trap more heat. Human-made GHGs work against us when they trap too much sunlight and block outward radiation. Scientists worry that the accumulation of these gases in the atmosphere has changed and will continue to change the climate.

The risk of climate change depends on the physical and socioeconomic implications of a changing climate. Climate change might have several effects:

- Reduced productivity of natural resources that humans use or extract from the natural environment (for example, lower agricultural yields, smaller timber harvests, and scarcer water resources).
- Damage to human-built environments (for example, coastal flooding from rising sea levels, incursion of saltwater into drinking water systems, and damage from increased storms and floods).
- Risks to life and limb (for example, more deaths from heat waves, storms, and contaminated water, and increased incidence of tropical diseases).
- Damage to less managed resources such as the natural conditions conducive to different landscapes, wilderness areas, natural habitats for scarce species, and biodiversity (for example, rising sea levels could inundate coastal wetlands, and increased inland aridity could destroy prairie wetlands).

All of these kinds of damage are posited to result from changes in long-term GHG concentrations in the atmosphere. Very rapid rates of climate change could exacerbate the damage. The adverse effects of climate change most likely will take decades or longer to materialize, however. Moreover, the odds that these events will come to pass are uncertain and not well understood. Numerical estimates of physical impacts are few, and confidence intervals are even harder to come by. The rise in sea level as a result of polar ice melting, for instance, is perhaps the best understood, and the current predicted range of change is still broad. For example, scenarios presented by the Intergovernmental Panel on Climate Change (IPCC) in *Climate Change 1995: The Science of Climate Change* indicate possible increases in sea level of less than 20 cm to almost 100 cm by 2100 as a result of a doubling of Earth's atmospheric GHG concentrations. The uncertainty in these estimates stems from not knowing how temperature will respond to increased GHG concentrations and how oceans and ice caps will respond to temperature change. The risks of catastrophic effects such as shifts in the Gulf Stream and the sudden collapse of polar ice caps are even harder to gauge.

Update

Since this piece was published in *Climate Change Economics and Policy: An RFF Anthology* in 2001, the economics and policy literatures on climate change have continued to grow astronomically. That year the Intergovernmental Panel on Climate Change (IPCC) issued its Third Assessment Report, which took a more definitive view on how human activities are changing the climate. It also provided much more detail on the anticipated adverse impacts of climate change, though the details and timing of these impacts remain uncertain.

These advances in knowledge have not yet led to a working consensus on how to answer the question posed in the title of this essay. There continues to be broad (though not universal) agreement that initial efforts are needed to rein in greenhouse gas (GHG) emissions, and mechanisms for doing so, involving many developed and developing countries, have gone into operation as the Kyoto Protocol went into force in February 2005. At the same time, however, developing countries have elevated their concerns about support for adaptation to what increasingly seem inevitable adverse impacts of unavoidable climate change. Discussion has begun as of this writing on the shape of the commitments to contain and eventually reduce global GHG emissions beyond the period of the Kyoto Protocol commitments (2012), but the end results of this process are impossible to envision.

Unknown physical risks are compounded by uncertain socioeconomic consequences. Cost estimates of potential impacts on market goods and services such as agricultural outputs can be made with some confidence, at least in developed countries. But cost estimates for nonmarket goods such as human and ecosystem health give rise to serious debate.

Moreover, existing estimates apply almost exclusively to industrial countries such as the United States. Less is known about the adverse socioeconomic consequences for poorer societies, even though these societies arguably are more vulnerable to climate change. Economic growth in developing countries presumably will lessen some of their vulnerability—for example, threats related to agricultural yields and basic sanitation services would decline. But economic growth in the long term could be imperiled in those regions whose economies depend on natural and ecological resources that would be adversely affected by climate change. Aggregate statistics mask considerable regional variation: Some areas probably will benefit from climate change while others lose.

In weighing the consequences of climate change, it is important to remember that humans adapt to risk to lower their losses. In general, the ability to adapt contributes to lowering the net risk of climate change more in situations where the human control over relevant natural systems and infrastructure is greater. Humans have more capacity to adapt in agricultural activities than in wilderness preservation, for example. The potential to adapt also depends on a society's wealth and the presence of various kinds of social infrastructure, such as educational and public health systems. As a result, richer countries probably will face less of a threat to human health from climate change than poorer societies that have less infrastructure. Beyond this general point, the potential for adaptation to reduce climate change risks continues to be debated.

GHGs remain in the atmosphere for tens or hundreds of years. GHG concentrations reflect longterm emissions; changes in any one year's emissions have a trivial effect on current overall concentrations. Even significant reductions in emissions made today will not be evident in atmospheric concentrations for decades or more. This point is important to keep in mind in deciding when to act—we do not have the luxury of waiting to see the full implications of climate change before taking ameliorative action. Many observers characterize responding to the risks of climate change as taking out insurance; nations try to reduce the odds of adverse events occurring through mitigation, and to reduce the severity of negative consequences by increasing the capacity for adaptation once climate change occurs. The insurance analogy underscores both the uncertainty that permeates how society and policymakers evaluate the issue and the need to respond to the risks in a timely way.

In constructing a viable and effective risk-reducing climate policy, policymakers must address hazy estimates of the risks, the benefits from taking action, and the potential for adaptation against the uncertain but also consequential cost of reducing GHGs. Costs of mitigation matter, as do costs of climate change itself. One must consider the consequences of committing resources to reducing climate change risks that could otherwise be used to meet other human interests, just as one must weigh the consequences of different climatic changes.

Why Consider the Costs and Benefits of Climate Policy?

Responding effectively to climate change risks requires society to consider the potential costs and benefits of various actions as well as inaction. By costs we mean the opportunity costs of GHG mitigation or adaptation—what society must forgo to pursue climate policy. Benefits are the gains from reducing climate change risks by lowering emissions or by enhancing the capacity for adaptation. An assessment of benefits and costs gives policymakers information they need to make educated decisions in setting the stringency of a mitigation policy (for example, how much GHG abatement to undertake, and when to do it) and deciding how much adaptation infrastructure to create.

It is important to consider the costs and the benefits of climate change policies because all resources—human, physical, and natural—are scarce. Policymakers must consider the benefits not obtained when resources are devoted to reducing climate change risks, just as they must consider the climate change risks incurred or avoided from different kinds and degrees of policy response.

Marginal benefits and costs reveal the gain from an incremental investment of time, talent, and other resources into mitigating climate risks, and the other opportunities forgone by using these resources for climate change risk mitigation. It is not a question of *whether* to address climate change but *how much* to address it.

Critics object to a benefit–cost approach to climate change policy assessment on several grounds. Their arguments include the following:

- The damages due to climate change, and thus the benefits of climate policies to mitigate these damages, are uncertain and thus inherently difficult to quantify given the current state of knowledge. Climate change also could cause large-scale irreversible effects that are hard to address in a simple benefit–cost framework. Therefore, the estimated benefits of action are biased downward.
- Climate mitigation costs are uncertain and could escalate rapidly from too-aggressive emission control policies. Proponents of this view are indicating a concern about the risk of underestimating mitigation costs.
- Climate change involves substantial equity issues—among current societies and between current and future generations—that are questions of morality, not economic efficiency. Policymakers should be concerned with more than benefit–cost analysis in judging the merits of climate policies.

As these arguments indicate, some critics worry that economic benefit–cost analysis gives short shrift to the need for climate protection, whereas others are concerned that the results of the analysis will call for unwarranted expensive mitigation.

Both groups of critics have proposed alternative criteria for evaluating climate policies, which can be seen as different methods of weighing the benefits and costs of policies given uncertainties, risks of irreversibility, the desire to avoid risk, and distributional concerns. For example, under the *precautionary principle*, which seeks to avoid "undue" harm to the climate system, cost considerations are absent or secondary. Typically, the idea is that climate change beyond a certain level simply involves too much risk, if one considers the distribution of benefits and costs over generations.

Knee-of-the-cost-curve analysis, in contrast, seeks to limit emission reductions to a point at which marginal costs increase rapidly. Benefit estimation is set aside in this approach because of uncertainty. The approach implicitly assumes that the marginal damages from climate change (which are the flip side of marginal benefits from climate change mitigation) do not increase much as climate change proceeds and that costs could escalate rapidly from a poor choice of emissions target.

The benefit–cost approach can address both uncertainty and irreversibility. We do not mean to imply that estimates in practice are always the best or that how one evaluates and acts on highly uncertain assessments will not be open to philosophical debate. For example, as people become more informed about climate change, it is safe to presume that the importance they attach to the issue will change. Critics of the economic methodology argue that this process reflects in part a change in preferences through various social processes, not only a change in information. Moreover, under conditions of great uncertainty, the legitimacy of a policy decision may depend even more than usual on whether the processes used to determine it are deemed inclusive and fair, as well as on the substantive evidence for the decision.

But it is fundamentally inaccurate to see analysis of economic benefits and costs from climate change policies as inherently biased because of uncertainty and irreversibility. Nor should benefit–cost analysis be seen as concerned only with market values accruing to developed countries. One of the great achievements in environmental economics over the past 40 years has been a clear demonstration of the importance of nonmarket benefits, which include benefits related to the development aspirations of poorer countries. These values can be given importance equal to that of market values in policy debates.

Our advocacy that benefits and costs be considered when judging climate change policies does not mean we advocate a simple, one-dimensional benefit–cost test for climate change policies. In practice, decisionmakers can, will, and should bring to the fore important considerations about the equity and fairness of climate change policies across space and time. Decisionmakers also will bring their own judgments about the relevance, credibility, and robustness of benefit and cost information and about the appropriate degree of climate change and other risks that society should bear. Our argument in favor of considering both benefits and costs is that policy deliberations will be better informed if good economic analysis is provided.

The alternative decision criteria advanced by critics also are problematic in practice. The definition of "undue" is usually heuristic or vague. The approach is equivalent to assuming a sharp spike, or peak, in damages caused by climate change beyond the proposed threshold. It may be the case, but not enough evidence yet exists to assume this property (let alone to indicate at what level of climate change such a spike would occur). On the other hand, with knee-of-the-curve analysis, benefits are ignored so there is no assurance of a sound decision either.

Benefits and costs are unavoidable. How their impacts are assessed is what differentiates one approach from another. We maintain throughout this discussion that the assessment and weighing of costs and benefits is an inherent part of any policy decision.

Equity and Fairness Issues

The fairness of climate change policies to today's societies and to future generations continues to be at the core of policy debates. These issues go beyond what economic benefit–cost analysis can resolve, though such analysis can help illustrate the possible distributional impacts of different climate policies. In this section, we focus first on intergenerational equity issues. Then, contemporaneous equity issues are addressed.

Advocates of more aggressive GHG abatement point to the potential adverse consequences of less aggressive abatement policies for the well-being of future generations as a moral rationale for their stance. They assert that conventional discounting—even at relatively low rates—may be inequitable to future generations by leaving them with unacceptable climate damages or high costs from the need to abate future emissions very quickly. Critics also have argued that conventional discounting underestimates costs in the face of persistent income differences between rich and poor countries. Essentially, the argument is that because developing countries probably will not close the income gap over the next several decades, and because people in those countries attach higher incremental value to additional well-being than people in rich countries, the effective discount rate used to evaluate reductions in future damages from climate change should be lower than that applied to richer countries.

Supporters of the conventional approach to discounting on grounds of economic efficiency argue just as vehemently that any evaluation of costs and benefits over time that understates the opportunity cost of forgone investment is a bad bargain for future generations because it distorts the distribution of investment resources over time. These supporters of standard discounting also argue that future generations are likely to be better off than the present generation, casting doubt on the basic premise of the critics' concerns.

Experts attempting to address this complex mixture of issues increasingly recognize the need to distinguish principles of equity and efficiency, even though there is as yet no consensus on the practical implications for climate policy. We can start with the observation that anything society's decision-makers do today—abating GHGs, investing in new seed varieties, expanding health and education facilities, and so on—should be evaluated in a way that reflects the real opportunity cost, that is, the options forgone both today and in the long term. This answer responds to the critics who fear a misallocation of investment resources if climate policies are not treated similarly to other uses of society's scarce resources.

Long-term uncertainty about the future growth of the economy provides a rationale for low discount rates on grounds of economic efficiency. The basic argument is that if everything goes well in the future, then the economy will be productive, the rate of return on investment will remain high, and the opportunity cost of displacing investment with policy today likewise also will be high. However, if things do not go so well and the rate of return on capital is low because of climate change or some other phenomenon, then the opportunity cost of our current investment in climate change mitigation versus other activities also will be low.

But economic efficiency only means a lack of waste given some initial distribution of resources. Specifically how much climate change mitigation to undertake is a different question, one that refers to the distribution of resources across generations. The answer depends on how concerned members of the current generation are about the future in general, how much they think climate change might imperil the well-being of their descendants, and the options at their disposal to mitigate unwelcome impacts on future generations. For example, one could be very concerned about the well-being of the future but also believe that other investments—such as health and education—would do more to enhance the well-being of future generations. Not surprisingly, experts and policymakers do not agree on these points.

We turn next to a brief discussion of international equity issues associated with climate change. The most immediate aspect of this debate involves the international distribution of responsibility for reducing GHGs and the associated costs. Developing nations have many pressing needs, such as potable water and stable food supplies, and less financial and technical capacity than rich countries have for mitigating GHGs. These nations have less incentive to agree to a policy that they see as imposing unacceptable costs.

Beyond this question are even more vexing issues surrounding the distribution of climate change risks. As already noted, it is likely that developing countries are both relatively more vulnerable to climate change than advanced industrialized countries and have less adaptive capacity; however, these disadvantages likely will be reduced in the future with further economic development. These differences are only beginning to be accounted for in climate change risks assessments. Analyses that consider only aggregate benefits and costs of climate change mitigation, without addressing the distribution of these benefits and costs, miss an important dimension of the policy problem. For example, the absolute magnitude of avoided costs from slowing climate change may be smaller in developing countries simply because per capita incomes are lower. But the implication that climate change mitigation should be given short shrift just because it mainly affects poorer people is ethically troubling.

Differences in perceptions about what constitutes equitable distributions of effort complicate any agreement. No standard exists for establishing the equity of any particular allocation of GHG control responsibility. Simple rules of thumb, such as allocating responsibility based on equal per capita rights to emit GHGs (advantageous to developing countries) and allocations that are positively correlated to past and current emissions (advantageous to developed countries) are unlikely to command broad political support internationally.

What Do Existing Economic Analyses Say?

Analyzing the benefits and costs of climate change mitigation requires understanding biophysical and economic systems as well as the interactions between them. Integrated assessment (IA) modeling combines the key elements of biophysical and economic systems into one integrated system (Figure 1). IA models strip down the laws of nature and human behavior to their essentials to depict how more GHGs in the atmosphere raise temperature and how temperature increase induces economic losses. The models also contain enough detail about the drivers of energy use and interactions between energy and economy that the economic costs of different constraints on CO_2 emissions can be determined.

Researchers often use IA models to simulate a path of carbon reductions over time that would maximize the present value of avoided damages

(that is, the benefits of a particular climate policy) less mitigation costs. As noted earlier, considerable controversy surrounds this criterion for evaluation.

A striking finding of many IA models is the apparent desirability of imposing only limited GHG controls over the next 20 or 30 years. According to the estimates in most IA models, the costs of sharply reducing GHG concentrations today are too high relative to the modest benefits the reductions are projected to bring.

The benefit of reducing GHG concentrations in the near term is estimated in many studies to be on the order of $5–25 per ton of carbon. Only after GHG concentrations have increased considerably do the impacts warrant more effort to taper off emissions, according to the models.

Even more striking is the finding of many IA models that emissions should rise well into this century. In comparison, the models indicate that policies pushing for substantial near-term control, such as the Kyoto Protocol, involve too much cost, too soon, relative to their projected benefits. Critics argue that IA models inadequately address several important elements of climate change risks: uncertainty, irreversibility, and risk of catastrophe. Assessing the weight of these criticisms requires us to explore the influences on the economic benefits and costs of climate protection.

Influences on the Benefits

The IPCC Second Assessment Report concluded that climate change could pose some serious risks. The IPCC presented results of studies showing that the damaging effects of a doubling of GHG concentrations in the atmosphere could cost on the order of 1.0–1.5% of gross domestic product (GDP) for developed countries and 2.0–9.0% of GDP for developing countries (see also Frankhauser and others in Suggested Reading). Reducing such losses is the benefit of protecting against the negative effects of climate change.

Several factors affect the potential magnitude of the benefits. One is the potential scale and timing of damages avoided. Although IA models differ greatly in detail, most have economic damage representa-

Figure 1. Climate Change and Its Interaction with Natural, Economic, and Social Processes

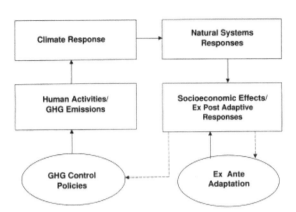

tions calibrated to produce damages resulting from a doubling of atmospheric GHG concentrations roughly of the same order as the IPCC Second Assessment Report. This point is worth keeping in mind when evaluating the results. The models increasingly contain separate damage functions for different regions. Generally, the effects in developing countries are presumed to be worse than those in the developed world, as in the IPCC Second Assessment Report. For the most part, these costs would be incurred decades into the future. Consequently, the present value of the costs would be relatively low today.

Assumptions about adaptation also affect estimates of potential benefits. Some critics of the earlier IPCC estimates argue that damages likely will be lower than predicted because expected temperature increases from a doubling of atmospheric GHG concentrations probably will be less than projected, ecosystems seem to be more resilient over the long term than the estimates suggest, human beings can adapt more than was supposed, and damages are not likely to increase proportionally with GDP. The implication is that the optimal path for GHG control (in a present value sense) should be even less aggressive than the IA results indicate. These new assessments remain controversial. One ongoing question concerns the cost of adjusting to a changing climate versus the long-term cost of a changed climate. Another is whether the effects of climate change (for example, in encouraging the spread of human illness through a greater incidence of tropical diseases, reducing river flows that concentrate pollutants, and increasing the incidence of heat stress) are being underestimated.

A third factor affects benefits: Damage costs not only are uncertain but also involve a chance of a catastrophe. However, a general finding from IA models is that GHG reductions should be gradual, even if damages are larger than conventionally assumed. A risk of catastrophe provides a rationale for more aggressive early actions to reduce GHG concentrations, but the risk has to be very large to rationalize near-term actions as aggressive as those envisioned in the Kyoto Protocol in a present-value IA framework. Part of the reason for this finding is that the outcome with the lowest cost also is the most likely to occur. IA models also do not incorporate risk-averse attitudes, which would provide a stronger rationale for avoiding large costs. Moreover, discounting in the models reduces the effective impact of all but the most catastrophic costs after a few decades.

Irreversibility of GHG emissions is yet another factor influencing the benefits of GHG abatement. Because GHG emissions persist in the atmosphere for decades, even centuries, the resulting long-term damages strengthen the rationale for early and aggressive GHG control. Moreover, given that some damage costs from adjusting to a changed climate

Table 1. Implications of a Carbon Tax for U.S. Gasoline and Coal Prices

	Price ($)		
Commodity	1997 average	With $100/ton carbon tax	With $400/ton carbon tax
Bituminous coal	26.16	87.94	273.28
Motor gasoline	1.29	1.53	2.26

Note: Coal price is national average annual delivered price per ton to electric utilities; gasoline price is national average annual retail price per gallon.

Sources: U.S. DOE (see Suggested Reading).

depend on the *rate* of climate change, immediate action also might be valuable. To date, however, the importance of this factor has not been conclusively demonstrated; the gradual abatement policies implied by the IA models do not seem likely to greatly increase the speed of further climate change.

Finally, policies that reduce CO_2 also can yield ancillary benefits in terms of local environmental quality improvement, such as fewer threats to human health and reduced damage to water bodies from nitrogen deposition. The magnitudes of these ancillary effects remain fairly uncertain. They are lower to the extent that more environmental improvement would occur anyway, in the absence of GHG policy. They also depend on how GHG policies are implemented (for example, a new boiler performance mandate that encouraged extending the lives of old, dirty boilers would detract from the environment).

Influences on the Costs

Estimates of the cost of mitigating GHG emissions vary widely. Some studies suggest that the United States could meet its Kyoto Protocol target at negligible cost; other studies claim that the United States would lose at least 1–2% of its GDP each year. A study by the Energy Modeling Forum helped explain the range of results in assessing the costs to meet the Kyoto Protocol policy targets (see Weyant and Hill in Suggested Reading). For example, the carbon price (carbon tax or emissions permit price) needed to achieve the Kyoto Protocol emissions target in the United States with domestic policies alone ranges from about $70 per metric ton of car-

bon to more than $400 per ton (in 1990 dollars) across the models. The corresponding GDP losses in 2010 range from less than 0.2% to 2.0% relative to baseline. (The percentages of GDP are not reported in Weyant and Hill but implied from graphs presented there.) Carbon prices are put in perspective by relating them to prices for common forms of energy, as listed in Table 1.

The results reported by Weyant and Hill and previous assessments of GHG control costs reflect different views about three key assumptions that drive the estimated costs of climate policy: stringency of the abatement policy, flexibility of policy instruments, and possibilities for development and diffusion of new technology. First, as one would expect, the greater the degree of CO_2 reduction required (because the target is ambitious, baseline emissions are high, or both), the greater the cost.

Costs of GHG control depend on the speed of control as well as its scale. Wigley and others (see Suggested Reading) showed that most long-term target GHG concentrations could be achieved at substantially lower present value costs if abatement were increased gradually over time, rather than rapidly, as envisaged under the Kyoto Protocol. Subsequent elaboration of this idea has shown that, in principle, cost savings well in excess of 50% could be achieved by using a cost-effective strategy for meeting a long-term concentration target versus an alternative path that mandates more aggressive early reductions (see the 1997 paper by Manne and Richels in Suggested Reading). These cost savings come about not only because costs that come later are discounted more but also because less existing capital becomes obsolete prematurely. There is an irreversibility problem associated with premature commitment to a form and scale of low-emissions capital, just as irreversibility is associated with climate change. The former irreversibility implies lower costs with a slower approach to mitigation.

Another important factor in assessing the costs of CO_2 control is the capacity and willingness of consumers and firms to substitute alternatives for existing high-carbon technologies. Substitution undertaken depends partly on the technological ease of substituting capital and technological inputs for energy inputs and partly on the cost of lower-carbon alternatives. Some engineering studies suggest that 20–25% of existing carbon emissions could be eliminated at low or negligible cost if people switched to new technologies such as compact fluorescent light bulbs, improved thermal insulation, efficient heating and cooling systems, and energy-efficient appliances. Economists counter that the choice of energy technology offers no free lunch. Even if new technologies are available, many people are unwilling to experiment with new devices at current prices. Factors other than energy efficiency also matter to consumers, such as quality, features, and the time and effort required to learn about a new technology and how it works. People behave as if their time horizons are short, perhaps reflecting their uncertainty about future energy prices and the reliability of the technology.

In addition, the unit cost of GHG control in the future may be lower than in the present, as a consequence of presumed continuation in trends toward greater energy efficiency in developed and developing countries (as well as some increased scarcity of fossil fuels). These trends will be enhanced by policies that provide economic incentives for GHG-reducing innovation. It is possible that the cost associated with premature commitment to irreversible long-lived investments in low-emissions technologies is more important in practice than climatic irreversibility, at least over the medium term. The reason is that sunk investments cannot be undone if climate change turns out to be less serious than might be expected, whereas society can accelerate GHG control if it learns that the danger is greater than estimated. The strength of this point depends in part on how irreversible low-GHG investment is and on the costs of irreversible climate change. In addition, critics of this view argue that without early action to reduce GHG emissions, markets for low-emission technologies would not develop and societies would lock in to continued use of fossil fuel–intensive energy systems.

Still another important factor is the flexibility and cost-effectiveness of the policy instruments imposed, both domestically and internationally. For example, Weyant and Hill's review showed that the flexibility to pursue CO_2 reductions anywhere in the Annex I countries (the industrialized countries that would cap their total emissions under the Kyoto Protocol) through some form of international emissions trading system could lower U.S. costs to meet the Kyoto Protocol target by roughly 30–50%. Less quantitative analysis has been done of alternative domestic policies. Nevertheless, it can be presumed from studies of the costs of abating other pollutants that cost-effective policies will lower the cost of GHG abatement, perhaps significantly. In contrast, constraints on the use of cost-effective policies—for example, the imposition of rigid technology mandates in lieu of more flexible performance standards—will raise costs, perhaps considerably. This factor often is neglected in analyses of domestic abatement activity that consider only the use of cost-effective policies such as emissions permit trading, although use of such policies is hardly foreordained. Ignoring this factor means that the costs reported in the economic models probably understate the costs societies will actually incur in GHG control. By the same token, studies of international policies that assume ideal conditions of implementation and compliance are overoptimistic.

A subtle but important influence on the cost of GHG control is whether emission-reducing policies also raise revenues (such as a carbon tax) and what is done with those revenues. When revenue generated by a carbon tax or other policy is used to reduce other taxes (a process commonly referred to as revenue recycling), some of the negative effect on incomes and labor force participation of the increased cost of energy is offset. However, it may be more effective at stimulating employment and economic activity in countries with chronically high unemployment than in the United States. The issue of revenue recycling applies also to policies that would reduce CO_2 through carbon permits or "caps." If CO_2 permits are auctioned, then the revenues can be recycled through cuts in existing taxes; freely offered CO_2 permits do not allow the possibility of revenue recycling. The difference in net social costs of GHG control in the two cases can be dramatic. Reducing CO_2 emissions with auctioned permits and revenue recycling can have net costs less than the benefits of GHG control indicated by the IA models. In contrast, with a system of freely provided CO_2 permits, *any* level of emissions reduction yields environmental benefits (according to the IA models) that fall short of society's costs of abatement.

Most cost analyses presume that the relevant energy and technology markets work reasonably efficiently (other than the commonly recognized failure of private markets to provide for all the basic R&D that society wants, because this is a kind of public good). This assumption is more or less reasonable for most developed industrial economies. Even in these countries, one can identify problems such as direct and indirect energy subsidies that encourage excessive GHG emissions. Problems of market inefficiency are far more commonplace in the developing countries and in countries in transition toward market systems; accordingly, one expects incremental CO_2 control costs to be lower (even negative) in those countries. However, the institutional barriers to accomplishing GHG control in these economic systems may negate the potential efficiency gains.

Thus far, our discussion had focused on CO_2 control. Because CO_2 is only one of several GHGs, and because CO_2 emissions can be sequestered or even eliminated by using certain technologies, emissions targets related to climate change can be met in several ways. Some recent analyses suggest that the costs of other options compare very favorably with the costs of CO_2 reduction. For example, counting the results of forest-based sequestration and the reduction of non-CO_2 gases toward total GHG reduction goals could lower the cost to the United States of meeting its Kyoto Protocol emissions target by roughly 60%. But care is needed in interpreting some of the cost estimates. In particular, low estimates for the cost of carbon sequestration may not adequately capture all the opportunity cost of different land uses.

Uncertainty, Learning, and the Value of New Information

Another key factor in choosing the timing and intensity of climate change mitigation is the opportunity to learn more about both the risks of climate change and the costs of mitigation. Several studies show that the value of more and better information about climate risks is substantial. This value arises because one would like to avoid putting lots of resources into mitigation in the short term, only to find out later that the problems related to climate change are not serious. However, one also would like to minimize the risk of doing too little mitigation in the short term, only to find out later that very serious consequences of climate change will cost much more to avert because of the delay.

Manne and Richels, as well as Kolstad, showed that it generally pays to do a little bit of abatement in the short run under these conditions—to hedge against the downside without making too rapid a commitment. One virtue of some delay in emissions control is that it allows us to learn more about the severity of the risk of climate change and the options for responding to it. If the risk turns out to be worse than expected, mitigation can be accelerated to make up for lost time. To be sure, the strength of this argument depends on how costly it is to accelerate mitigation and on the degree of irreversibility of climate change. Analysts will continue to debate these points for some time to come.

Concluding Remarks

In this chapter, we have explained that benefits and costs matter, for reasons of both efficiency and equity, and that benefits and costs must and can be considered in the context of the uncertainties that surround climate change. Economic analyses provide several rationales for pursuing only gradual abatement of GHG emissions. Because damages accrue gradually, catastrophes are uncertain and far off in the future, and unit mitigation costs are likely to fall over time (especially with well-designed climate policies), it makes sense to proceed slowly. To the extent that innovation is slower than desired with this approach, government programs targeted at basic R&D can help. The IA models indicate that rapid abatement does not maximize the present value of all society's resources.

We have not argued that current benefit–cost analyses are the last word on the subject. Opportunities certainly exist to improve the measurement of benefits and costs and to track the incidence of costs and risks across groups and over time. In practice, policy decisions will turn on a broader set of considerations than a single expected benefit–cost ratio. However, the arguments in favor of purposeful but gradual reduction in GHGs seem strong.

Economic analysis also could be used to justify not only a slower approach to GHG mitigation but also a less stringent long-term target. Here is where the potential conflict can arise between individuals' narrower economic self-interests and their concern for the well-being of future generations. Determining the right long-term policy goals ultimately requires us to address our attitudes toward intergenerational equity as well as to better understand the scale of environmental and economic risks that different climate policies imply for future generations. A more gradual GHG policy over the next 10–20 years does not preclude any but the most environmentally stringent targets, while potentially increasing the political acceptability of increasingly demanding mitigation measures. These considerations warrant renewed attention as the international community continues to grapple with the problem of finding a climate policy it can really live with.

Suggested Reading

IPCC Third Assessment Report: Climate Change 2001 (available at www.ipcc.ch).

Kolstad, Charles D. 1996. Learning and Stock Effects in Environmental Regulation: The Case of Greenhouse Gas Emissions. *Journal of Environmental Economics and Management* 31: 1–18.

Kopp, R. Near-Term Greenhouse Gas Emissions Targets. Discussion Paper 04-41. Washington, DC: Resources for the Future.

Manne, Alan S., and Richard Richels. 1992. *Buying Greenhouse Insurance: The Economic Costs of CO_2 Emission Limits.* Cambridge, MA: MIT Press.

———. 1997. On Stabilizing CO_2 Concentrations—Cost-Effective Emission Reduction Strategies. *Environmental Modeling and Assessment* 2(4): 251–65.

Mendelsohn, Robert, and James E. Neuman (eds.). 1999. *The Impact of Climate Change on the United States Economy.* Cambridge, U.K.: Cambridge University Press.

Shogren, Jason F., and Michael A. Toman. 2000. Climate Change Policy. In *Public Policies for Environmental Protection*, edited by P. Portney and R. Stavins. Washington, DC: Resources for the Future, 125–168.

Weyant, John P., and Jennifer N. Hill. 1999. Introduction and Overview. *The Energy Journal*, Special Issue (*The Costs of the Kyoto Protocol: A Multi-Model Evaluation*): vi-xiiv.

Wigley, Thomas M.L., Richard Richels, and James A. Edmonds. 1996. Economic and Environmental Choices in the Stabilization of Atmospheric CO_2 Concentrations. *Nature* 379 (6562): 240–43.

See also The Pew Center on Global Climate Change for a number of useful papers: www.pewclimate.org

38 Choosing Price or Quantity Controls for Greenhouse Gases

William A. Pizer

To control emissions of greenhouse gases (GHG), there are two basic forms of economic incentives that can be used: taxes on emissions of these gases (a price instrument) or an emissions-trading system (a quantity instrument). The analysis suggests that, in principle, the tax instrument is much superior to an emissions trading system for controlling GHG. However, existing policies are leaning strongly towards systems of emissions trading. Such systems can be made more efficient by allowing a "safety valve" in the form of a relatively high tax that can be paid on excess emissions in the event that allowance prices in the trading market become very high.

Much of the debate surrounding climate change has centered on verifying the threat of climate change and deciding the magnitude of an appropriate response. After years of negotiation, this effort led to the 1997 signing of the Kyoto Protocol, a binding commitment by industrialized countries to reduce their emissions of carbon dioxide (CO_2) to slightly below 1990 recorded levels. Without approving or disapproving of the response effort embodied in the Kyoto Protocol, I believe that an important element has been ignored. Namely, should we specify our response to climate change in terms of a quantitative target?

The appeal of a quantitative target is obvious. A commitment to a particular emissions level provides a straightforward measure of environmental progress as well as compliance. Commitment to an emissions tax, for example, offers neither a guarantee that emissions will be limited to a certain level nor an obvious way to measure a country's compliance (when other taxes and subsidies already exist). Yet, this concern points to an important observation.

Quantity targets guarantee a fixed level of emissions. Emission taxes guarantee a fixed financial incentive to reduce emissions. Both can be set at aggressive or modest levels. Aside from the appeal of the known and verifiable emissions levels that quantity targets can ensure, might there be other important differences between price and quantity controls? Economists

Originally published in Michael Toman (ed.), *Climate Change Economics and Policy: An RFF Anthology* (Washington, DC: Resources for the Future, 2001).

would say "Yes." With uncertain outcomes and policies that are fixed for many years, it is important to carefully consider both the costs and benefits of alternate price and quantity controls to judge which is best. My own analysis of the two approaches indicates that price-based greenhouse gas (GHG) controls are much more desirable than quantity targets, taking into account both the potential long-term damages of climate change and the costs of GHG control. This can be argued on the basis of both theory and numerical simulations. On the basis of the latter, I find that price mechanisms produce expected net gains five times higher than even the most favorably designed quantity target.

To explain this conclusion, I first characterize the differences between price and quantity controls for GHGs. I then present both theoretical and empirical evidence that price-based controls are preferable to quantity targets on the basis of these differences. Finally, I discuss how price controls can be implemented without a general carbon tax. This point is particularly salient for the United States, where taxes are generally unpopular. The "safety valve," as it is often called, involves a cap-and-trade GHG system accompanied by a specified fee or penalty for emissions beyond the initial cap.

How Do Quantity- and Price-Based Mechanisms Work?

A quantity mechanism—usually referred to as a permit or cap-and-trade system—works by first requiring individuals to obtain a permit for each ton of CO_2 they emit, and then limiting the number of permits to a fixed level. (CO_2 emissions from fossil fuel sources constitute the bulk of GHG emissions and are the general focus of most policy discussions. However, the arguments made in this context apply equally well to the regulation of GHG emissions more broadly defined.) This kind of system has been used with considerable success in the United States to regulate sulfur dioxide and lead. The permit requirement could be imposed on the individuals who release CO_2 into the atmosphere by burning coal, petroleum products, or natural gas. However, unlike the emissions of conventional

pollutants, which depend on various other factors, CO_2 emissions can be determined very accurately by the volume of fuel being used. Rather than requiring *users* of fossil fuels to obtain permits, we could therefore require *producers* to obtain the same permits. This method has the advantage of involving far fewer individuals in the regulatory process, thereby reducing both monitoring and enforcement costs.

One key element in a permit system is that individuals are free to buy and sell existing permits in an effort to obtain the lowest cost of compliance for themselves, which in turn leads to the lowest cost of compliance for society. In particular, when individuals observe a market price for permits, those who can reduce emissions more cheaply will do so to sell excess permits or to avoid having to buy additional ones. Similarly, those who face higher reduction costs will avoid reductions by buying permits or by keeping those they already possess. In this way, total emissions will exactly equal the number of permits, and only the cheapest reductions are undertaken.

A price mechanism—usually referred to as a carbon tax or emissions fee—requires the payment of a fixed fee for every ton of CO_2 emitted. Like the permit system, this fee could be levied upstream on fossil fuel producers or downstream on fossil fuel consumers. Either way, we associate a positive cost with CO_2 emissions and create a fixed monetary incentive to reduce emissions. Such price-based systems have been used in Europe to regulate a wide range of pollutants (although the focus is usually revenue generation rather than substantial emissions reductions).

Like a tradable permit system, price mechanisms are cost-effective. Only those emitters who can reduce emissions at a cost below the fixed fee or tax will choose to do so. Because only the cheapest reductions are undertaken, we are guaranteed that the resulting emission level is obtained at the lowest possible cost.

The important distinction between these two systems is how they adjust when costs change unexpectedly. A quantity or permit system adjusts by allowing the permit price to rise or fall while

Update

Much has happened in the arena of climate change policy since this article was first published in 2001. The Kyoto Protocol entered into force at the beginning of 2005, despite the withdrawal of the United States and Australia. More importantly, various nations have begun designing and implementing domestic policies using both price and quantity instruments. The European Union has already begun its greenhouse gas Emissions Trading Scheme (ETS), New Zealand has introduced a carbon tax starting in 2007, and Canada is establishing a trading program for Large Final Emitters (LFE) of carbon dioxide. While the E.U. ETS is a pure quantity policy and the New Zealand carbon tax a pure price policy, the Canadian LFE is a hybrid with a targeted reduction through the trading program and a safety valve to limit costs. In the United States, a group of nine northeastern states are negotiating the Regional Greenhouse Gas Initiative to cap and reduce emissions from power plants through a quantity-based regional trading program. At the national level, Senators McCain and Lieberman have introduced a proposal for a trading program with a further proposal from Senator Bingaman for a safety valve to supplement trading. In the midst of all these domestic developments, international discussions are commencing on what to do "post-Kyoto" from 2013 onward. In particular, these negotiations must focus on whether there will be a continued international focus on quantitative targets.

holding the emissions level constant. A price or tax system adjusts by allowing the level of total emissions to rise or fall while holding the price associated with emissions constant. Ignoring uncertainty and assuming that we know the costs of controlling CO2, both policies can be used with the same results. Consider the following example:

Suppose we know that with a comprehensive domestic CO_2 trading system in place in the United States by 2010, a permit volume of 1.2 gigatons (billion tons) of carbon equivalent emissions (GtC) will lead to a $100 permit price per ton of carbon. (U.S. emissions of carbon from fossil fuels were estimated at 1.5 GtC for 1998.) In other words, faced with a price incentive of $100 per ton to reduce emissions, regulated firms in the United States will find ways to reduce emissions to 1.2 GtC. Then, the same outcome can be obtained by imposing a $100 per ton carbon tax.

Uncertainty about Costs

In reality, we have only a vague idea about the permit price that would occur with emissions of 1.2 GtC or any other emission target. These costs are hard to pin down for three reasons. First, little evidence exists concerning reduction costs. There are no recent examples of carbon reductions on a substantial scale from which to base estimates. In the 1970s, energy prices doubled and encouraged increased energy efficiency, but these events occurred in a context of considerable uncertainty about the future and alongside many other confounding factors (such as increased environmental regulation). Alternatively, engineering studies provide a bottom-up approach to estimating costs. However, comparisons of past engineering forecasts with actual implementation costs suggest that forecasts are inaccurate at best.

A second source of uncertainty arises because we need to forecast compliance costs in the future. This task involves difficult predictions about the evolution of new technologies. Proponents of aggressive policy argue that reductions will be cheap as new low-carbon or carbon-free energy technologies become available. Proponents of more modest policies argue that these are unproven, pie-in-the-sky technologies that may never be practical.

Finally, it is impossible to know how uncontrolled emission levels will change in the future. That is, to achieve 1990 emission levels in 2010, it is unclear whether reductions of 5%, 25%, or even 50% will be necessary. The Intergovernmental

Figure 1: Distribution of Emissions in 2010

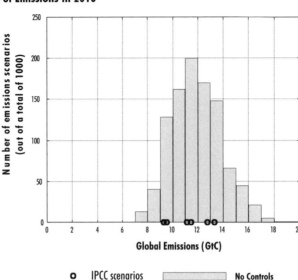

Panel on Climate Change (IPCC), the international agency charged with studying climate change, gives a range of six possible global emission scenarios in 2010 that include a low of 9 GtC and a high of 13 GtC. My own simulations suggest a broader possible range, 7–18 GtC.

The low end of both ranges reflects the possibility that population and economic growth may slow in the future and the energy intensity of production may fall. The high end reflects the opposite possibility, that growth remains high and energy intensity rises. Figure 1 shows the distribution of uncontrolled emissions arising from my simulations of 1,000 possible outcomes in 2010 alongside the six IPCC scenarios. (For details about the model, see Pizer in Suggested Reading.)

In summary, we have only vague ideas about the cost of alternative emission targets for two important reasons. First, there is little historic evidence about costs. Second, as we examine policies 10 or more years in the future, it is unclear how baseline emissions and available technologies will change between now and then. Figure 1 indicates that global emissions could be anywhere from 7 to 18 GtC in 2010. The cost associated with a target of

8.5 GtC (1990 level) will be uncertain because the necessary reduction is uncertain—somewhere between 0 and 10 GtC—and because costs are difficult to estimate, even knowing the reduction level.

Effects of Price and Quantity Controls with Cost Uncertainty

When the cost of a particular emission target is uncertain, price and quantity controls will have distinctly different consequences for the actual level of emissions as well as the overall cost of a climate policy. Even if both policies are designed to deliver the same results under a best-guess scenario, they will necessarily behave differently when control costs deviate from this best guess. These differences arise because a price policy provides a fixed incentive (dollars per ton of CO_2 emissions), regardless of the emission level, and a quantity policy generates whatever incentive is necessary to strictly limit emissions to a specified level.

Figure 2 illustrates these differences by showing the emission consequences in 2010 associated with two policies that are roughly equivalent under a best-guess scenario: a quantity target of 8.5 GtC

Figure 2: Effect of Price and Quantity Controls on Emissions in 2010

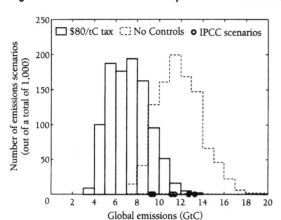
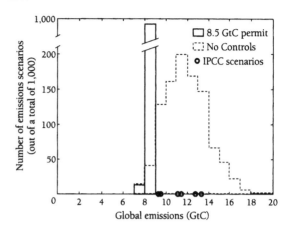

Two policies are roughly equivalent under a best-guess scenario: a carbon tax of $80/ton (left) and a quantity target of 8.5 GtC (right).

and a carbon tax of $80/ton. Using the same 1,000 emission scenarios shown in Figure 1, simulations are used to calculate the effect of these two policies for each outcome. With a carbon tax, emissions are below 8.5 GtC in more than 75% of the outcomes. In other words, on average the carbon tax achieves more reductions than a quantity target of 8.5 GtC. Sometimes, the reductions are much more; emissions may be as low as 3 GtC. Yet, the carbon tax fails to guarantee that emissions will always be below any particular threshold.

The quantity target, in contrast, never results in emission levels above 8.5 GtC. Because some emission outcomes in the absence of controls were rather high, on the order of 18 GtC, we would expect that the cost of this policy could be quite high. At the other extreme, the quantity policy would be costless if uncontrolled emissions were unexpectedly low.

These data suggest that the cost associated with quantity controls will be high or low depending on future reduction costs as well as the future level of uncontrolled emissions. In contrast, price controls create a fixed incentive to reduce each ton of CO_2 regardless of the uncontrolled emission level. Therefore, costs under a carbon tax should fluctuate much less than costs under a quantity control.

Figure 3 shows the estimated cost consequences of both policies. The range of costs associated with the quantity target is quite wide, as we suspected. The estimates extend from 0 to 2.2% of global gross domestic product (GDP), almost four times higher than the highest cost outcome under the carbon tax. In fact, the cost associated with emission reductions under a carbon tax are concentrated entirely in the range 0.2–0.6% of GDP. Because the carbon tax always applies the same per ton incentive to reduce emissions, the cost outcomes are more narrowly distributed than those occurring under a quantity target.

Choosing between Price and Quantity Controls

So far, the discussion has been limited to the different emission and cost consequences of alternative price and quantity controls. Choosing between them, as well as choosing the appropriate stringency of either policy, requires making judgments about climate change consequences as well as control costs. To understand when one policy instrument probably will be preferred to the other, it is useful to consider two extreme cases.

First, imagine that there is a known climate

Figure 3: Distribution of 2010 Costs associated with Price and Quantity Controls

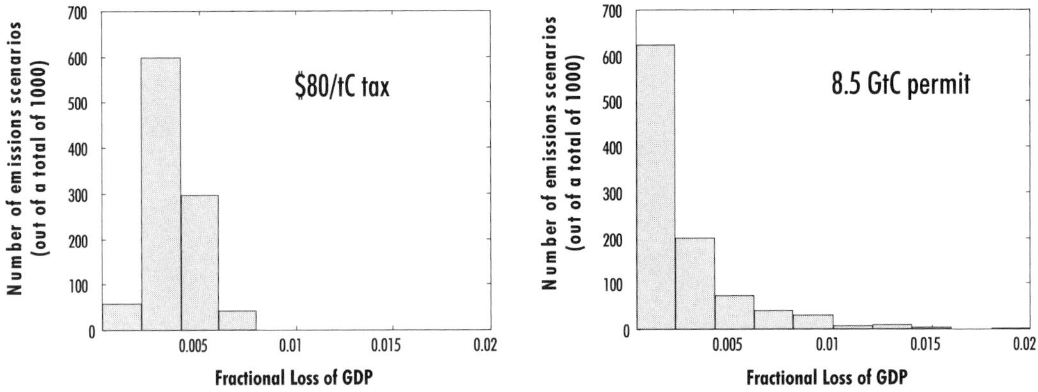

change threshold. When CO_2 emissions are below this threshold, the consequences are negligible. Above this threshold, however, damages are potentially catastrophic. For example, research suggests that the process by which CO_2 is absorbed at the surface of the oceans and circulated downward could change dramatically under certain circumstances. If we also believe that these changes will have severe consequences and that we can identify a safe emission threshold for avoiding them, then quantity controls seem preferable. Quantity controls can be used to avoid crossing the threshold, and in this case, large expenditures to meet the target are justified by the dire consequences of missing it.

Now, imagine instead that every ton of CO_2 emitted causes the same incremental amount of damage. These damages might be very high or low, but the key is that each ton of emissions is just as bad as the next. Such a scenario is also plausible, as indicated by a survey of experts including both natural and social scientists who do research on global warming. Their beliefs suggest that the damage caused by each ton of emitted CO_2 may be quite high but that there is no threshold: Damages are essentially proportional to emissions. Each additional ton is equally damaging, whether it is the first ton emitted or the last.

In this case, it makes sense to use a price instrument. Specifically, a carbon tax equal to the damage per ton of CO_2 will lead to exactly the right balance between the cost of reducing emissions and the resulting benefits of less global warming. Every time a firm decides to emit CO_2, it will be confronted with an added financial burden equal to the resulting damage. It will lead to reduction efforts as well as investments in new technology that are commensurate with the alternative of climate change damage. In this scenario, little emphasis is placed on reaching a particular emission target because there is no obvious quantity target to choose. This argument applies even if we are uncertain about the magnitude of climate damage per unit of CO_2.

Arguments for Price Policies

Given this characterization of circumstances under which alternative price and quantity mechanisms are preferred, we can make the argument for price controls. This argument hinges on two basic points. The first point is that climate change consequences generally depend on the stock of GHGs in the atmosphere, rather than annual emissions. GHGs emitted today may remain in the atmosphere for hundreds of years. It is not the level of annual emissions that matters for climate change but the total amount of CO_2 and other GHGs that have accumulated in the atmosphere. The second point is that although scientists continue to argue over a wide range of climate change consequences, few advocate an immediate halt to emissions. For example, the most aggressive stabilization target discussed by the IPCC is a 450-ppm concentration in the atmosphere (roughly 1,035 GtC), a level that we will not reach before 2030, even in the absence of emission controls.

If only the stock of atmospheric GHGs matters for climate change, and if experts agree that the stock will grow at least in the immediate future, then there is almost no rationale for quantity controls. The fact that only the stock matters should first draw our attention away from short-term quantity controls for emissions and toward long-term quantity controls for the stock. It cannot matter whether a ton of CO_2 is emitted this year, next year, or in 10 years if all we care about is the total amount in the atmosphere. Taking the next step and presuming that the stock will grow over the next few decades, this approach suggests that there is some room to rearrange emissions over time and that a short-term quantity control on emissions is unnecessary.

Quantity controls derive their desirability from situations where strict limits are important, when dire consequences occur beyond a certain threshold. Such policies trade off low expected costs in favor of strict control of emissions in all possible outcomes. However, under the assumption that it is acceptable to allow the stock of GHGs to grow in the interim, there is no advantage to such strict control. We give up the flexible response of price controls without the benefit of an avoided catastrophe.

Even for those who believe the consequences of global warming will be dire and that current emission targets are not aggressive enough, price policies are still better. An aggressive policy designed to stabilize the stock eventually does not demand a strict limit on emissions before stabilization becomes necessary. Additional emissions this year are no worse than emissions next year. Why not abate more when costs are low, less when costs are high—exactly the outcome under a price mech-

anism? When we eventually move closer to a point where the stock must be stabilized, a switch to quantity controls will be appropriate.

In addition to these theoretical arguments, integrated assessment models can provide support. To this end, I have constructed an integrated model of the world economy and climate based on the dynamic integrated climate-economy (DICE) model developed by William Nordhaus (see Suggested Reading). In contrast to the DICE model, I simultaneously incorporate uncertainty about everything from growth in population and energy efficiency to the cost of emission reductions, the sensitivity of the environment to atmospheric CO_2, and the damages arising from global warming.

The results of these simulations indicate the price-based mechanisms can generate overall economic gains (expected benefits minus expected costs) that are *five times higher* than even the most prudent quantity-based mechanism. These results are robust. Even allowing for catastrophic damages beyond 3°C of warming, price mechanisms continue to perform better. This robustness can be explained in two ways. First, the catastrophe—if it exists—is in the future. Before we reach that point, it is desirable to have some flexibility in emission reductions. Specifically, we will want to delay those reductions if the costs are unexpectedly high in the short run, provided those reductions can be obtained more cheaply in the future but before the catastrophe.

Second, unlike the stylized description in which climate consequences depended directly on CO_2 concentrations presented earlier, in this model, damages depend on temperature change. In reality, damages probably depend on an even more complex climatic response. Either way, the links between CO_2 emissions, concentrations, temperature change, and other climatic effects are not precisely known. Therefore, a quantity control on *emissions* is not equivalent to a quantity control on *climate change*. Both price and quantity controls will lead to uncertain climate consequences. Therefore, the advantage of the quantity control—namely, its ability to avoid with certainty the threat of climate catastrophe—is substantially weakened.

Combined Price and Quantity Mechanisms

Even if a carbon tax is preferable to a cap-and-trade approach in terms of social costs and benefits, this policy obviously faces steep political opposition in the United States. Businesses oppose carbon taxes because of the transfer of revenue to the government. Under a permit system, there is a hope that some, if not all, permits would be given away for free. Environmental groups oppose carbon taxes for an entirely different reason: They are unsatisfied with the prospect that a carbon tax, unlike a permit system, fails to guarantee a particular level of emissions. Such antagonism from both sides of the debate makes it unlikely that a carbon tax will become part of the U.S. response to the Kyoto Protocol.

However, the advantages of a carbon tax can be achieved without the baggage accompanying an actual tax. In particular, a combined mechanism (often referred to as a hybrid, or a safety valve) can obtain the economic advantages of a tax while preserving at least some of the political advantages of a permit system.

In such a scheme, the government first distributes a fixed number of tradable permits—freely, by auction, or both. The government then provides additional permits to anyone willing to pay a fixed ceiling or "trigger" price. The initial distribution of permits allows the government the flexibility to give away a portion of the right to emit CO_2, thereby satisfying concerns of businesses about government revenue increases. The sale of additional permits at a fixed price then gives the permit system the same compliance flexibility associated with a carbon tax.

With a combined price/quantity mechanism, it will be necessary to consider how both the trigger price and the quantity target should evolve over time. One possibility is to raise the trigger price over time to guarantee that the quantity target is eventually reached. A second possibility is to carefully choose future trigger prices as a measure of how much we are willing to pay to limit climate change. As we learn more about the costs of future emission reductions, however, this distinction between price and quantity controls will diminish.

That is, after uncertainty about future compliance costs is reduced through experience, then price and quantity controls can be used to obtain similar cost and emission outcomes.

Operationally, when this safety valve is used in conjunction with international emissions trading, as the Kyoto Protocol allows, problems potentially arise. In general, there would be a need for either harmonization of the trigger price across countries or restrictions on the sale of permits from those countries with low trigger prices. Otherwise, there would be an incentive for countries with a low trigger price to simply print and export permits to countries with higher permit prices. This action would not only effectively create low trigger prices everywhere; it also would create large international capital flows to the governments of countries with the low trigger prices.

Instead of harmonizing trigger prices, the trigger price could be set low enough to avoid the need for international GHG trades. This may be a desirable end in light of concerns about the indirect economic consequences of large volumes of international GHG trade flows.

Finally, if we find it desirable to raise the trigger price rapidly, it will be necessary to limit the possibility that permits can be purchased now and held for long periods of time. Otherwise, there will be a strong incentive to buy large volumes of cheap permits now to sell them at high prices in the future. This problem is easily addressed by assigning an expiration date for permits as they are issued, for perhaps one or two years in the future.

Building Domestic and International Support for a Price-Based Approach

Although the safety valve approach is potentially appealing to businesses concerned about the uncertainty surrounding future permit prices, environmental groups will be wary of giving up the commitment to a fixed emission target. Such a commitment is already an integral part of the Kyoto Protocol. However, a strict target policy ultimately may lack political credibility and viability. Although a low trigger price would clearly rankle environmentalists as an undesirable loosening of the commitment to reduce emissions, a higher trigger price could allay those fears while still providing insurance against high costs.

Perhaps more controversial than the concept of a safety valve is the fact that a hybrid policy requires setting a trigger price. It extends the debate over targets and timetables to include perceived benefits on the basis of the trigger price. Business interests undoubtedly will seek a low trigger price and environmental groups a high trigger price. I believe this conflict is desirable. The debate will focus on the source of disagreement between different groups—namely, the value placed on reduced emissions. Rather than leaning on rhetoric that casts reduction commitments as either the source of the next global recession (according to businesses) or the costless ushering in of a new age of cheaper and more energy-efficient living (according to environmentalists), it will be necessary to decide how much we are realistically willing to spend to deal with the problem.

Although seemingly provocative in its challenge of the core concept of targets and timetables embedded in the Kyoto Protocol, some concept of the safety valve is already part of many countries' notion of their commitments to the protocol. European countries that are likely to implement carbon taxes must have some idea how they will handle target violations if their tax proposals fail to sufficiently reduce emissions before the end of the first commitment period. Likewise, other countries that are considering either a quantity or command-and-control approach must envision a way out if their actual costs begin to surpass their political will to reduce emissions.

Among the many implicit safety valve possibilities, one could imagine a more flexible interpretation of existing provisions, such as the clean development mechanism or the use of carbon sinks. Alternatively, Article 27 specifies that parties can withdraw from the protocol by giving notice one year in advance. A country that foresaw difficulty in meeting its target in the first commitment period could serve notice that it wished to withdraw before the commitment period ended.

Therefore, flexibility in meeting current commitments already exists implicitly. Countries can choose to massage their commitments using existing provisions, violate their targets and risk penalties (which have yet to be defined), or simply withdraw. In these cases, however, the outcome and consequence are unclear. The advantage of a price mechanism is that it makes the safety valve concept explicit and transparent. Establishing a price trigger for additional emissions allows countries, and private economic decisionmakers in turn, to approach their reduction commitments with greater certainty about the future. This method not only improves the credibility of the protocol but also its prospects for future success in reducing GHG emissions.

Conclusions

The considerable uncertainty surrounding the cost of international GHG emission targets means that price- and quantity-based policy instruments cannot be viewed as alternative mechanisms for obtaining the same outcome. Price mechanisms will lead to uncertain emission consequences, and quantity mechanisms will lead to uncertain cost consequences. Economic theory as well as numerical simulations indicate that the price approach is preferable for GHG control, generating five times the net expected benefit associated with even the most prudent quantity control. The essence of this result is that a rigid quantity target over the next decade is indefensible at high costs when the stock of GHGs is allowed to increase over the same horizon.

Importantly, a price mechanism need not take the form of carbon tax. The key feature of the price policy is its ability to relax the stringency of the target if control costs turn out to be higher than expected. Such a feature can be implemented in conjunction with a quantity-based mechanism as a safety valve. A quantity target is still set, but with the understanding that additional emissions (beyond the target) will be permitted only if the regulated entities are willing to pay an agreed-upon trigger price.

This approach can improve the credibility of the protocol and its prospects for successful GHG emission reductions. The last point is particularly relevant for ongoing climate negotiations. Should the emission incentives and consequences remain ambiguous and uncertain, or should they be made explicit and transparent? Specifying a price at which additional, above-target emissions rights can be purchased provides a transparent incentive; the current approach does not. Although ambiguity may prove to be the easier negotiating route, it also may be a disincentive for true action.

Suggested Reading

Fischer, Carolyn, Suzi Kerr, and Michael Toman. 1998. Using Emissions Trading to Regulate U.S. Greenhouse Gas Emissions. RFF Climate Issues Brief 10-11. Washington, DC: Resources for the Future.

Fisher, B.S., et al. 1996. An Economic Assessment of Policy Instruments for Combating Climate Change. In *Economic and Social Dimensions of Climate Change*, edited by J. Bruce et al. Cambridge, U.K.: Cambridge University Press.

Hahn, R., and R. Stavins. 1995. Trading in Greenhouse Permits: A Critical Examination of Design and Implementation Issues. In *Shaping National Responses to Climate Change*, edited by H. Lee. Washington, DC: Island Press.

Nordhaus, William D. 1992. The "DICE" Model: Background and Structure of a Dynamic Integrated Climate-Economy Model of the Economics of Global Warming. Discussion Paper 1009. New Haven, CT: Yale University, Cowles Foundation for Research in Economics.

———. 1994. *Managing the Global Common*. Cambridge, MA: MIT Press.

Pizer, W.A. 1998. Prices versus Quantity Revisited: The Case of Climate Change. RFF Discussion Paper 98-02. Washington, DC: Resources for the Future.

39 Rethinking Fossil Fuels
The Necessary Step Toward Practical Climate Policy

Raymond J. Kopp

To curtail greenhouse gas emissions, we will have to make fundamental changes in the ways in which we generate electricity and power transportation. The real issue is how this can be done in ways that are technically feasible, and economically and politically viable. This must involve a long-term effort to develop and implement new technologies. An integral part of this process will be efficient policies that raise the price of carbon-emitting energy sources and thereby encourage the development and use of carbon-free sources of energy.

If we are to stabilize global greenhouse gas (GHG) emissions, the world's largest current emitter—the United States—along with other industrialized countries must radically curtail the emission of GHGs, most notably carbon dioxide (CO_2). This means that our current primary energy sources—which drive both the electricity and transport sectors in the United States, and are responsible for carbon dioxide emissions—must change. But change to what and how?

These two questions, what and how, are at the core of the domestic and, to a large extent, international policy debate about climate policy. What changes to the electricity and transport sectors are practical, technically feasible, and economically and politically viable over the next several decades? Given we can answer the first question, how do we formulate a constellation of domestic and international policies that will provide the proper incentives to undertake the desired changes?

These questions are quite a bit different from the issues surrounding international discussions of the Kyoto Protocol or domestic debates over tailpipe standards in California and federal greenhouse gas legislation like that introduced by Senators John McCain and Joseph Lieberman. While these other discussions concern near-term reductions in greenhouse gas emissions, the questions posed here relate to the fundamental problem of providing the incentives and resources necessary to transform the global energy system over the next half century.

Originally published in *Resources*, No. 154, Summer 2004.

Rethinking Coal

For many the answer to the "what" question is obvious. We must dramatically reduce reliance on fossil fuels, especially coal and petroleum, and expend the resources necessary to develop and deploy non-carbon energy technologies, namely renewables. As the old saying goes, "easier said than done." Current reliance on coal and oil is the result of past and present economics and technology, and the answer to the "what" question requires us to rethink our approach to these fuels.

Rethinking fossil fuels begins with coal and recognition of these facts: there is a lot of coal, it is spread widely around the globe, and it's cheap. Add to this the political clout that coal interests have in the United States—and a perception that the Chinese are anxious to use their coal reserves and fear growing dependence on foreign oil—and you have a picture of two very important CO_2 emitters that are wedded strongly to coal. While renewables such as wind, solar, and biomass will likely be a part of the energy future, the coal reality—its abundance, price, and political power—suggests that it is it naïve to think of a world where coal is not a large part of the energy system.

The problem with coal is the harmful by-products released into the atmosphere when it is burned. Of these, CO_2 is the most significant global pollutant, but mercury, sulfur dioxide, nitrogen oxides, and particulates cause serious problems as well. A sensible solution exists: avoid burning coal directly and convert it to natural gas instead, using oxygen and steam, then clean it of impurities such as mercury and sulfur. The natural gas can be further processed to increase its hydrogen content and convert carbon monoxide to carbon dioxide. This CO_2 can then be separated and stored underground while the remaining hydrogen is used for carbon-free electricity generation.

At present, there are no actual demonstrations of this process, and it sounds a bit like science fiction. But we do know how to make this work. What we don't know is how to do this on a grand scale and do it cheaply. If the technology—everything from the gasification, to the CO_2 separation

Update

At the time this article was written in summer 2004, low sulfur Central Appalachia coal was selling for approximately $60 per ton (with lower grade coal selling for lower prices). In the summer of 2005, this price had risen only very slightly to $61 per ton. However, over this same period, the price of natural gas had risen from $6 per million BTU to about $12 per million BTU, and the price of crude petroleum was up from about $35 per barrel to almost $70 per barrel; in both cases, prices roughly doubled.

These large price movements have two implications for the arguments advanced in this article. First, the economic advantage of coal as an energy source for electricity generation has increased significantly as a result of the rise in the price of the main competing fuel (natural gas). Thus, the economics of power generation has tipped more favorably to coal. Second, the rapid rise in the price of petroleum—and thereby gasoline—provides enhanced incentives for research and investment in alternative fuels. While the price rise is not sufficient to make hydrogen competitive at this juncture, it has brought increased attention to alternative fuels.

and storage, to hydrogen production—could be commercialized and available at a competitive cost, coal can be an environmentally benign and important part of the world's energy future. Of course, simultaneous research would have to be done to improve the environmental impact and safety of coal mining itself.

Rethinking Petroleum

Transportation and petroleum use poses a number of important public policy questions ranging from local issues such as traffic congestion, land use, and highway safety to international ones such as foreign policy and of course global climate change. When rethinking petroleum, three facts must be kept front and center.

First, from the perspective of climate policy, all issues concerning transportation revolve around fuels, and, at present, that fuel is petroleum. It is the dependence of the world's transportation system on petroleum that causes the greatest local and global environmental impacts.

Second, the transportation sector is growing in virtually every corner of the world and with it, the combustion of petroleum and resultant emissions. Without a doubt, the greatest growth is in the developing world, with China leading the way. In the decade of the 1990s, China doubled its petroleum consumption. Even in the developed countries of the European Union, where gasoline is highly taxed, aggressive programs to increase fuel efficiency and subsidize public transportation are, at best, flat with respect to petroleum consumption during the same decade. In some countries, like France and Germany, consumption has in fact increased.

Third, one does not simply switch quickly to another more environmentally friendly transportation fuel. Both vehicles and the necessary infrastructures are designed solely around petroleum and internal combustion engines using gasoline and diesel. From a technical point of view, there is no alternative fuel ready for prime time. The hope that battery-powered electric vehicles would be the wave of the future seems to have faded, along with the visions of a vehicle fleet powered by compressed natural gas. The jury is still out with respect to the feasibility of large-scale transformation of waste products and crops other than corn into fuels, also called second-generation biomass. Even if vehicles can be created that use an alternative fuel in a reasonable amount of time, the petroleum infrastructure (oil fields, pipelines, refineries, and retail outlets) represents an enormous amount of very long-lived capital.

What is on the table are alternatives, like hydrogen, fuel cells, and electric motors. Existing electric motor technology is sufficient for our needs. If we have hydrogen, the missing piece of technology is the fuel cell that turns hydrogen into

electricity to drive the motors, along with a cost-effective way to store hydrogen on vehicles. Fortunately, fuel cell research is on an upswing. To be sure, there are difficult technical issues to master, but the future seems bright. It appears the truly difficult technical part will be the development of safe and low-cost on-board hydrogen storage.

The next technological challenge will be bringing hydrogen to the marketplace. Moving and storing hydrogen is not as easy as petroleum. We may need entirely new pipeline systems (although some portions of the existing natural gas transmission system may work), distribution networks, and of course hydrogen stations. To complicate matters, hydrogen will displace petroleum as a transport fuel slowly as the internal combustion engine stock (namely, billions of cars and trucks) rolls over to fuel cell electric motors and as the hydrogen transportation system develops. So, at the same time we will have petroleum competing with hydrogen in the marketplace.

There may be as many answers to the "what" question as there are technologists. For example, there is much talk about production of hydrogen by electrolysis using nuclear power as the ultimate hydrogen production method. Similarly, second generation biomass liquid fuels hold real potential. There are those who argue that conservation combined with enhanced end-use efficiency will reduce energy needs enough that they can be met by wind and solar. All of these areas deserve further investigation. But the facts surrounding two of the three dominant fossil fuels remain. Coal's low market cost and abundance make it difficult if not impossible to ignore, and petroleum's current preeminent position as the only viable transport fuel suggests its environmental problems will remain with us (and perhaps get worse) unless we find a suitable alternative.

The Obstacles Ahead

Moving the U.S. energy system (that is, the electric power and transport sectors) to a coal-driven hydrogen future faces a series of obstacles. The country has a sizable investment in carbon-based energy technologies. Scrapping that capital precipitously and replacing it with something else will cost a significant amount of money. And, with a few exceptions, a good portion of the hydrogen technology needed is still on the drawing board. Thus, the dilemma is twofold: how to not only bring down the price of delivered hydrogen to a point where it is competitive with natural gas in electricity generation and with petroleum in transport, but also how to bring the cost so low that it will cause a rapid turnover in the existing energy-related capital stock.

To accomplish this feat, a new, integrated energy policy must do the following: hasten the development and commercialization of the necessary technologies, reduce their cost, and provide incentives to the private sector to replace the current carbon-based capital with the new hydrogen capital before the existing stock is fully depreciated. Funds will have to be directed to research and development (R&D) and spread across a number of different technologies ranging from coal gasification to hydrogen generation, among others. Each technology will have a different pace of development, so funding policies should be targeted differently to each technology or class of technologies.

Spurring R&D

While government may in fact undertake some of this research (perhaps through the system of National Laboratories run by the Department of Energy), the bulk of the R&D must be accomplished by the private sector. One might then ask why government policies are necessary for the private sector to undertake this work. The answer is quite straightforward. Absent any government policy, there are simply no incentives for the private sector to make the large investments needed to transform the energy system. Environmental problems aside, from a business perspective the current carbon-based energy system functions fairly well, and few firms would make massive investments to change it without external pressure.

Some would argue that all that is needed is for the government to "get the prices right," that is, set a price on carbon emissions either through a carbon tax or a cap-and-trade permit system and let the private sector react to the price signal. This view makes sense and should be part of a portfolio, but if a single, price-based policy had to do all the "heavy lifting" the resulting high price would be politically unacceptable.

It is common knowledge that commercial firms under-invest in R&D because they are never able to appropriate all the benefits for themselves. This "market failure" is the primary justification for government policy to stimulate R&D in private markets through such things as tax incentives, grants, and private–public partnerships, as well as government support of research at universities and other public institutions. While firms would no doubt increase their R&D expenditures in carbon-free technologies in response, say, to a $100/ ton carbon tax, they would be investing less than the socially optimal amount. Rather than raise the tax to, say, $200 to bring forth more R&D, it is more efficient from a policy perspective to augment the tax with other policies that induce firms to invest more in R&D.

Learning by Doing

Once nearly commercialized technologies are available, a second set of policies is needed to buy down their cost. History has revealed that a great deal of learning takes place as a technology moves to the plant floor and production begins, and this learning lowers cost. This phenomenon, called learning-by-doing, causes the cost of producing a new technology (say a fuel cell) or the cost of production from a new technology (for example, tons of hydrogen from coal gasification) to fall as the number of units produced increases. A variety of these buy-down policies can be put in place to expand the production rates for new technologies, thereby lowering their cost and hastening the time when they can compete in the marketplace. These policies usually take the form of government purchase commitments or a subsidy, where the subsidy is used to lower the cost of production in the future.

Perhaps the most important part of this three-policy trilogy in getting the "prices right" is a carbon cap and tradable permit system. This policy acts to alter the economic playing field by disadvantaging carbon-emitting energy sources, causing their cost to rise, and thereby promoting carbon-free sources. Lowering the cost of the latter will help "pull" the nascent coal-to-hydrogen technology and all its other components (carbon separation and storage, fuel cells, and hydrogen-based transportation infrastructure) into the commercial marketplace.

Also, a system in which carbon permits are auctioned by the government provides the crucial revenues to fund the likely expensive and long-lived R&D and cost buy-down policies discussed above. The importance of a dedicated revenue stream to the funding of long-term basic research cannot be overstated.

The pull policy alluded to above can act like an accelerator pedal in a car. If new climate science suggests that we must move to a noncarbon energy system more quickly than anticipated, the number of permits is reduced (causing auction revenues to rise). The rising permit price further advantages the commercial adoption of the noncarbon technology while at the same time generating more revenue that can be used to hasten technology development.

If stabilizing greenhouse gas emissions is our goal—and it is a worthy one to strive for—we had better be prepared for the long haul. New technologies such as electric- and hydrogen-powered motors will eventually be readily available. New techniques for deriving energy from coal and other fossil fuels will eventually be standardized and in greater use. Since the transformation of the global energy system to one that emits zero greenhouse gases can be expected to take half a century or more, the sooner the transformation begins, the lower the accumulated greenhouse gases in the atmosphere will be.

Suggested Reading

Energy Information Administration. 2003. *Annual Energy Outlook 2003*. Washington, DC: Office of Integrated Analysis and Forecasting, U.S. Department of Energy.

Environmental Protection Agency. 2002. *Inventory of U.S. Greenhouse Gas Emissions and Sinks: 1990–2000*. Washington, DC.

National Research Council. 2004. *The Hydrogen Economy: Opportunities, Costs, Barriers, and R&D Needs*. Washington, DC: National Academy of Engineering, The National Academies.

Forest 'Sinks' as a Tool for Climate-Change Policymaking
A Look at the Advantages and Challenges

Roger A. Sedjo

Forests can trap or "sink" carbon. This provides a potentially low-cost mitigation strategy for helping to contain global climate change. But, as a policy tool, forest sinks pose formidable implementation challenges that will require planning and diplomacy to resolve.

The degree to which natural processes can mitigate the build-up of atmospheric carbon has generated considerable debate among the countries that have been drafting the detailed rules to implement the Kyoto Protocol, the international climate-change treaty. While the Kyoto process may now collapse following the withdrawal of support by the United States, the concept of forest "sinks" offers advantages that are likely to make it important in any successor policy to address climate change. Since President Bush has also moved away from support of caps on carbon dioxide (CO_2) emissions because of his concerns about energy supply, while acknowledging that climate change is a "real problem," this could mean that sinks are all the more important, particularly in the early phases of any long-term comprehensive carbon mitigation plan.

The fundamental science of carbon sinks is well understood—biological growth binds carbon in the cells of trees and other plants while releasing oxygen into the atmosphere, through the process of photosynthesis. Ecosystems with greater biomass divert more CO_2 from Earth's atmosphere and sequester it; forests in particular can absorb large amounts of carbon. Under the Kyoto Protocol, a forest is a carbon sink and a new or expanded (through human effort) forest is allowed to generate credits for removing carbon from the atmosphere.

The most recent round of Kyoto Protocol negotiations, held last November in The Hague, came to a standstill in part because a compromise over carbon sinks failed. American and European negotiators could not reach agreement on the extent

Originally published in *Resources*, No. 143, Spring 2001.

to which carbon captured in biological sinks, would be given credit in meeting country carbon-reduction targets as agreed to earlier at Kyoto.

At first glance, the idea of providing carbon credits for forest sinks sounds easy to implement, but a number of questions have been raised:

- Should existing forests count?
- Is there an agreed measure of absorption?
- How long will it take for a newly planted forest to start absorbing CO_2 and at what rate?
- Should a country receive CO_2 credits if it develops forests in a country other than its own?
- What are the politics of sinks?
- What are the economics of sinks?

Some experts claim that there seem to be no precise answers to these and other questions about forest sinks. (It's important to point out that a substantial amount of carbon is sequestered in the oceans as well as modest amounts in soil.) So, in the context of strategy to control climate change, how important is the sink issue and what compromises may be necessary to prevent sinks from fouling up the grand design?

Let us address these questions one at a time.

Should existing forests count?

In general, the view is that existing forests have inadvertently served as sinks and thus should not count under the Kyoto Protocol. However, there may be some exceptions to this rule. For example, it may be sensible to provide carbon credits for protecting forests that would otherwise be converted to other uses, such as agriculture. In many cases, the value of the carbon credits would exceed the value of the land in nonforest uses. In addition, if existing forests continue to grow and sequester additional carbon, particularly as a result of forest management, then one can argue that credits should be provided for the additional carbon. This is sometimes referred to as a "baseline" problem—deciding which measures are considered over and above what would have happened anyway.

Is there an agreed measure of absorption?

Yes. The amount of carbon held in the forest depends on the amount of dry biomass there. Most developed countries have accurate forest inventories that can provide the baseline for estimating forest biomass. About 50% of the dry weight of the biomass will be carbon. Different tree and plant species have different densities, but these differences are well known, and forest biomass is easy to estimate by using sampling techniques.

How long will it take for a newly planted forest to start absorbing CO_2 and at what rate?

The rate of carbon absorption depends on the amount of dry biomass in the forest. Trees typically grow slowly at first, then at an increasing rate until growth begins to level off as they approach maturity. The growth pattern depends on species, climatic conditions, soil fertility, and other factors. In some parts of the world, certain species grow quickly and can accumulate substantial biomass in less than a decade.

Should a country receive CO_2 credits if it develops forests in a country other than its own?

Forest growth is much more rapid in some regions than in others. Resource conservation would dictate that most of the carbon-sequestering forests should be located in regions where carbon can be absorbed efficiently. Thus, it is sensible for one country to invest in the forests of another—with permission, of course—as a way to earn carbon credits. Additionally, such an approach may transfer large amounts of capital from developed countries to developing countries, thus promoting their economic development.

What are the politics of sinks?

Forest sinks appear to offer the potential of low-cost carbon absorption. However, not all countries are equally blessed with these resources. Much of Europe consists of even-aged growth in what are called "regulated" forests. The expected potential for additional forest growth there to absorb carbon is limited. In fact, many observers argue that European forests are likely to experience some decline over the early decades of the twenty-first century. Thus, it is not surprising that European countries would resist the inclusion of forest sinks for carbon monitoring under the Kyoto Protocol.

By contrast, many countries outside of Europe, including the United States, expect their stock of managed forests to increase during the first decades of the twenty-first century. The United States, Australia, Canada, and Japan are keen to use forest sinks to meet any climate treaty obligations. Many environmental groups appear to believe that meeting carbon targets should be painful and thus view forest sinks as insufficiently austere. However, other environmental groups view carbon credits from forests as offering the potential to help protect tropical forests from destruction and from forestland conversion to agricultural and other uses.

What are the economics of forest sinks?

Most studies indicate that the costs associated with sinks appear to be modest compared with the costs of making the necessary changes in the energy sector. Forest sinks often have other associated benefits, such as erosion reduction, watershed protection, and biodiversity protection of existing native forests. However, their potential to offset carbon emissions is limited. At best, the potential of forests and other terrestrial systems that act as carbon sinks to offset emissions is probably not more than one-third of current net emissions.

Additionally, as the volume of forest sinks increases across the globe, their costs will rise and their additional potential will decline. Thus, perhaps the best way to view sinks is as a temporary low-cost mitigation strategy that can buy humanity three to five decades to make more fundamental adjustments.

Looking Ahead

Carbon sinks appear to offer substantial potential to assist humankind in addressing the challenge posed by climate change but they are more than just forest ecosystems. Grasslands, wetlands, and agriculture all offer the potential to absorb carbon. Although grasslands do not build up a large aboveground mass like forests do, they are effective in the sequestration of carbon into the soil. Wetlands, too, hold

---

Update

The Kyoto Protocol contains provisions for credits for forestry carbon sinks. It allows Annex 1 countries (industrialized nations) to meet part of their carbon-containment commitments through land-use change and forestry sinks. These include a range of activities that result in carbon sequestration in managed forests and agricultural lands. In addition, Annex 1 countries can obtain carbon credits by underwriting these kinds of programs in other nations. After some controversy over the allowable extent of such credits, it was agreed that a cap of 15 percent would apply to carbon sequestration totals that are attributable to forest management (with a few exceptions for some countries).

---

large amounts of carbon in storage. Agricultural lands can contribute to carbon absorption if proper management is followed. No-tillage agriculture offers the potential to restore large volumes of carbon to agricultural soils and contribute to the absorption of carbon from the atmosphere.

Forests appear to offer the greatest potential because they can absorb large volumes of carbon both above and below the ground. Furthermore, the measurement and monitoring of aboveground forest carbon is reasonably simple. The condition of the forest can readily be ascertained visually and with standard forest inventorying procedures, which have been used for decades—indeed, centuries. Carbon can be estimated from the standard forest inventories with only modest additional data requirements. Furthermore, if payments are made for carbon absorbed in the forest biomass, they typically do not reflect the true values because the forest soils also sequester carbon.

As a policy tool, forest sinks pose some distinct challenges. Suppose that a huge reforestation effort is driven by the desire to absorb carbon and that many of these trees would also be suitable as timber. Timber producers, which annually plant an estimated several million hectares of trees for industrial wood purposes, are going to reconsider their tree-growing investments. After all, with all of these new forests being created, the outlook for future timber prices must appear to be bleak. Thus, many timber producers may decide to reduce their own investments in timber growing. The net effect will be to offset some of the increased planting for carbon purposes with the reduction in industrial forest-growing investments. This reduction—that is, the impacts that are precipitated by carbon-absorbing forest projects but are external to those projects—is called *leakage*.

A second form of leakage is associated with protecting threatened forests, as often is proposed for the tropics. Suppose that a particularly valuable forest is threatened with conversion to agriculture. Intervention may be able to save this forest and thus claim credits for the carbon that is prevented from being emitted. However, such an action might simply deflect the deforestation pressure from one forest to another, with no net reduction in carbon emissions.

It should be noted that leakage is not unique to forest sinks. Potential leakage is pervasive throughout many of the proposed climate remedies. Consider the proposal to tax carbon emissions from fossil fuels in developed countries as a way to provide financial incentives to assist developed countries in meeting their emissions reduction targets. Such a policy would increase energy prices in the developed world and energy-intensive industries would have incentives to move to the developing world, where no emissions targets or carbon taxes exist, and hence energy is cheaper. The net effect could be the transfer of emissions from developed countries to developing countries without a significant reduction in global emissions.

This leakage in the energy sector could be substantial and could have significant implications for the world economy.

Can such an outcome be avoided for both carbon sinks and energy? Yes, but it would require implementing similar rules across countries so that leakage is not created through circumvention outside a project or outside a particular country. One step would be to allow sink credits only on a country's net carbon sink increases, and debits for net sink reduction.

Overall, forest sinks have the potential to play a valuable role in carbon sequestration. Although sinks are only a partial solution to anticipated global warming, they do appear to have the potential to sequester 10 to 20% of the anticipated build-up of atmospheric carbon over the next 50 years. Furthermore, sinks can accomplish the task at relatively low costs compared to many other approaches.

Suggested Reading

Sedjo, R., B. Sohngen, and P. Jagger. 2001. Carbon Sinks in the Post Kyoto World. In *Climate Change Economics and Policy: An RFF Anthology*, edited by M. Toman. Washington, DC: Resources for the Future.

Shugart, Herman, Roger Sedjo, and Brent Sohngen. 2003. *Forests and Global Change: Potential Impacts on U.S. Forest Reserves*. Report prepared for the PEW Center for Climate Change.

Sohngen, Brent, and Roger Sedjo. 2000. Potential Carbon Flux from Timber Harvests and Management in the Context of a Global Carbon Flux. *Climate Change* 44: 151–172.

Part 9

Thinking about Sustainable Development

41 The Difficulty in Defining Sustainability

Michael A. Toman

The term "sustainability" has inherent ambiguities. For ecologists, the term connotes the preservation and maintenance of ecological systems. For an economist, the term suggests maintaining or improving the standard of living for human beings. In spite of these ambiguities, there may be ways to reconcile at least partially the implications of these different views.

"Sustainability" has become a new watchword by which individuals, organizations, and nations are to assess human impacts on the natural environment and resource base. A concern that economic development, exploitation of natural resources, and infringement on environmental resources are not sustainable is expressed more and more frequently in analytical studies, conferences, and policy debates. This concern is a central theme in the international deliberations leading up to the United Nations Conference on Environment and Development in June 1992. To identify what may be required to achieve sustainability, it is necessary to have a clear understanding of what sustainability means.

Like many evocative terms, the word sustainability (or the phrase "sustainable development," which more strongly connotes concerns of particular importance to developing countries) means many things to different people and can be used in reference to a number of important issues. The term inherently evokes a concept of preservation and nurturing over time. The World Commission on Environment and Development (known popularly as the Brundtland Commission) labeled sustainable development in its 1987 report *Our Common Future* as "development that meets the needs of the present without compromising the ability of future generations to meet their own needs." Thus sustainability involves some notion of respect for the interests of our descendants. Beyond this point, however, uncertainty and disagreement are rife.

In scholarly usage, the term sustainability originally referred to a harvesting regimen for specific reproducible nat-

Originally published in *Resources*, No. 106, Winter 1992.

ural resources that could be maintained over time (for example, sustained-yield fishing). That meaning has been considerably broadened by ecologists in order to express concerns about preserving the status and function of entire ecological systems (the Chesapeake Bay, the biosphere as a whole). Economists, on the other hand, usually have emphasized the maintenance and improvement of human living standards, in which natural resources and the environment may be important but represent only part of the story. And other disciplines (notably geography and anthropology) bring in concerns about the condition of social and cultural systems (for example, preservation of aboriginal knowledge and skills).

Beyond ambiguity of meaning there also is disagreement about the prospects for achieving sustainability. The Brundtland Report foresees "the possibility for a new era of economic growth, one that must be based on policies that sustain and expand the environmental resource base." Some scholars, notably the economist Julian Simon, question whether sustainability is a significant issue, pointing out that humankind consistently has managed in the past to avoid the specter of Malthusian scarcity through resource substitution and technical ingenuity. Others, notably the ecologists Paul and Anne Ehrlich and the economist Herman Daly, believe that the scale of human pressure on natural systems already is well past a sustainable level. They point out that the world's human population likely will at least double before stabilizing, and that to achieve any semblance of a decent living standard for the majority of people the current level of world economic activity must grow, perhaps fivefold to tenfold. They cannot conceive of already stressed ecological systems tolerating the intense flows of materials use and waste discharge that presumably would be required to accomplish this growth.

Ascertaining more clearly where the facts lie in this debate and determining appropriate response strategies are difficult problems—perhaps among the most difficult faced by all who are concerned with human advance and sound natural resource management. Progress on these fronts is hampered by continued disagreements about basic concepts

and terms of reference. To narrow the gaps, it may be helpful first to identify salient elements of the sustainability concept about which there are contrasts in view between economists and resource planners on the one hand, and ecologists and environmental ethicists on the other.

Key Conceptual Issues

As noted above, intergenerational fairness is a key component of sustainability. The standard approach to intergenerational tradeoffs in economics involves assigning benefits and costs according to some representative set of individual preferences, and discounting costs and benefits accruing to future generations just as future receipts and burdens experienced by members of the current generation are discounted. The justifications for discounting over time are first, that people prefer current benefits over future benefits (and weight current costs more heavily than future costs); and second, that receipts in the future are less valuable than current receipts from the standpoint of the current decisionmaker, because current receipts can be invested to increase capital and future income.

Critics of the standard approach take issue with both rationales for unfettered application of discounting in an intergenerational context. They maintain that invoking impatience entails the exercise of the current generation's influence over future generations in ways that are ethically questionable. The capital growth argument for intergenerational discounting also is suspect, critics argue, because in many cases the environmental resources at issue—for example, the capacity of the atmosphere to absorb greenhouse gases or the extent of biological diversity—are seen to be inherently limited in supply.

These criticisms do not imply that discounting should be abolished (especially since this could increase current exploitation of natural and environmental capital), but they do suggest that discounting might best be applied in tandem with safeguards on the integrity of key resources like ecological life-support systems. Critics also question whether the preferences of an "average" member of the current generation should be the sole or even primary guide to intergenerational resource tradeoffs, particularly if some resource uses threaten the future wellbeing of the entire species but are only dimly experienced by current individuals. Adherents of "deep ecology" even take issue with putting human values at the center of the debate, arguing instead that other elements of the global ecological system have equal moral claims to be sustained. Even if one accepts that human values should occupy center stage, it is difficult to gauge what the values held by future generations might be.

A second key component of sustainability involves the specification of what is to be sustained. If one accepts that there is some collective responsibility of stewardship owed to future generations, what kind of "social capital" needs to be intergenerationally transferred to meet that obligation? One view, to which many economists would be inclined, is that all resources—the natural endowment, physical capital, human knowledge and abilities—are relatively fungible sources of well-being. Thus large-scale damages to ecosystems such as degradation of environmental quality, loss of species diversity, widespread deforestation, or global warming are not intrinsically unacceptable from this point of view; the question is whether compensatory investments for future generations in other forms of capital are possible and are undertaken. Investments in human knowledge, technique, and social organization are especially pertinent in evaluating these issues.

An alternative view, embraced by many ecologists and some economists, is that such compensatory investments often are infeasible as well as ethically indefensible. Physical laws are seen as limiting the extent to which other resources can be substituted for ecological degradation. Healthy ecosystems, including those that provide genetic diversity in relatively unmanaged environments, are seen as offering resilience against unexpected changes and preserving options for future generations. For natural life-support systems, no practical substitutes are possible, and degradation may be irreversible. In such cases (and perhaps in others as well), compensation cannot be meaningfully specified. In addition,

in this view environmental quality may complement capital growth as a source of economic progress, particularly for poorer countries. Such complementarity also would limit the substitution of capital accumulation for natural degradation.

In considering resource substitutability, economists and ecologists often also differ on the appropriate level of geographical scale. On the one hand, opportunities for resource tradeoffs generally are greater at the level of the nation or the globe than at the level of the individual community or regional ecosystem. On the other hand, a concern only with aggregates overlooks unique attributes of particular ecosystems or local constraints on resource substitution and systemic adaptation.

A third key component of sustainability is the scale of human impact relative to global carrying capacity. As already noted, there is sharp disagreement on this issue. As a crude caricature, it is generally true that economists are less inclined than ecologists to see this as a serious problem, putting more faith in the capacities of resource substitution (including substitution of knowledge for materials) and technical innovation to ameliorate scarcity. Rather than viewing it as an immutable constraint, economists regard carrying capacity as endogenous and dynamic.

The Safe Minimum Standard

Concerns over intergenerational fairness, resource constraints, and human scale provide a rationale for some form of intergenerational social contract (though such a device can function only as a "thought experiment" for developing our own moral precepts, since members of future and preceding generations cannot actually be parties to a contract). One way to give shape to such a contract is to apply the concept of a safe minimum standard, an idea that has been advanced (sometimes with another nomenclature) by a number of economists, ecologists, philosophers, and other scholars.

To simplify somewhat, suppose that damages to some natural system or systems can be entirely characterized by the size of their cost and degree of irreversibility. Since ecologists do not view all the effects of irreversibility as readily monetizable, these two attributes of damages are treated separately (see Figure 1). The magnitude of cost can be interpreted in terms of opportunity cost by economists or as a physical measure of ecosystem performance by ecologists.

Irreversibility reflects uncertainty about system performance and the resulting human consequences. At one extreme, very large and irreversible effects may threaten the function of an entire ecosystem. At a global level, the threat could be to the cultural if not the physical survival of the human species. In Figure 1, this extreme is represented at the upper lefthand corner. At the other extreme (the lower righthand corner), small and readily reversible effects are relatively easily mediated by private market transactions or by corrective government policies based on comparisons of benefits and costs.

There is uncertainty about how rapidly the threat to current and future human welfare grows as damages become costlier and irreversibility becomes more likely. The safe minimum standard posits a socially determined dividing line between moral imperatives to preserve and enhance natural resource systems and the free play of resource tradeoffs. To satisfy the intergenerational social con-

tract, the current generation would rule out in advance actions that could result in natural impacts beyond a certain threshold of cost and irreversibility. Rather than depending on a comparison of expected benefits and costs from increased pressure on the natural system, such proscriptions would reflect society's value judgment that the cost of risking these impacts is too large. Possible resources for which society would not risk damages beyond a certain cost and degree of irreversibility include wetlands, other sources of genetic diversity, the climate, wilderness areas, Antarctica, and other ecosystems with unique functional or aesthetic values (like the Grand Canyon).

There is a distinct difference between the safe minimum standard approach and the standard prescriptions of environmental economics, which involve obtaining accurate valuations of resources in benefit-cost assessments and using economic incentives to achieve efficient resource allocation given these valuations. Whether a resource-protection criterion is established by imperatives through an application of the safe minimum standard concept or by tradeoffs through cost-benefit analyses, that criterion can be cost-effectively achieved by using economic incentives. However, for impacts on the natural environment that are uncertain but

---

Figure 1. Diagram of the safe minimum standard for balancing natural resource tradeoffs and imperatives for preservation.

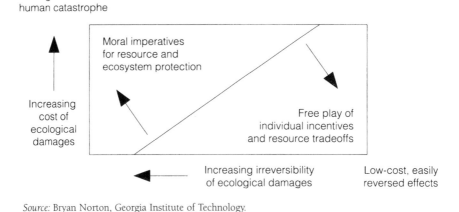

Source: Bryan Norton, Georgia Institute of Technology.

Update

Since I wrote this piece in 1992, the term "sustainability" has been put to an increasing variety of uses in economics and policy analysis. Within the mainstream economics literature on sustainable growth and the environment, there has been continued refinement of the concept in the context of the feasibility of non-decreasing well-being in the longer term (across generations). The literature has advanced a long way theoretically, but we still have far to go in developing empirical insights into long-term sustainability.

In a policy context, there are continuing debates surrounding how physical limits on resource depletion and environmental degradation might supplement or replace more conventional analyses of economic tradeoffs. Much of this debate has been stimulated by the analysis of global climate change, which is a canonical case of sustainability issues. (See chapter 37 in this volume—my essay with Jason Shogren on "How Much Climate Change is Too Much?"). In addition, policy debates on sustainability in general and on climate change in particular have increasingly incorporated concerns about poverty and equity, as well as the more intergenerational focus in my paper here.

may be large and irreversible, the safe minimum standard posits an alternative to comparisons of economic benefits and costs for developing resource-protection criteria. It places greater emphasis on potential damages to the natural system than on the sacrifices experienced from curbing ecological impacts. The latter are seen as likely to be smaller and more readily reversible. In addition, the safe minimum standard invokes a wider, possibly less individualistic set of values in assessing impacts. Since societal value judgments determine the level of safeguards, public decisionmaking and the formation of social values are explicit parts of the safe minimum standard approach.

This illustrative discussion of course provides no actual guidance on where and how (if at all) such a dividing line between imperatives and trade-offs should be drawn. The location of the line will depend on the range of individual beliefs in society and available knowledge about human impacts on ecosystems. For example, ecologists who are concerned mainly about irreversibility and believe that ecological systems are fragile might draw an essentially vertical line, with a large area covered by moral imperatives for ecosystem protection; economists who are concerned mainly about expected cost and believe that the well-being of future generations should be highly discounted might draw an essentially horizontal line, with little (or no) scope for moral imperatives. Acquisition of additional knowledge also will alter the relative weight given to imperatives and tradeoffs for specific ecosystems or the environment as a whole. In addition, how the delineation would be made depends on complex social decision processes, some of which probably have not yet been constructed.

The safe minimum standard thus does not provide an instant common rallying point for resolving the disagreements discussed here. However, this concept does seem to provide a frame of reference and a vocabulary for productive discussion of such disagreements. Such discussion would refine understanding of what sustainability means and the steps that should be taken to enhance prospects for achieving it.

Research Needs

There is a need for much additional interdisciplinary work to refine the concept of sustainability. Along with basic concept definitions, extensions of economic and ecological theory to more fully account for the objectives and constraints of sustainability would be useful. To clarify some of the points of disagreement already outlined, substantial interdisciplinary data-gathering and analysis also would be required. This empirical work should address issues in developing countries and in developed countries, and those relevant to the entire world.

The tension between ecological and economic perspectives on sustainability suggests several ways in which both economists and ecologists could adapt their research emphases and methodologies to make the best use of interdisciplinary contributions. Economists could usefully expand analyses of resource values to consider the function and value of ecological systems as a whole, making greater use of ecological information in the process. Economic theory and practice also could be extended to consider more fully the implications of physical resource limits that often are not reflected in more stylized economic constructs. In addition, research by economists and other social scientists (psychologists and anthropologists) could help to improve understanding of how future generations might value different attributes of natural environments. Finally, the sustainability debate should remind economists to carefully distinguish between efficient allocations of resources—the standard focus of economic theory—and socially optimal allocations, which may include intergenerational (as well as intragenerational) equity concerns.

For ecologists, the challenges include providing information on ecological conditions in a form that could be used in economic valuation. Ecologists also must recognize the importance of human behavior, particularly behavior in response to economic incentives—a factor often given short shrift in ecological impact analyses. Finally, it must be recognized that human behavior and social decision processes are complex, just as ecological processes are. What may appear as self-evident to the student of natural environment need not seem so for the student of human society, and vice versa.

Suggested Reading

Norton, B. and M. Toman. 1997. Sustainability: Ecological and Economic Perspectives. *Land Economics* 73: 553–568.

Pezzey, J. and M. Toman. 2005. Sustainability and Its Economic Interpretation. In *Scarcity and Growth Revisited: Natural Resources and the Environment in the New Millennium*, edited by R. David Simpson et al. Washington, DC: Resources for the Future, 121–141.

Portney, P. and J. Weyant (eds.). 1999. *Discounting and Intergenerational Equity*. Washington, DC: Resources for the Future.

Toman, M. 1994. Economics and 'Sustainability': Balancing Tradeoffs and Imperatives. *Land Economics* 70: 399–413.

———. 1999. Sustainable Decision-Making: The State of the Art from an Economic Perspective. In *Valuation and the Environment: Theory, Method, and Practice*, edited by M. O'Connor and C. Splash. Cheltenham, U.K.: Edward Elgar.

Toman, M., R. Lile, and D. King. 1999. Assessing Sustainability in a Social Context: Some Conceptual and Empirical Challenges. *International Journal of Environment and Pollution* 12 (4): 369–382.

42 An Almost Practical Step Toward Sustainability

Robert Solow

In this classic 1992 essay, Robert Solow explores the meaning of sustainable development in the context of economic theory and its implications for social accounting practices.

You may be relieved to know that this talk will not be a harangue about the intrinsic incompatibility of economic growth and concern for the natural environment. Nor will it be a plea for the strict conservation of nonrenewable resources, even if that were to mean dramatic reductions in production and consumption. On the other hand, neither will you hear mindless wish fulfillment about how ingenuity and enterprise can be counted on to save us from the consequences of consuming too much and preserving too little, as they have always done in the past.

Actually, the argument I want to make seems to be particularly appropriate on the occasion of the fortieth anniversary of Resources for the Future; it is precisely about resources for the future. And it is even more appropriate for a research organization: I hope to show how some fairly interesting pure economic theory can offer a hint—though only a hint—about a possible improvement in the way we talk about and think about our economy in relation to its endowment of natural resources. The theoretical insight that I will present suggests a potentially important line of empirical research and a possible guideline for long-term economic policy. Then I will make a naive leap and suggest that, if we talked about the economy in a more sensible and accurate way, we might actually be better able to conduct a rational policy in practice with respect to natural and environmental resources. That is probably foolishness, but I hope you will find it a disarming sort of foolishness.

An invited lecture on the occasion of the fortieth anniversary of Resources for the Future; originally published as a pamphlet in 1992.

Previewing the Arguments

It will be useful if I tell you in advance where the argument is leading. It is a commonplace thought that the national income and product accounts, as currently laid out, give a misleading picture of the value of a nation's economic activity to the people concerned. The conventional totals, gross domestic product (GDP) or gross national product (GNP) or national income, are not so bad for studying fluctuations in employment or analyzing the demand for goods and services. When it comes to measuring the economy's contribution to the well-being of the country's inhabitants, however, the conventional measures are incomplete. The most obvious omission is the depreciation of fixed capital assets. If two economies produce the same real GDP but one of them does so wastefully by wearing out half of its stock of plant and equipment while the other does so thriftily and holds depreciation to 10 percent of its stock of capital, it is pretty obvious which one is doing a better job for its citizens. Of course the national income accounts have always recognized this point, and they construct net aggregates, like net national product (NNP), to give an appropriate answer. Depreciation of fixed capital may be badly measured, and the error affects net product, but the effort is made.

The same principle should hold for stocks of nonrenewable resources and for environmental assets like clean air and water. Suppose two economies produce the same real net national product, with due allowance for depreciation of fixed capital, but one of them is wasteful of natural resources and casually allows its environment to deteriorate, while the other conserves resources and preserves the natural environment. In such a case we have no trouble seeing that the first is providing less amply for its citizens than the second. So far, however, the proper adjustments needed to measure the stocks and flows of our natural resources and environmental assets are not being made in the published national accounts. (The United Nations has been working in this direction for some years, so the situation may change, although only with respect to environmental accounting.) The nature of this problem has been understood for some time, and individual scholars, beginning with William D. Nordhaus and James Tobin in 1972, have made occasional passes at estimating the required corrections.

That is hardly news. The additional insight that I want to explain is that there is a "right" way to make that correction—not perhaps the easiest or most direct way, but the way that properly charges the economy for the consumption of its resource endowment. The same principle can be extended to define the right adjustment that must be made to allow for the degradation or improvement of environmental assets in the course of a year's economic activity. The properly adjusted net national product would give a more meaningful indicator of the annual contribution to economic well-being.

The corrections are more easily defined than performed. The necessary calculations would undoubtedly be more error-prone than those the U.S. Department of Commerce already does with respect to the depreciation of fixed capital. Nevertheless, I would suggest that talk without measurement is cheap. If we—the country, the government, the research community—are serious about doing the right thing for the resource endowment and the environment, then the proper measurement of stocks and flows ought to be high on the list of steps toward intelligent and foresighted decisions.

The second and last step in my argument is more abstract. It turns out that the measurements I have just been discussing play a central role in the only logically sound approach to the issue of sustainability that I know. If "sustainability" is anything more than a slogan or expression of emotion, it must amount to an injunction to preserve productive capacity for the indefinite future. That is compatible with the use of nonrenewable resources only if society as a whole replaces used-up resources with something else. As you will see when I return to this point for a full exposition, the very same calculation that is required to construct an adjusted net national product for current evaluation of economic benefit is also essential for the construction of a strategy aimed at sustainability. This conclusion confirms the importance of a serious effort to dig out the relevant facts.

That is a brief preview of what I intend to say, but before going on to say it, I would like to mention the names of the economists who have contributed most to this line of thought. They include Professors John Hartwick of Queen's University in Canada, Partha Dasgupta of the University of Cambridge, England, and Karl-Göran Mäler of the Stockholm School of Economics; my sometime colleague Martin L. Weitzman, now of Harvard University; and, more on the practical side, Robert Repetto of the World Resources Institute. I have already mentioned the early work of Nordhaus and Tobin; Nordhaus has continued to contribute common sense, realism, and rigorous economic analysis. Finally, I should confess that I have contributed to this literature myself. My idea of heaven is an occasion when a piece of pretty economic theory turns out to suggest a program of empirical research and to have implications for the formulation of public policy.

Finding the True Net Product of Our Economy

Now I go back to the beginning and make my case in more detail. Suppose we adopt a simplified picture of an economy living in some kind of long run. What I mean by that awkward phrase is that we are going to ignore all those business-cycle problems connected with unemployment and excess capacity or overheating and inflation. From quarter to quarter and year to year this economy fully exploits the resources of labor, plant, and equipment that are available to it.

To take the easiest case—that of natural resources—first, imagine that this economy starts with a fixed stock of nonrenewable resources that are essential for further production. This is an over-simplification, of course. Even apart from the possibility of exploration and discovery, the stock of nonrenewable resources is not a pre-existing lump of given size, but a vast quantity of raw materials of varying grade, location, and ease of extraction. Those complications are not of the essence, so I ignore them.

It is of the essence that production cannot take place without some use of natural resources. But I shall also assume that it is always possible to substitute greater inputs of labor, reproducible capital, and renewable resources for smaller direct inputs of the fixed resource. Substitution can take place on reasonable terms, although we can agree that it gets more and more costly as the process of substitution goes on. Without this minimal degree of optimism, the conclusion might be that this economy is like a watch that can be wound only once: it has only a finite number of ticks, after which it stops. In that case there is no point in talking about sustainability, because it is ruled out by assumption; the only choice is between a short happy life and a longer unhappy one.

Life for this economy consists of using all of its labor and capital and depleting some of its remaining stock of resources in the production of a year's output (GDP approximately). Part of each year's output is consumed, and that gives pleasure to current consumers; the rest is invested in reproducible capital to be used for production in the future. There are various assumptions one could make about the evolution of the population and employment. I will assume them to have stabilized, since I want to talk about the very long run anyway. Next year is a lot like this year, except that there will be more plant and equipment, if net investment was positive this year, and there will be less of the stock of resources left.

Each year there are two new decisions: how much to save and invest, and how much of the remaining stock of nonrenewable resources to use up. There is a sense in which we can say that this year's consumers have made a trade with posterity. They have used up some of the stock of irreplaceable natural resources; in exchange they have saved and invested, so that posterity will inherit a larger stock of reproducible capital.

This intergenerational trade-off can be managed well or badly, equitably or inequitably. I want to suppose that it is done well and equitably. That means two things. First, nothing is simply wasted; production is carried on efficiently. Second, although the notion of intergenerational equity is much more complicated and I cannot hope to explain it fully here, the idea is that each generation is allowed to favor itself over the future, but not too much. Each generation can, in turn, discount the welfare of all future generations, and each successive generation applies the same discount rate to the welfare of its successors. To make conservation an interesting proposition at all, the common discount rate should not be too large.

You may wonder why I allow discounting at all. I wonder, too: no generation "should" be favored over any other. The usual scholarly excuse—which relies on the idea that there is a small fixed probability that civilization will end during any little interval of time—sounds farfetched. We can think of intergenerational discounting as a concession to human weakness or as a technical assumption of convenience (which it is). Luckily, very little of what I want to say depends on the rate of discount, which we can just imagine to be very small.

Given this discounting of future consumption, we have to imagine that our toy economy makes its investment and resource-depletion decisions so as to generate the largest possible sum of satisfactions over all future time. The limits to this optimization process are imposed by the pre-existing stock of resources, the initial stock of reproducible capital, the size of the labor force, and the technology of production.

This assumption of optimality is an embarrassing load to carry around. Its function is primarily to allow the semi-fiction that market prices accurately reflect scarcities. A similar assumption is implicit whenever we use ordinary GDP as a measure of economic well-being. In practice, no doubt, prices reflect all sorts of distortions arising from monopoly, taxation, poor information, and other market imperfections. In practice one can try to make adjustments to market prices to correct for the worst distortions. The conceptual points I want to make would survive. They are not to be taken literally in any case, but more as indicators of the sort of measurements we should be aiming at in principle.

Properly Charging the Economy for the Consumption of Its Resource Endowment

Now I come to the first major analytical step in my

argument. If you look carefully at the solution to the problem of intergenerational resource allocation I have just sketched, you see that an excellent approximation of each single period's contribution to social welfare emerges quite naturally from the calculations. It is, in fact, a corrected version of net domestic product. The new feature is precisely a deduction for the net depletion of exhaustible resources. (I use the phrase "net depletion" because it is possible to extend this reasoning to allow for some discovery and development of new resources. In the pure case, where all discovery and development have already taken place, net and gross depletion coincide.)

The correct charge for depletion should value each unit of resource extracted at its net price, namely, its real value as input to production minus the *marginal* cost of extraction. As Hartwick has pointed out, if the marginal cost of mining exceeds average cost, which is what one would expect in an extractive industry, then the simple procedure of deducting the gross margin in mining (that is, the value of sales less the cost of extraction) will overstate the proper deduction and thus understate net product in the economy. If I may use the jargon of resource economics for a moment, the correct measure of depletion for social accounting prices is just the aggregate of Hotelling rents in the mining industry. That is the appropriate way to put a figure on what is taken from the ground in any given year, that year's withdrawal from the original endowment of nonrenewable resources.

This proposal presents two practical difficulties for national income accounting. The first is that observed market prices have to be corrected for the worst of the distortions I have just listed (that is, the distortion that would result from deducting the gross margin in mining—overstatement of the proper deduction and understatement of the net product in the economy). Making adjustments to market prices to correct for distortions is attempted routinely by the World Bank and other agencies in making project evaluations in developing countries. We seem to ignore the problem of such distortions when we use our own national income accounts to study and judge the economies of advanced countries. If we are justified in that practice, the same casual treatment may be satisfactory in this context. (Not always, however: the large observed fluctuations in the price of oil cannot be accepted as indicating "true" values.) Either way, this is a surmountable problem.

I am not sure whether it is safe to be so casual about the second practical difficulty that my proposal for deducting net depletion of exhaustible resources presents for national income accounting. In principle, the proper measurement of resource rents requires the use of a numerical approximation to the marginal cost of mining. As I said, if marginal cost exceeds average cost by a lot, then taking the easy way out (just deducting the gross margin in mining) would entail a large error by overstating the depreciation of the resource stock. It seems to me that this is exactly where the fund of knowledge embodied in an organization like RFF can find its application. Tentative calculations for the main extractive industries would tell us something important about the true net product of our own economy. That would be important not merely because it would allow a more accurate evaluation of the path the economy has been following, but also, as you will see, because the measurement of resource rents should be an input into policy decisions with a view to sustainability.

Correcting National Accounts to Reflect Environmental Amenities

Pretty clearly, similar ideas should apply to a program of correcting the conventional national accounts to reflect environmental amenities. Much more attention has been lavished on environmental accounting than on resource accounting, and I have very little to add. Henry M. Peskin's work (much of which was done here at RFF) goes back to the early 1970s, and the Organisation for Economic Co-operation and Development, the World Bank, and the U.S. Department of Commerce are preparing a framework for integrating national income and environmental accounts. The sooner it happens the better. My only comment is a theoretical one. Without too much strain, it may be possible to treat environmental quality as a stock, a kind

of capital that is "depreciated" by the addition of pollutants and "invested in" by abatement activities. In such cases the same general principles apply as to other forms of capital. The same intellectual framework will cover reproducible capital, renewable and nonrenewable resources, and environmental "capital."

The data problems may be altogether different, of course, especially when it comes to the measurement of benefits, a nicety that does not arise in the case of resource depletion. But the underlying treatment will follow the same rules. This counts for more than fastidiousness, I think. It would be a real achievement if it were to become a commonplace that capital assets, natural assets, and environmental assets were equally "real" and subject to the same scale of values, indeed the same bookkeeping conventions. Deeper ways of thinking might be affected.

That completes the first phase of my argument, so I will summarize briefly. The very logic of the economic theory of capital tells us how to construct a net national product concept that allows properly for the depletion of nonrenewable resources, and also for other forms of natural capital. Carrying out those instructions is far from easy, but that only makes the process more interesting. The importance of doing the work and doing it right is that theory underlines the basic similarity among all forms of capital, and that is a lesson worth learning. It will be reinforced by routine embodiment in the national accounts. Perhaps RFF could take the lead, as it has done with respect to environmental costs and benefits.

Analyzing Sustainable Paths for a Modern Industrial Society

Now I want to start down an apparently quite different path, but I promise that it will eventually link up with the unromantic measurement issues I have discussed so far, and will even reinforce the argument I have made.

I do not have to remind you that "sustainability" has become a hot topic in the last few years, beginning, I suppose, with the publication of the Brundtland Commission's report, *Our Common Future*, in 1987. As far as I can tell, however, discussion of sustainability has been mainly an occasion for the expression of emotions and attitudes. There has been very little analysis of sustainable paths for a modern industrial economy, so that we have little idea of what would be required in the way of policy and what sorts of outcomes could be expected. As things stand, if I express a commitment to sustainability, all that tells you is that I am unhappy with the modern consumerist life-style. If I pooh-pooh the whole thing, on the other hand, all you can deduce is that I am for business as usual. It is not a very satisfactory state of affairs.

Understanding What It Is That Must Be Conserved

If sustainability means anything more than a vague emotional commitment, it must require that something be conserved for the very long run. It is very important to understand what that something is: I think it has to be a generalized capacity to produce economic well-being.

It makes perfectly good sense to insist that certain unique and irreplaceable assets should be preserved for their own sake; nearly everyone would feel that way about Yosemite or, for that matter, about the Lincoln Memorial, I imagine. But that sort of situation cannot be universalized: it would be neither possible nor desirable to "leave the world as we found it" in every particular.

Most routine natural resources are desirable for what they do, not for what they are. It is their capacity to provide usable goods and services that we value. Once that principle is accepted, we are in the everyday world of substitutions and trade-offs.

For the rest of this talk, I will assume that a sustainable path for the national economy is one that allows every future generation the option of being as well off as its predecessors. The duty imposed by sustainability is to bequeath to posterity not any particular thing—with the sort of rare exception I have mentioned—but rather to endow them with whatever it takes to achieve a standard of living at least as good as our own and to look after

their next generation similarly. We are not to consume humanity's capital, in the broadest sense. Sustainability is not always compatible with discounting the well-being of future generations if there is no continuing technological progress. But I will slide over this potential contradiction because discount rates should be small and, after all, there is technological progress.

All that sounds bland, but it has some content. The standard of living achievable in the future depends on a bundle of endowments, in principle on everything that could limit the economy's capacity to produce economic well-being. That includes nonrenewable resources, of course, but it also includes the stock of plant and equipment, the inventory of technological knowledge, and even the general level of education and supply of skills. A sustainable path for the economy is thus not necessarily one that conserves every single thing or any single thing. It is one that replaces whatever it takes from its inherited natural and produced endowment, its material and intellectual endowment. What matters is not the particular form that the replacement takes, but only its capacity to produce the things that posterity will enjoy. Those depletion and investment decisions are the proper focus.

Outlining Two Key Propositions

Now it is time to go back to the toy economy I described earlier and to bring some serious economic theory to bear. There are two closely related logical propositions that can be shown to hold for such an economy. The first tells us something about the properly defined net national product, calculated with the aid of the right prices. At each instant, net national product indicates the largest consumption level that can be allowed this year if future consumption is never to be allowed to decrease.

To put it a little more precisely: net national product measures the maximum current level of consumer satisfaction that can be sustained forever. It is, therefore, a measure of sustainable income given the state of the economy—capital, resources, and so on—at that very instant.

This is important enough and strange enough to be worth a little explanation. How can this year's

NNP "know" about anything that will or can happen in the future? The theorist's answer goes something like this. The economy's net product in any year consists of public and private consumption and public and private investment. (I am ignoring foreign trade altogether. Think of the economy as representing the world.) The components of investment, including the depletion of natural resources, have to be valued. That is where the "rightness" of the prices comes in. If the economy or its participants are forward-looking and far-seeing, the prices of investment goods will reflect the market's evaluation of their future productivity, including the productivity of the future investments they will make possible. The right prices will make full allowance even for the distant future, and will even take account of how each future generation will look at its future.

This story makes it obvious that everyday market prices can make no claim to embody that kind of foreknowledge. Least of all could the prices of natural resource products, which are famous for their volatility, have this property; but one could entertain legitimate doubts about other prices, too. The hope has to be that a careful attempt to average out speculative movements and to correct for the other imperfections I listed earlier would yield adjusted prices that might serve as a rough approximation to the theoretically correct ones. We act as if that were true in other contexts. The important hedge is not to claim too much.

While it is closely related to the proposition that NNP measures the maximum current level of consumer satisfaction that can be sustained forever, the second theoretical proposition I need is considerably more intuitive, although it may sound a little mysterious, too. Properly defined and properly calculated, this year's net national product can always be regarded as this year's interest on society's total stock of capital. It is absolutely vital that "capital" be interpreted in the broadest sense to include everything, tangible and intangible, in which the economy can invest or disinvest, including knowledge. Of course this stock of capital must be evaluated at the right prices. And the interest rate that capitalizes the net national product will generally

be the real discount rate implicit in the whole story. Investment and depletion decisions determine the real wealth of the economy, and each instant's NNP appears as the return to society on the wealth it has accumulated in all forms. There are some tricky questions about wage incomes, but they are off the main track and I shall leave them unanswered.

Maintaining the Broad Stock of Society's Capital Intact

Something interesting happens when these two propositions are put together. One of them tells us that NNP at any instant is a measure of the highest sustainable income achievable, given the total stock of capital available at that instant. The other proposition tells us that NNP at any instant can be represented as that same stock of capital multiplied by an unchanging discount rate. Suppose that one goal of economic policy is to make investment and depletion decisions this year in a way that does not erode sustainable income. Then those same decisions must not allow the aggregate capital stock to fall. To use a Victorian phrase, preserving sustainability amounts to maintaining society's capital intact.

Let me say that in a slightly different way, speaking more picturesquely of generations rather than of instants or years. Each generation inherits a capital stock in the very broad and inclusive sense that matters. In turn, each generation makes consumption, investment, and depletion decisions. It enjoys its own consumption and leaves a stock of capital for the next generation. Of course, generations do not make decisions; families, firms, and governments do. Still, if all those decisions eventuate in a very large amount of current consumption, clearly the next generation might be forced to start with a lower stock of capital than its parents did. We now know that this is equivalent to saying that the new sustainable level of income is lower than the old one. The high-consumption generation has not lived up to the ethic of sustainability.

In the opposite case, consider a generation that consumes very little and leaves behind it a larger stock of capital than it inherited. That generation will have increased the sustainable level of income, and done so at the expense of its own consumption. Obviously that is what most past generations in the

United States have done. Equally obviously, they were helped by ongoing technological progress. I have left that factor out of account, because it makes things too easy. It could probably be accommodated in the theoretical picture by imagining that there is a stock of technological knowledge that is built up by scientific and engineering research and depreciates through obsolescence. We know so little about that process that the formalization seems almost misleading. But the fact is very important.

A concern for sustainability implies a bias toward investment. That does not mean investment *über alles;* it means just enough investment to maintain the broad stock of capital intact. It does not mean maintaining intact the stock of every single thing; trade-offs and substitutions are not only permissible, they are essential. Unfortunately I have to make the limp statement that the terms on which one form of capital should be traded off against another are given by those adjusted prices— "shadow prices" we call them—and they involve a certain amount of guesswork. The guesswork has to be done; it cannot be avoided by defining the problem away. It is better that the guesswork be based on careful research than that the decision be fudged.

Connecting Up the Arguments

Knowing What and How Much Should Be Replaced

Now I can connect up the two halves of my argument. Every generation uses up some part of the earth's original endowment of nonrenewable resources. There is no alternative. Not now anyway. Maybe eventually our economy will be based entirely on renewables. (The theory I have been using can be applied then too, with routine modifications.) Even so, there will be a long meanwhile. What should each generation give back in exchange for depleted resources if it wishes to abide by the ethic of sustainability? We now have an answer in principle. It should add to the social capital in other forms, enough to maintain the aggregate social capital intact. In other words, it should replace the used-up resources with other assets of equal value, or equal shadow value. How much is that? The shadow value of resource depletion is exactly the

aggregate of Hotelling rents. It is exactly the quantity that should be deducted from conventional net national product to give a truer NNP that takes account of the depletion of resources. A research project aimed at estimating that deduction would also be estimating the amount of investment in other forms that would just replace the productive capacity dissipated in resource depletion. This is sometimes known as Hartwick's rule: a society that invests aggregate resource rents in reproducible capital is preserving its capacity to sustain a constant level of consumption.

Once again, I should mention that the same approach can be applied to environmental assets— the most complete treatment is by Karl-Göran Mäler—and to renewable resources—as in the work of John Hartwick. The environmental case is more complex, because even a stylized model of environmental degradation and rehabilitation is more complex than a model of resource depletion. The principle is the same, but the execution is even more difficult. Remember that even the simplest case offers daunting measurement problems.

Translating Sustainability into Policy

It is possible that the clarity brought to the idea of sustainability by this approach could lift the policy debate to a more pragmatic, less emotional level. But I am inclined to think that a few numbers, even approximate numbers, would be much more effective in turning discussion toward concrete proposals and away from pronunciamentos.

Suppose that the Department of Commerce published routinely a reasonable approximation to the "true" value of each year's depletion of nonrenewable resources. We could then say to ourselves: we owe to the future a volume of investment that will compensate for this year's withdrawal from the inherited stock. We know the rough magnitude of this requirement. The appropriate policy is to generate an economically equivalent amount of net investment, enough to maintain society's broadly defined stock of capital intact. Of course, there may be other reasons for adding to (or subtracting from) this level of investment. The point is only that a commitment to sustainability is translated into a commitment to a specifiable amount of productive investment.

By the way, the same sort of calculation should have a very high priority in primary producing countries, the ones that supply the advanced industrial world with mineral products. They should also be directing their—rather large—Hotelling rents into productive investment. They will presumably want to invest more than that, because sustainability is hardly an adequate goal in poor countries. In this perspective, the cardinal sin is not mining; it is consuming the rents from mining.

It goes without saying that this concrete translation of sustainability into policy leaves a lot of questions unanswered. The split between private and public investment has to be made in essentially political ways, like the split between private and public saving. There are other reasons for public policy to encourage or discourage investment, because there are social goals other than sustainability. One could hope for more focused debate as trade-offs are made more explicit.

I want to remind you again that environmental preservation can be handled in much the same way. It is a more difficult context, however, for several reasons. Many, though not all, environmental assets have a claim to intrinsic value. That is the case of the Grand Canyon or Yosemite National Park, as noted earlier. The claim that a feature of the environment is irreplaceable, that is, not open to substitution by something equivalent but different, can be contested in any particular case, but no doubt it is sometimes true. Then the calculus of trade-offs does not apply. Useful minerals are in a more utilitarian category, and that is why I dealt with them explicitly.

Yet another difficulty is the deeper uncertainty about environmental benefits and costs. Marketed commodities, like minerals or renewable natural resources, are much simpler. I have admitted, fairly and squarely, how much of my argument depends on getting the shadow prices approximately right. Ordinary transaction prices are clearly not the whole answer; but they are a place to start. With environmental assets, not even that benchmark is available. I do not need to con-

vince this audience that the difficulty of doing better does not make zero a defensible approximation for the shadow price of environmental amenity. I think the correct conclusion is the one stated by Karl-Göran Mäler: that we are going to have to keep depending on physical and other special indicators in order to judge the economy's performance with regard to the use of environmental resources. Even so, the conceptual framework should be an aid to clear thinking in the environmental field as well.

Maybe this way of thinking about environmental matters offers a way out of a dilemma facing less developed countries. The dilemma arises because they sometimes find that the adoption of developed-country environmental standards makes local industries uncompetitive in world markets. The poor countries then seem to have a choice between cooperating in the degradation of their own environment or acquiescing in their own poverty. At least when pollution is localized, the resolution of the dilemma appears to be a controlled trade-off between an immediate loss of environmental amenity and a gain in future economic well-being. Temporary acceptance of less-than-the-best environmental conditions can be made more palatable if the "rents" from doing so are translated into productive investment. Higher incomes in the future could be spent in part on environmental repair, of course, but it is general well-being that counts ultimately.

Notice that I have limited this suggestion to the case of localized pollution. When poor countries in search of their own economic goals contribute to global environmental damage, much more difficult policy questions arise. Their solution is not so hard to see in principle, but the practical obstacles are enormous. In any case, I leave those problems aside.

Concluding Comments

That brings me to the end of my story. I have suggested that an innovation in social accounting practice could contribute to more rational debate and possibly to more rational action in the economics of nonrenewable resources and the approach to a sustainable economy. There is a trick involved here, and I guess I should confess what it is. In a complex world, populated by people with diverse interests and tastes, and enmeshed in uncertainty about the future (not to mention the past), there is a lot to be gained by transforming questions of yes-or-no into questions of more-or-less. Yes-or-no lends itself to stalemate and confrontation; more-or-less lends itself to trade-offs. The trick is to understand more of what and less of what. This lecture was intended to make a step in that direction.

Part 10

Environmental Policy in Developing and Transitional Countries

Are Market-Based Instruments the Right First Choice for Countries in Transition?

Ruth Greenspan Bell

Emissions-trading systems have become popular and widely-used instruments for environmental regulation in the West. But such systems depend on markets in which emissions allowances can be effectively traded. In settings where market institutions are not well developed, as in the former socialist nations of Eastern Europe (the so-called transition countries) and in some developing countries, the use of this policy instrument becomes problematic. Environmental polices must be crafted to the specific conditions in individual countries.

It has now been more than 10 years since the fall of the Iron Curtain. Those of us who closely watched the transition can vividly recall the excitement and sense of possibility of those early days.

Environmental activism appeared to be an integral part of the systemic changes occurring throughout Central and Eastern Europe. Severe environmental degradation throughout the region had been an early rallying point for the democratic opposition, which used it to demonstrate the failures of state socialism. It seemed logical that these concerns would translate into a commitment by the new governments to strengthening environmental protection and cleaning up the mess left behind. In 1989, President George H.W. Bush gave a speech in Budapest that became famous in the environmental community, as it pledged U.S. help for efforts at environmental rebuilding.

Resources were brought to bear on the environment from a number of sources, including the European Union's PHARE program (which assists the applicant countries of central Europe in their preparations for joining the European Union), western European countries, the international financial institutions, Japan, and the U.S. Agency for International Development (AID). William K. Reilly, then-U.S. Environmental Protection Agency administrator, was instrumental in writing a commitment to protect the environment into the charter of the European Bank for Reconstruction and Development. Bedrich Moldan, board chair of the Regional Environmental Center for Central and Eastern Europe (REC), an organization originally

Originally published in *Resources*, No. 146, Winter 2002.

set up with AID funding to support regionwide environmental reform, characterized these contributions as efforts to introduce the best environmental practices and ideas from the West.[1]

One issue ripe for examination is the quality and impact of the environmental assistance that started flowing to the countries of the former Soviet bloc after 1989. Many western observers and some central European experts apparently envisioned a tabula rasa that would support leapfrogging over the mistakes committed in the name of environmental protection in the west. (Their aspirations were much like those who apparently thought that markets in all their aspects would magically appear once communism was removed.) This hope was expressed, in part, through a push for the development of new ways to control pollution.

Much attention focused on the development of efficient regulatory instruments and attempts to avoid the mistakes of environmental regulation in the west. Many donors and advisors—including the Organisation for Economic Co-operation and Development (OECD), AID, World Bank, and the European environment ministers themselves— pushed and continue to push for the development of economic instruments, such as pollution taxes, marketable permits, and the like. Whether deliberate or not, the language used to discuss these environmental tools frequently obscured complex issues. The most notable example is repeatedly characterizing traditional approaches as "command and control" and contrasting them with "markets," for an audience reacting to years of hated central planning. Some advisors flatly promised that economic instruments would have lower institutional and human resource requirements than command and control, a glittering and ultimately incorrect promise in countries with small and underfunded environment ministries.

Ten years offers time for reflection. With a few interesting exceptions, the principal environmental improvements in the former Soviet empire have been not the result of improved regulatory tools, but a consequence of the collapse of unproductive state-owned industries and decreased reliance on heavy industry. Meanwhile, overall environmental

institutions remain weak and most of the new ideas proposed after 1989 have not been implemented. Poland's substantial domestic investment in environmental improvement and Hungary's gains in energy efficiency are illuminating exceptions of gains made without great social costs.

Despite donor enthusiasm, most countries in the region were not ready to take on the challenge of environmental reform, for two primary reasons. The environmental movement no longer played the catalytic role it had before 1989. When Communist Party dominance ended, opposition leaders did not need environmental camouflage and could move into more direct roles in political life. The smaller number of people and organizations that continued to focus on environmental issues were pushed to the margin. The groups that remained tended to be top-heavy with technical experts and scientists, who were not very good at communicating with the broader audiences necessary to change policy.

Even more fundamentally, exhilaration was eclipsed by the enormity of the challenges on every possible front—depressed economies, badly frayed social safety nets, and widespread concerns about social unrest. The extraordinary difficulty of doing everything at once (including instituting environmental change) in a time of intense social and economic change was not the "most conducive...to furthering the huge constructive and cooperative effort of institution building that society [was] now challenged to perform."[2]

Introducing Market Mechanisms

Even as it became apparent that most countries in transition did not have the resources, motivation or public support to pursue environmental reform, donors continued, nevertheless, to push them toward the adoption of sophisticated tools. The effort to move directly to market-based instruments is worth examining as an illustration of the disconnect between hopes and expectations and on-the-ground conditions. It is a classic case in which optimism overtook good sense and little attention was given to institutional and social constraints.

What was overlooked was that markets do not

act in a vacuum; institutions do matter for economic instruments, as they do for all tools of environmental protection. The example of emissions trading, which was pursued in a number of countries and was the subject of several regionally based efforts, illustrates the gulf between advice and implementation.

One of the key motivations for industry to want emissions trading has been the economic pain firms have suffered from investing in compliance, which in turn is at least partly related to a clear expectation of consistent and reliable enforcement. When firms have to grapple with the reality—rather than the theory—of environmental regulation, they develop a good grasp of what are the real costs to them of regulation, and of what it takes, at a practical level, to achieve compliance. There is little evidence of industries theoretically coming to the conclusion that emissions trading will be a cheaper way of achieving compliance than directed regulation. Why try to save money on regulation if you are not expending any to begin with and don't expect to in the future?

However, the environmental regulatory systems of the Soviet bloc countries were weak institutions. Laws were not the most important motivator of firm behavior and in any case were riddled with formal and informal exceptions. The state controlled everything and rewarded production over other values. Industry had not been hit on the head with the hard realities of environmental compliance. This is beginning to change in a few of the countries in transition. But even today in most of these countries, environmental enforcement is no more rigorous than it was during the Soviet period, and likely weaker because of the general confusion.

A second institutional requirement for emissions trading to work is very clear knowledge—not guesses—of what pollution each plant is discharging to the environment. Believable end-of-pipe monitoring assures that real, not imaginary, pollution reductions are being traded. But monitoring throughout the former Soviet bloc most often emphasized ambient measurements over end of pipe, and, in any case, was not consistent. In truth, no one could be sure what particular factories were emitting and

whether they were meeting their discharge requirements. One could make estimations using the sulfur content of coal, but the accuracy of the estimations would depend on a number of assumptions, including that the control equipment had been turned on and had been maintained—not a trivial issue in the countries in question. The environmental equivalent of "trust but verify" was missing.

Lack of Transparency

A whole series of measures and institutions are necessary to keep emissions trading honest. One of the most important in the United States is transparency. Permit requirements, emissions data, and the transactions themselves are all available for inspection by the public, including the firm's competitors. In the United States, where environmental regulation is a very contentious subject, this has helped to create a level of trust, a necessary predicate if government regulators, economic competitors, nongovernmental organizations, and the public interest community are going to go along with unconventional programs.

Government transparency was not a hallmark of the Soviet bloc governments, nor is it particularly a European tradition. Nine years into the transition, some of the countries in transition signed the Aarhus Convention in 1998, agreeing to increase their citizens' opportunities to obtain environmental information on demand. These countries are struggling with the nuts and bolts of implementation. It may be that emissions trading programs can work without as much transparency as the United States demands; in many countries, the public is more tolerant when industry and government sit down to negotiate. But it is clearly an issue that architects of any trading program must consider.

The connection between transparency and emissions trading is a particularly sensitive issue in the former Soviet bloc. Trading is, in some ways, a recognition that one party will be held to stricter standards than another similarly situated. When arrangements are made in the sunshine, there are fewer reasons to be concerned that these differentials will creep toward corruption. However, given their experience over the past 45 years, most citizens in the countries in transition are acutely aware how quickly this can happen.

In the mid-1990s, a project in Poland developed a legal basis for granting compliance schedules—essentially an alternative environmental compliance tool that involves grants of discretion. The Polish Ministry participants spent a considerable amount of time and energy devising safeguards to be sure that discretion would not be highjacked to serve the purposes of people in power, rather than the environment.

Need for Legal Systems and Institutions

No firm with any degree of sophistication is likely to participate in emissions trading programs if transactions are not backed up by disinterested mediating institutions available to act in a timely manner to protect a wronged party. Emissions rights are complex intangible property rights and sometimes involve future rights. Buying and selling them is not the same as buying and selling apples in a local market. Emissions trading and other complex market-based mechanisms need a viable, reliable legal system or some analogous set of institutions to ensure the integrity of trades and protect everyone involved.

These certainly did not exist in the early days of the transition. Some of the westernmost countries in transition were only beginning to reestablish a European legal system free of the political and economic "'safety valves'—the legal means of last resort by which Party and state authorities could avoid their own rules"—that existed throughout the period of Soviet dominance.[3] Other countries, particularly Russia and the other parts of the Union of Soviet Socialist Republics had never really been subject to the reliable rule of law. While some countries have made progress in this regard, donor advice on emissions trading did not distinguish between countries with working legal systems and those without.

Trading systems are based on a real, rather than theoretical, understanding of how markets work and of how transactions are constructed, recorded, and policed—the very institutions of capitalism. Complex market transactions don't just happen; the

actors must have considerable skills. Before 1989, scholars throughout the bloc studied non-Marxist economics, but the actual economy was structured under the rules of state socialism.

Industrial managers had been tutored in the old systems. They were not motivated by profit and loss, not held to western accounting principles, and not responsive to shareholders or the stock market. In general, they lacked the kinds of skills normally applied in complex emissions trading systems. The last 10 years have introduced masters of business administration programs and practical market experience, but not without a great deal of pain. A few countries retained a trading mentality that was reflected in small businesses, but, in general, industry has faced a steep learning curve and was not ready, certainly in the early part of the transition, to take on market-based environmental responsibilities.

In sum, not only was there no tabula rasa, but there also were considerable although varied histories to overcome and institutions to build throughout the countries of the former Soviet bloc. The key elements—monitoring, transparency, a working legal system, and a realistic incentive to trade— were nowhere to be found. Scholars can debate whether the single-minded push of some donors to concentrate such intense efforts on developing sophisticated, market-based environmental regulation was a disservice, diverting energy from efforts that might have been more productive.

In any case, pushing inexperienced governments prematurely toward highly sophisticated environmental policy tools was not the only miscalculation by the donor community. Another was the emphasis placed on drafting new state-of-the-art environmental laws without apparently giving much thought to the existing laws and how they operated, much less to what the countries could actually manage in terms of the resources and experience they could bring to bear in implementing the new laws.

My purpose here is not to argue against the use of market-based instruments. I don't advocate throwing the baby out with the bath water. Rather, I am arguing that market-based instruments, and specifically emissions trading, were done a disservice when the OECD, World Bank, and others pushed these tools too hard and too fast in countries that were institutionally unprepared to implement them. The power of these instruments may have been trivialized when the experts were less than candid about the total package.

If environmental professionals in the countries in transition were led to believe that they could make this leap without at the same time constructing supporting institutions, the cause of environmental protection itself may have been dealt a blow by the disappointments that followed.

If the notion of a great environmental leap forward was not sensible, what approach might have worked better? Certainly, the old system could not be left in place. An emphasis on incremental improvements in pursuit of pragmatic goals might have been smarter, particularly one that helped to build a transitional system that would have taken account of existing capabilities and institutions. This might have resulted in real, although small, initial environmental gains, and might have been accomplished without losing sight of the ultimate goal of developing the most efficient ways to manage the environment. Another constructive approach would have elevated the importance of institutional reform in the advice rendered on economic instruments.

The donor community also needs to rethink its way of doing business. Formulaic advice should be replaced with crafted responses that explicitly recognize the varied conditions in each country that would support reform. Donors need to do their homework, as well, which means getting to know each country in a very different way than they have in the past.

The importance of examining these issues today is not only a question of historical review. Many of the same countries that were the subject of environmental assistance efforts are trying to enter the European Community. They will be required to incorporate into their environmental practices many highly sophisticated tools, layering them on to still-weak, thinly staffed domestic environmental institutions. Moreover, the same donors continue to urge countries with weak institutions in other parts of the world to adopt highly sophisticated tools for environmental protection. It would be wise to consider the lessons of environmental assistance in the countries in transition, as others embark on these new challenges.

Suggested Reading

Bell, Ruth Greenspan. 1992. "Environmental Law Drafting in Central and Eastern Europe", XXII *Environmental Law Reporter*, News & Analysis (September).

Cole, Daniel H. 1998. *Instituting Environmental Protection: From Red to Green in Poland*. New York: St. Martin's Press.

Cole, Daniel H., and Peter Z. Grossman. 1999. "When is Command-and-Control Efficient? Institutions, Technology, and the Comparative Efficiency of Alternative Regulatory Regimes for Environmental Protection." *Wisconsin Law Review* 5 (1999): 887–938.

Elster, Jon, Claus Offe, and Ulrich K. Preuss. 1998. *Institutional Design in Post-Communist Societies: Rebuilding the Ship at Sea (Theories of Institutional Design)*. Cambridge; New York: Cambridge University Press.

Panayotou, T. 1994. Economic Instruments for Environmental Management and Sustainable Development. Paper prepared for the United Nations Environment Programme, Environmental Economics Series Paper No. 16

Stiglitz, Joseph. 1997. Interview in World Bank *Transition Newsletter: The Newsletter about Reforming Economies*. Vol. 8, No. 6, December. http://www.worldbank.org/html/prddr/trans/d ec97/pgs1-3.htm (accessed Oct. 4, 2005).

Notes

1. Europe After 10 Years of Transition, Speech at the REC on June 18, 2000. http://www.rec.org/REC/Programs/10th_anniversary/Speech.html

2 Elster, Offe, and Preuss, Institutional Design in Post-Communist Societies.

3 The phrase is Daniel H. Cole's in Instituting Environmental Protection: From Red to Green in Poland.

44 Demonstrating Emissions Trading in Taiyuan, China

Richard Morgenstern, Robert Anderson, Ruth Greenspan Bell, Alan Krupnick, and Xuehua Zhang

Can emissions-trading systems function effectively in a country where monitoring and enforcement systems are still largely untested and where state enterprises are the largest polluters? An experiment in a city in the People's Republic of China is trying to answer this question.

Can market-based instruments (MBIs) be effective tools for improving environmental quality in the People's Republic of China (PRC)? Can such instruments really reduce emissions at lower costs than conventional command-and-control approaches in a planned-market economy where monitoring and enforcement systems are still in their infancy and where state-owned enterprises are the dominant polluters? No one knows for sure whether MBIs are suitable for such an application. But a number of senior Chinese officials, along with experts from the Asian Development Bank, are betting that the time is ripe to test out these ideas in a full-scale experiment. Toward this end, they have recruited a team of international and domestic experts, led by researchers at Resources for the Future, to try to demonstrate the feasibility of emissions trading in Taiyuan, a heavily polluted industrial city in northern China.

Since spring 2001, the RFF team has been assessing the local situation and, most recently, designing a program for emissions trading among large emitters in Taiyuan, the capitol of Shanxi Province. A formal regulation that would implement the trading system was formally adopted by the city government in late 2002. How the system will actually work, whether the design will prove viable in Taiyuan, whether tangible environmental improvements can be obtained at reasonable cost, and what modifications might be necessary to improve the system are all unknown at this time. What is known is that there is strong interest in trying to adapt the western-style emissions trading experience to the real-world conditions in China—and

Originally published in *Resources*, No. 148, Summer 2002.

a major effort is underway to demonstrate the viability of such an approach.

Background

Heavy reliance on relatively uncontrolled coal combustion as a source of heat and power has created serious environmental problems in China—particularly in its coal-rich northern provinces. Particulate matter (PM) and sulfur dioxide (SO_2) are the major pollutants of concern, although with recent progress in reducing PM emissions, attention is increasingly shifting to the control of SO_2. In many urban areas, high SO_2 concentrations–along with fine particles created by the atmospheric transformations of SO_2 into sulfates—represent a serious public health threat.

Situated about 500 kilometers southwest of Beijing, Taiyuan has a population of 2.7 million and covers an area of almost 7,000 square kilometers. Topographically, the city is surrounded by mountains on three sides, resulting in a Los Angeles-type of smog trap in which air pollutants tend to accumulate. A 1998 report by the World Bank identified Taiyuan as among the most polluted cities in the world. SO_2 concentrations—which have been relatively flat over the past decade—averaged 200 parts per million (ppm) in 2000, more than three times the PRC's Class II annual standard (60 ppm). With recent economic growth averaging 10% per year, the reported absence of deterioration in air quality reflects the considerable effort that has already been devoted to environmental improvement.

In China, local governments generally take their cues for controlling pollution from the central government. Currently, the national control policy relies on emissions standards based on the concentration of SO_2 in the boiler's stack at a specific point in time. Pollutant concentrations are based on self-reported data from the enterprises and periodic stack testing by the local Environmental Protection Bureaus (EPBs). These estimated concentrations are combined with limited data on pollutant flows to calculate mass emissions from the enterprises. The calculated mass emissions, in turn, form the basis of a small emissions levy ($25/ton), which is used to support local EPB activities with the balance returned to individual enterprises to finance their pollution control investments. It is widely recognized that the current approach to calculating mass emissions is not robust enough to support more aggressive efforts to reduce emissions—including via emissions trading.

A new, more sophisticated mass-based system—"total emissions control" (TEC)—is being implemented in China as a supplement to the existing stack-gas concentration standards. The TEC system is similar in many ways to individual facility-level caps on SO_2 emissions imposed under Title IV of the U.S. Clean Air Act. Once it is fully implemented and enforced, the mass-based TEC system will be able to serve as a key building block of an emissions trading system.

China typically experiments with new pollution control programs through pilots or demonstrations. The Taiyuan city government began experimenting with emissions permits and earlier pilot versions of the TEC system in the 1980s. As early as 1985, emissions controls were introduced in local regulations issued by the Taiyuan government. The city conducted two experiments with emissions offsets and administratively determined trading in the mid-1990s. For instance, in one "trade," the buyer paid the Taiyuan EPB for an addition to its emissions quota. The EPB took this money to finance reductions in emissions elsewhere in the city. In 1998, the Taiyuan city government issued TEC "management rules", including a provision for "permit exchange," a form of emissions permit trading. The Taiyuan EPB has begun to issue updated permits with TEC-based limits to large enterprises. So far, more than three dozen permits have been issued—all of them to large enterprises. Although widely seen as extremely ambitious, the Tenth Five-Year Plan for Taiyuan calls for 2005 SO_2 emissions to be reduced by about 50% below 2000 levels.

Key Questions for Using MBIs in Taiyuan

Since early 2001 members of the RFF team have made numerous trips to Taiyuan to collect data and work with local officials and enterprise managers

Recent Transactions in the Emissions Trading Program in Taiyuan

Through May 2005 the Taiyuan EPB reports a total of 12 transactions of emissions allowances in Taiyuan. Unlike the textbook type of trading, the government has been directly involved in most of the transactions, including establishing the prices. Reportedly, prices are on the order of 1500 Yuan (slightly below $200 per ton of SO₂). In contrast to the experimental trades in the 1990s, however, the government is now only the middleman between firms buying and selling emissions allowances.

The buyers have all been new sources (for old sources it is cheaper to pay the levy). Although many of the transactions are quite small (10 tons or less), several were in the range of 40-70 tons. The largest transaction involved the use of offsets in Guijiao City (part of the Taiyuan airshed). In that case the newly built Coal Slack Power Plant in Guijiao City was emitting 2000 tons of SO₂ even after coal-washing. To support this level of emissions, the power plant was directed by the provincial government to fund, at a cost of 60 million Yuan, a government-owned central heating system in the inner city, thereby allowing the closure of more than 50 boilers, whose emission rights were then transferred to the plant owner.

on design issues. Several critical questions were considered at the outset:

Do market-based instruments really have the potential to reduce environmental control costs in Taiyuan?

The RFF team conducted an assessment of expected SO_2 control costs for different enterprises in the city. The marginal costs of abatement, as shown in the accompanying table, suggests reasons for at least a *prima facie* case for MBIs. Estimates of these costs range from $60–$1160 per ton of SO_2. In a western context, this wide range of abatement costs would certainly provide an incentive for both government and industry to look favorably on MBIs. The alternative—a traditional command-and-control approach—would likely force some firms to undertake unnecessarily expensive mitigation options to achieve the same emission reductions available at lower cost.

If conditions are ripe for MBIs, why not simply increase the existing pollution levy?

The current levy is clearly not high enough to create incentives to change behavior. Thus, substantially increasing the levy, as was recently done in Beijing, is a potential option. After extensive discussions with provincial and local officials, however, it became clear that increasing the levy rate to create significant new abatement incentives, especially at a time when reforms in emissions measurement were being introduced, would not be politically acceptable in Taiyuan. Introducing an MBI in the context of the newly established TEC was seen as being more acceptable, especially to the enterprises, which feared the imposition of higher taxes in the form of levies.

Could an emissions trading program survive in a place that historically lacks strong enforcement and compliance systems?

Our purpose is to translate the international experience with emissions trading to the real-world conditions in Taiyuan, a resolve that required us to consider China's unique historical, institutional, and technological context. Emissions trading works in the United States and elsewhere when certain conditions—often the same conditions necessary to support traditional methods of environmental control—are present. But Chinese environmental institutions are still in their infancy in the development of their compliance and enforcement systems. What is the incentive for industry to adopt more efficient methods of regulation when enterprises are expending only limited resources on environmental protection to begin with or if the system is so riddled with exceptions that enforcement is not sure?

The courts generally cannot be relied on for an independent source of compliance oversight, as

Table 1. Cost-Effectiveness of SO₂ Control Measures in Taiyuan

Control Measure	Where Applied	$/ton
Treat post-combustion gas	Taiyuan District Heating	($60)*
Flue-gas desulfurization (FGD)	Eastern Mountain Plant	($80)
Lower sulfur coal (~1.3%)	Taiyuan #1 & #2, Taiyuan Iron & Steel	($100)
FGD (simplified)	Taiyuan #1	($240)**
FGD	Taiyuan #1(Planned)	($180)*
Add limestone to fuel	Coal gasification plant (Planned)	($130)
Coal washing	Future Sites (Possible)	($1,160)

* as reported by plant officials to RFF team, March 2001.

** plus unspecified investment costs paid for by the Government of Japan.

judges owe their appointments, salaries and social benefits to the provincial and local governments, the same institutions that own and control most heavy industry, and pay the salaries of the local EPB officials. Similarly, the penalties for violating pollution standards are inadequate to support an enhanced environmental management system of any type (see below).

Is the current monitoring system suitable for emissions trading?

The dearth of experience with consistent monitoring raises questions about whether the appropriate incentives exist for industry to participate in a trading program. Whatever environmental management system is ultimately adopted in Taiyuan, the basic tracking, recordkeeping, and enforcement functions will need extensive upgrading and renewed compliance commitments. The monitoring system is a case in point: continuous emissions monitors (CEMs) of the type used in the U.S. acid rain program were installed in only a handful of units in Taiyuan. Although there were ambitious plans to add several hundred more CEMs by the end of 2002, there had already been significant delays and alternative and generally less accurate methods of estimating emissions were necessary during the demonstration. However, significant numbers of CEMs have now been added, primarily on the largest enterprises.

If emissions trading is to be introduced in Taiyuan, which sources would participate in the program and how would the permits be allocated?

The obvious choice is to expand the ongoing efforts to introduce the TEC mass-based permit system among large emitters. The possibility of allocating permits via auction was discussed early on but it was clear that existing financial and political constraints would stymie such an effort. However, the regulation does allow the city government to retain a small portion of the permits to be auctioned to the highest bidder.

To improve the local capacity to manage the demonstration program, the U.S. Environmental Protection Agency (EPA), which has extensive experience in developing and managing trading programs, has helped provide training under a parallel agreement with the Asian Development Bank. To date, the training has focused on the technical aspects of MBIs, U.S. and other international experiences with emissions trading, design elements of a trading program, measurement, emission and allowance tracking, and emission verification. In addition, the RFF team has formed a highly successful alliance with the Chinese Research Academy of Environmental Sciences, which has provided technical assistance on a wide range of issues.

The Current Situation

Twenty-six of the largest enterprises—representing

about half the reported emissions in Taiyuan —are slated to participate in the new trading system. Apart from a few small adjustments to reflect recent investments and expansions, the baseline for individual enterprises is the actual reported SO_2 emissions in 2000–2001. Although discussions are still continuing on this point, the 50% reduction mandated in the Tenth Five-Year Plan is reflected in the 2005 initial permit allocations slated for individual enterprises. Targets for interim years also have been established. New sources may purchase permits from existing sources or from the small set-aside held by the local EPB. Banking of permits is allowed.

The largest sources are required to install CEMs. For others, a new emissions tracking system has been developed to integrate into a single data system all the relevant information reported by the government as well as by the enterprises concerning emissions, fuel purchases (including sulfur content), output and other factors. A parallel allowance-tracking system has also been developed by the U.S. EPA to monitor actual trades. It is anticipated that the operation of both the emissions and the allowance-tracking systems, in combination with the available CEMs, will provide a credible basis for emissions trading.

Establishing the appropriate penalty for violations is an issue that still must be resolved. Based on international experience, it is clear that the penalty for exceeding permitted emissions must be high enough to ensure that sources have adequate incentives to either control their emissions within permitted levels or to trade with other sources in order to satisfy their obligations. However, pursuant to a regulation previously issued by the provincial government, the maximum penalty the local EPB can impose on a single enterprise is less than $4000 per violation, which is generally interpreted as an annual maximum. Clearly, even if fully

enforced, a penalty of this form creates a limited incentive to adhere to the standards, especially for large emitters. Despite extensive discussions with local officials on options for revising this penalty, it was not possible to obtain the necessary agreements at the local and provincial levels to modify this penalty limit. Discussions are continuing.

As trading gets underway, further training on data reporting, emissions and allowance tracking, and other topics has been conducted. Depending on the results of trading in Taiyuan, officials may try to expand the Taiyuan emissions trading system to other cities in the province and beyond.

Overall, there is widespread interest, both among the Chinese and the international community, in the SO_2 trading demonstration project and subsequent trading experiences in Taiyuan. Although the operational success of this program is still unknown, the issuance of a formal regulation issued by the city government is a major milestone. Regardless of the ultimate outcome, important lessons will be learned about the suitability of MBIs for developing countries, particularly for planned-market systems such as the People's Republic of China. Further, the city of Taiyuan will have in place a much-improved system for administering whatever type of regulatory system it ultimately implements.

Suggested Reading

Bell, Ruth Greenspan, and Clifford Russell. 2002. Environmental Policy for Developing Countries, Issues. *Science and Technology* (Spring).

Cole, Daniel H., and Peter Z. Grossman. 2002. Toward a Total-Cost Approach to Environmental Instrument Choice. In R. Zerbe and T. Swanson (eds.), *Research in Law and Economics*.

45

Saving the Trees by Helping the Poor
A Look at Small Producers along Brazil's Transamazon Highway

Charles Wood and Robert Walker

Conserving tropical forests over the longer term requires policies that address the needs of the rural poor. Land tenure security, ensured by an effective system of property rights, is an important part of such a policy, as it provides incentives for care and investment in reforestation. But as this research in Brazil finds, it also requires protection from damaging, contagious fires, and this involves the development of local organizations that transform isolated farmers into civic partners.

The inexorable drive of subsistence farmers to clear tropical forest for an eked-out living presents a major environmental threat, nowhere more than in the frontier areas of Brazil where the rate of deforestation is among the highest in the world. Government responses have taken a number of forms, including setting land aside in nature preserves.

Desirable as this approach may seem, policies to establish conservation forests often founder on the *social* problem of rural poverty. Although small holders and shifting cultivators can be kept out of well-protected areas, on a regional scale it is unrealistic to assume that enforcement could ever be entirely effective. Apart from the high cost of monitoring large tracts of land, there remains the moral issue of depriving communities of needed land for agriculture. Large-scale enterprises that might create jobs for the rural poor while alleviating pressure on forest resources sound good in theory, but settlement frontiers offer few locational advantages for large-scale capital investments. The approach of greatest social viability, at least in the short run, is to create incentives for farming in ways that conserve natural resources.

Resource economists have long argued that subsistence farmers are more likely to conserve if they can establish property rights that guarantee legal ownership of land. Only then can they be sure of reaping the benefits of restraint and investment. Throughout the world such security is often viewed as synonymous with the receipt of land *title*, a legal document conferred by government agencies or obtained through sales transactions.

Originally published in *Resources*, No. 136, Summer 1999.

Conversely, these economists maintain, farmers without secure title are more likely to opt for immediate consumption over long-term investment. They tend to rapidly exploit land and timber resources rather than engage in sustainable production strategies.

Conserving Nature for Profit

If advocating property rights for purposes of conservation has enjoyed something of a renaissance over the past twenty years, resource economists have only recently expanded the concept to include wealth creation and economic development more generally. They point out that having no title to land *and* being poor encourages rural people not only to clear forest but to mine soil nutrients, an agricultural practice that rapidly exhausts soil fertility and degrades its structure.

By contrast, the economists say, freedom from eviction and the rights of individuals to monopolize land not only for personal consumption but for profit are powerful stimuli to economic activity and investment. The result can be conservation of valuable soils and timber resources. Granted, physical growth rates of commercial hardwoods are probably too slow to protect them from liquidation by individuals intent on short-term profit maximization. But where long-term investments are likely to pay off, conservation is bound to benefit. A farmer whose title to land is secure may, for example, leave a forest bequest to children. A relatively predictable future also lessens the need to clear noncommercial trees so as to make *de facto* claims on agricultural plots.

Although the theoretical reasoning that leads to these conclusions is cogent, in fact the predictions have rarely been subjected to empirical test, especially in rural Brazil. Thus, we participated in the study described later in this article whose results provide empirical support for theory.

Elsewhere, studies in Thailand and Africa are beginning to substantiate beneficial soil conservation effects related to land tenure security. A rise in secure property rights among poor farmers is expected to bring about land improvements, and

the switch to farming systems based on perennials or tree crops that are much less consumptive of soil nutrients. (See "Suggested Reading," page 278.)

Networking to Fight Fire

Whether deforestation can be slowed depends, then, on whether mitigation makes sense economically at the individual farm level. However, even where it does make sense, the forest resource base still faces substantial threats from outside forces over which the farmer profiting from conservation exercises no control.

Fire is a particularly acute threat, as recent events in Indonesia, Mexico, and Brazil dramatically underscore. In northern Brazil, one million hectares of forest burned in early 1998 before the onset of the rainy season. It was the first time in recorded history that an extensive area of forest closed to development burned in Amazonia. Drought conditions associated with El Niño, and possibly exacerbated by greenhouse gas buildup, will in all likelihood continue to bring the moist forest of the Amazon Basin to the point of flammability.

The fire threat that drought poses makes the prevailing land use and agricultural practices of Amazonia all the more a concern.

New studies of the effects of "surface" fires used to facilitate selective logging in the Amazon show that the measures of deforestation that we have come to rely on as environmental indicators vastly underestimate the magnitude of the damage done. In particular, surface fires unleash a cycle of increasing flammability and forest degradation, with effects that do not become visible on satellite images for years.

Research by Dan Nepstad and his colleagues calls attention to the previously unnoticed effects of these fires. Once out of control, they escape into standing primary or logged forest. While they burn with less intensity than the fires associated with agricultural use, surface fires nonetheless cause severe damage to the understory, and to tree species with fire-sensitive outer barks. Because they are slow-burning, surface fires also ignite a vicious

Table 1. Titling status of survey lots on the Transamazon Highway

| | Universe | Titled lots | Provisional title type | |
			Authorization	Recognition
Untitled	145 (42%)			
Titled	202 (58%)			
Definitive title		135 (67%)		
Provisional title		67 (33%)	45	21

Note: Title is definitive when a government document is in evidence. Provisional title is a step toward definitive title, and is indicated by the posession of an "authorization" or a "recognition" by the titling agency. Of the 67 provisional titles, 1 did not report provisional title type.

positive-feedback effect by increasing the subsequent flammability of the landscape. Thus far, these surface fires have affected one and a half times more forest than the fires that small farmers set.

Still, these latter "deforestation" fires can be contagious. Initially, farmers set them to clear land for planting, usually of rice and pasture. Later, they set them as part of the crop rotation process and to clear secondary growth. Farmers also burn pastures to keep out invasive plant species.

One response to the threat of spreading fire that these agricultural practices pose might be to consolidate individual land parcels into large farms. Unfortunately, however, such an approach flies in the face of other objectives, such as alleviating rural poverty through land reform, a pressing concern in Brazil, and one that calls for more land parcels, not fewer. Thus, alternatives must be sought.

As recent research in Brazil shows, institutions do exist to promote cooperative relations between small holders in forest frontiers. These community organizations with local bases of participation facilitate access to financial credit, ensure reasonable prices for raw resources and finished goods, and provide a political voice for poor farmers. They also provide a forum for farmers to unite against the spread of fire.

Interviewing in the Amazon

To better understand the connections among land tenure security, logging, and fire contagion, researchers at the University of Florida, Florida State University, and the Brazilian Agricultural Research Agency (EMBRAPA/CPATU) undertook a collaborative study among poor farmers in the Brazilian Amazon. The research team conducted interviews with 261 small producers on the Transamazon Highway, whose land possessions covered 347 lots of 100 hectares each, the original size of land grants to families in a colonization project that began with the highway's construction in the 1970s. The hope of the colonization scheme was to settle an empty region, thereby "bringing people without land to land without people."

Among other things, the interviews allowed us to collect extensive information on the farming households themselves, their farming systems, and their use of the forest. We also were able to obtain the land titling status of the individual lots (see the table).

In addition to the survey of properties, we interviewed individuals involved in the region's political organizations, such as the rural union, a number of cooperatives, and several other groups that facilitate access to financial credit. On the basis of these interviews, we were able to determine which communities within the study area were well organized to prevent the spread of fire and which ones were not. It was the research team's hypothesis that social ties among farmers would create a basis for both the formal and informal regulation of fires during the burning season.

Statistical results from the research show that having title to land does indeed influence the way that small land holders manage tropical hardwoods.

In particular, possession of title encouraged the longterm maintenance of valuable wood and reforestation activities. Results from logistic regression show that the frequency of forest conservation and reforestation was much higher among individuals with title than without, even after controlling for important determinants of land use such as family size and availability of economic resources. Indeed, the relative frequency of reforestation among those with title was about fifteen times higher than among those without. Likewise, possession of a title tended to discourage participation in timber markets. Although the statistical effect was not as strong as observed for forest conservation and reforestation, poor farmers holding title to land were less likely to have recently sold trees than those without title. The research findings thus offer empirical support for the predictions derived from the property rights paradigm.

With respect to fire contagion, the results are more complex. When the probability is high that fires will spread from one property to another, the very kind of security presumably afforded by property rights is eroded. The possession of title, and the associated right to do with a piece of property as one sees fit, provide no protection against the economic behavior of one's neighbors.

Results from logistic regression show that possession of land title did not lower the risk of damage caused by a neighbor's activity. The relative frequency of individuals experiencing fire contagion was about the same among individuals with and without title.

What appeared to make some difference, however, was whether the property was located in a well-organized part of the study area. Lots located in the vicinity of an effective credit organization, cooperative, or union representative tended to suffer less fire contagion than lots found in unorganized areas. The relative frequency of fire in a well-organized area was about 60 percent lower than elsewhere.

Our argument is that organizational effectiveness creates community cooperation, which in turn provides household incentives to control and manage the use of fire. People in well-organized areas were found to work with their neighbors to take such preventive measures as constructing firebreaks, coordinating the timing of their burns, and generally keeping each other informed of their fire-related plans.

Setting Policy

Based on the research done to date, conserving tropical forests over the long run will require setting policies that effectively address the needs of the rural poor. Land tenure security associated with land titling is important in this regard, as it reduces the rate of hardwood depletion by small holders and encourages their efforts at reforestation.

It is important, however, to recognize that the notion of land tenure security extends beyond the right to private property. Findings from the Brazilian research suggest that another dimension of tenure security rights should be considered, namely the right to remain free from the damage caused by fire contagion.

Promoting land tenure security thus involves attention not only to land titling but to the development and support of social organizations that transform isolated farmers into civic partners.

Suggested Reading

Alston, L.J., G.D. Libecap, and R. Schneider. 1995. Property Rights and the Preconditions for Markets: The Case of the Amazon Frontier. *Journal of International and Theoretical Economics* 151(1): 80–107.

Beaumont, P., and R. Walker. 1996. Land Degradation and Property Regimes. *Ecological Economics* 18: 55–66.

Feder, G., and D. Feeny. 1991. Land Tenure and Property Rights: Theory and Implications for Development Policy. *The World Bank Economic Review* 5(1): 135–153.

Schmink, M., and C. Wood. 1992. *Contested Frontiers in Amazonia*. New York: Columbia University Press.

46 Small Is Not Necessarily Beautiful
Coping with Dirty Microenterprises in Developing Countries

Allen Blackman

Urban clusters of small firms in developing countries can generate enormous amounts of pollution. As this study of two cities in Mexico finds, policies to address this problem may require a number of practical and informal strategies.

Sizable cities in developing countries typically host thousands of small manufacturers engaged in pollution-intensive activities such as leather tanning, metalworking, ceramics, textiles, and food processing. Collectively, these firms, which are often located in poor, densely populated neighborhoods, can have devastating environmental impacts. Nevertheless, pollution control efforts in developing countries have generally focused on large industrial sources. This bias stems in part from an enduring misperception that small-scale sources are relatively unimportant. It also is a matter of practicality— small firms are difficult to regulate by conventional means. Numerous, cash-strapped, and frequently unlicensed, they are often hard for regulators to identify, much less monitor and sanction.

Yet recent RFF case studies of small-scale brick kilns and leather tanneries in Mexico demonstrate not only that cutting small manufacturers' pollution yields large benefits, but also that the barriers to pollution control are far from insurmountable. Policymakers have at their disposal a number of cost-effective and practical—if unconventional—pollution control strategies.

Brick Kilns in Ciudad Juárez

In Mexico, small-scale traditional brick kilns fired with cheap, highly polluting fuels like scrap wood, plastic refuse, and used tires are notorious sources of urban air pollution. Ciudad Juárez—a sprawling industrial city on the U.S.-Mexico border

Originally published in *Resources*, No. 141, Fall 2000.

279

with some of the worst air pollution in North America—is home to approximately 350 such kilns clustered in eight brickyards.

A study of the benefits and costs of controlling emissions of particulate matter smaller than 10 microns (PM10)—a pollutant responsible for a large share of the noncarcinogenic adverse health impacts from air pollution—clearly demonstrates that the Ciudad Juárez's brick kilns inflict significant harm. Because these kilns do not have smokestacks, over 90% of their PM10 emissions are deposited less than a third of a mile away, a critical problem since most kilns are situated in residential neighborhoods. Thus, the kilns are partly responsible for over a dozen cases of premature mortality and hundreds of cases of respiratory illness each year in that city (see Table 1), damages that are valued at between $20 million and $90 million. By contrast, the annual cost of pollution control programs that would virtually eliminate these impacts is estimated at less than $300,000.

Although brick kilns are widely recognized to be a leading source of air pollution in Ciudad Juárez and its sister city, El Paso, Texas, a number of factors make it politically difficult to require brickmakers to bear the full costs of pollution control: brickmaking provides over 2,000 jobs, most brickmakers are impoverished (profits per kiln average $100 per month), and most belong to a trade association or other local organization that can lobby against pollution control efforts. Despite these obstacles, efforts to control brick kiln emissions had considerable success in the early 1990s. In 1989, the municipal environmental authority initiated a "clean technology" project aimed at substituting propane for dirty fuels. The next year, the propane initiative was handed off to the *Federación Mexicana de Asociaciones Privadas de Salud y Desarrollo Comunitario* (FEMAP), a private, nonprofit social services organization. FEMAP was able to attract considerable funding and participants from both sides of the border including local propane companies, universities, and Los Alamos National Laboratory in New Mexico.

Leaders of the propane initiative used a number of carrots and sticks to encourage participation.

First, they subsidized various costs associated with adopting propane. Mexican propane companies provided most of the requisite equipment and training free of charge. To make propane more attractive, despite the fact that it was more expensive than traditional fuels, engineers at Los Alamos National Laboratories designed new energy-efficient kilns. Unfortunately these kilns proved prohibitively expensive and complicated.

Second, the initiative's leaders worked to ratchet up pressures on brickmakers to adopt propane. Most importantly, they convinced brickmaker trade unions in several brickyards to prohibit their members from using dirty fuels. Also, the city government banned the use of particularly dirty fuels and set up a telephone hotline to register complaints about brickmakers violating the ban. Enforcement teams with the power to jail and fine violators were dispatched in response to complaints.

Third, FEMAP initiated an educational campaign to raise brickmakers' awareness of the health hazards associated with dirty fuels. Finally, project leaders tried to reduce competitive pressures for brickmakers to use cheap dirty fuels by organizing a boycott of bricks fired with dirty fuels. However, the boycott was quickly undone by rampant cheating.

By the end of 1993, over half of the Ciudad Juárez brickmakers were using propane. Unfortunately, this success proved short-lived. In the early 1990s, as part of a nationwide economic liberalization program, the federal government was phasing out long-standing propane subsides. As propane prices continued to rise in 1994, propane users switched back to dirty fuels. By 1995, only a handful of brickmakers were still using propane. However, the propane initiative has had some lasting impacts: local organizations and city officials continue to enforce a ban on particularly dirty fuels such as tires and plastics.

Leather Tanneries in León

The city of León in north-central Mexico produces about two-thirds of the country's leather. Of the 1,200 tanneries in the city, about half employ fewer than 20 workers. Virtually all of León's tanneries

Table 1. Estimated annual health effects of PM10 emissions from traditional brick kilns in Ciudad Juárez (number of cases above baseline = zero kiln emissions; mean values and 95% confidence intervals)

Health endpoint (number of cases)	Ciudad Juárez			El Paso		
	Low	Mean	High	Low	Mean	High
Mortality	2.5	14.1	31.0	0.5	2.6	5.8
Respiratory hospital admissions	0	262	770	0	37	107
Emergency room visits	0	607	1,719	0	85	240
Work loss days	0	3,216	8,500	0	448	1,185
Adult respiratory symptom days	91,610	376,600	794,000	14,430	59,300	125,000
Adult restricted activity days	2,704	138,000	349,100	377	19,240	48,670
Asthma attacks	180	42,680	108,600	25	5,950	15,130
Children's chronic bronchitis	0	1,637	4,416	0	184	497
Children's chronic cough	0	1,878	5,017	0	211	564
Adult chronic bronchitis	0	93	242	0	15	38

Source: Blackman, Newbold, Shih, and Cook (2000).

dump untreated toxic effluents directly into municipal sewers where they flow untreated into the Turbio River. The resulting pollution has contaminated ground water, destroyed irrigated agricultural land, and caused serious health problems. Regulations governing water pollution have been on the books for decades, but most are simply not enforced. By all accounts, the main reason is that, as one of the city's principal employers, tanners have considerable political power.

Concerted efforts to control pollution from tanneries in León began in 1986. Tannery representatives signed a *convenio* (voluntary agreement) with regulators in which they agreed to comply with written regulations within four years. But when it became apparent in 1990 that the tanners had not taken any action, they were given a second four-year grace period. At the end of this second grace period, tanneries still had made no progress, so they were granted yet another grace period. This cycle has continued until today.

In addition to the succession of *convenios*, there has also been an attempt to control pollution by relocating the tanneries. In the early 1990s, the city built the infrastructure for a tannery industrial park with a common wastewater treatment facility. Tanners were required to purchase land and build new facilities in the park. In consideration, they were to be provided with subsidized loans and tax credits. However, tanners ultimately refused to foot these costs, and today the industrial park stands empty.

Perhaps the most successful means of controlling tannery emissions in León has been the largely voluntary adoption of clean tanning technologies, including sedimentation tanks, that allow particulate matter to settle out of waste streams; low-chemical tanning recipes; enzymes that substitute for sulfur compounds used to rid hides of hair; recycling tanning baths; and chrome recovery. Estimates of the percentage of tanneries using these technologies range from 90% for sedimentation tanks, which are required by law (this law appears to be enforced), to less than 5% for chrome recovery.

In view of the proven political and economic constraints on relocation and conventional regulation, clean technologies currently represent the best hope for controlling tannery pollution. To assess the barriers to and incentives for adoption of clean technologies, a team of researchers from RFF and the University of Guanjuato in Mexico recently administered a detailed survey to the owners and managers of about 170 tanneries. Preliminary analysis of the survey data suggests that for most technologies, firm size is not determinative: small tanneries are just as likely to adopt cleaner methods as large ones. Rather, access to information about the technology as well as the education of tannery managers appears to be critical. Many clean technologies are simply not well known or well understood by tanners. Surprisingly, key sources of information and assistance are private sector entities, including chemical supply companies, fellow tanners, and the tannery trade associations.

Another finding is that adopters of certain technologies tend to be spatially clustered, suggesting that demonstration effects are important. In the case of one technology—sedimentation tanks—laws requiring installation are clearly driving adoption. Finally, even though most of these technologies lower materials costs (and sometimes labor costs), several entail significant setup costs that act as a significant barrier to adoption.

Lessons Learned

Political constraints. In both Ciudad Juárez and León, city regulatory authorities have been able to enforce regulations that impose minimal costs on polluters. For example in León, they have compelled tanneries to install sedimentation tanks. But they have not been able to consistently enforce more burdensome regulations. A key reason is that brickmakers in Ciudad Juárez and tanners in León are numerous and well-organized and, as a result, have the power to block enforcement. In general, when dealing with severe pollution problems created by small firms, political considerations are likely to be quite important, if not paramount. Severe pollution problems arise when small polluters are numerous, and when polluters are numerous, they are bound to have political power. Consequently, unless the victims of small manufacturers' pollution can be galvanized into action, successful policies will need to accommodate polluters' concerns about the costs of pollution control.

Informal regulation. The case studies suggest that so-called "informal" regulation—pressure generated by private sector actors that have day-to-day contact with polluters—is probably the most important ingredient of a successful pollution control program for small firms. In Ciudad Juárez,

brickmakers trade unions played a critical role in promoting propane by enforcing prohibitions on the use of traditional fuels in some brickyards. Also, citizen complaints facilitated enforcement of a municipal ban on dirty fuels. In León, pollution control efforts have been stymied by the absence of such private sector pressure.

There are a number of explanations for the lack of informal regulation in León. Most households in the city depend on the leather industry for their livelihood so it is difficult to generate public support for measures that raise tanners' production costs. Also, tannery emissions are less noticeable than those of brick kilns, and their health and environmental impacts are less immediate. Hence, informal regulation would seem to be easier to generate when polluters are not the mainstay of the local economy and when emissions are easy to detect and have immediate adverse health impacts.

Clean technological change. Given the barriers to regulating small firms by conventional means, clean technological change represents a particularly promising pollution control strategy. In both cities, technical assistance spurred clean technology adoption, and in Ciudad Juárez, a campaign to educate brickmakers about the health impacts of dirty fuels also appears to have had some impact. Surprisingly, both case studies suggest that private- sector organizations, such as equipment suppliers and trade associations, can be the principal purveyors of technical information, an encouraging prospect given chronic constraints on resources available for public sector initiatives.

The Ciudad Juárez case study also demonstrates that clean technologies need not be "win-win" propositions; that is, they need not reduce production costs as well as polluting emissions. The majority of brickmakers adopted propane and continued to use it for over a year even though it significantly increased production costs. Part of the explanation may have to do with the interplay between competition and informal regulation. The market for bricks is highly competitive, and as a result, brickmakers that use high-cost clean fuels are liable to be undercut by competitors using dirty fuels. Thus, competition in the market for bricks discouraged adoption initially. But once diffusion has progressed past a certain stage, competition appears to have worked *in favor* of adoption because those who had adopted have an incentive to ensure that their competitors adopt as well. Moreover, adopters generally had some leverage over competitors who were neighbors and/or fellow union members. As a result, once an initial cadre of brickmakers adopted, neighbors and fellow union members quickly followed suit. This suggests that if a critical mass of small firms can be convinced by hook or by crook to adopt a cost-increasing clean technology, eventually diffusion can become self-perpetuating.

Finally, the failure of efforts to successfully diffuse costly, complicated energy-efficient kilns in Ciudad Juárez demonstrates the well-established principle that in developing countries, new technologies must be appropriate, that is, both affordable and consistent with existing levels of technology.

The promise of private-sector-led environmental initiatives. The relatively successful Ciudad Juárez experience suggests that private- sector-led initiatives hold considerable promise as a means of addressing small-firm pollution problems. These initiatives would seem to enjoy a number of advantages over state-run programs. First, the willingness of the majority of the Ciudad Juárez brickmakers to cooperate with the project suggests that privatesector-led initiatives may be best suited to engage unlicensed firms that, by their nature, are bound to be wary of sustained contact with regulatory authorities. Second, the public enthusiasm that the propane initiative generated suggests that private-sector-led projects may be able to draw more freely on public sympathy for environmentalism than top-down bureaucratic initiatives. And finally, the projects' success at consensus building among a diverse set of stakeholders suggests that private-sector-led-initiatives may be better able to sidestep the politics and bureaucracy that often plague public-sector-led initiatives.

By contrast, the León efforts to establish *convenios* have been rife with such problems. The qualified success of the propane initiative, however, does not imply that small-scale pollution problems are best left to private sector organizers. In all likelihood, the propane initiative would not have had as much success without unusually strong public sector support.

Boycotts. In Ciudad Juárez, the attempt to organize a boycott of the brickmakers still using dirty fuels was an utter failure. Buyers simply continued to buy bricks from whoever was selling at the best price. This experience suggests that in most cases, contravening market forces—especially in informal or lightly regulated markets—simply does not work. Monitoring is too difficult and cheating is too easy.

Suggested Reading

Blackman, A. (ed.). 2006. *Small Firms and the Environment in Developing Countries: Collective Impacts, Collective Action.* Washington, DC: Resources for the Future.

Blackman, A., et al. Forthcoming. The Benefits and Costs of Informal Sector Pollution Control: Traditional Mexican Brick Kilns. *Environment and Development Economics.*

Blackman, A., and A. Kildegaard. 2003. Clean Technological Change in Developing Country Industrial Clusters: Mexican Leather Tanneries. Discussion Paper 03-12. Washington, DC: Resources for the Future.

Blackman, A. 2000. Informal Sector Pollution Control: What Options Do We Have? *World Development* 28: 2067-82.

Blackman A., and G.J. Bannister. 1998. Community Pressure and Clean Technology in the Informal Sector: An Econometric Analysis of the Adoption of Propane by Traditional Mexican Brickmakers. *Journal of Environmental Economics and Management* 35: 1-21.

47 New Investment Abroad
Can it Reduce Chinese Greenhouse Gas Emissions?

Allen Blackman

Foreign direct investment by multinational corporations in developing countries has the potential to introduce new technologies that increase efficiency and reduce polluting waste emissions. The recent experience in China seems to provide some support for this proposition. Data from an RFF survey of twenty American-owned or joint-venture power plants in China indicate that these plants are significantly more efficient than Chinese plants of similar scale.

In the next thirty years, developing countries will become the leading source of greenhouse gas emissions thought to cause global warming. Most of these emissions result from burning fossil fuels like coal and oil. Yet countries where weak public sectors and widespread poverty are pressing concerns are unlikely to be willing or able to undertake costly measures to lower emissions. Energy efficiency improvements may help resolve the dilemma. They not only reduce emissions of greenhouse gases but in some cases also significantly cut operating costs. Conceivably, firms in developing countries could be induced to invest in energy efficiency with minimal prodding. Foreign direct investment (FDI) by multinational corporations may be a principal means of transferring both the technology and the financial capital needed for such investments.

The Chinese electricity generating sector is an important test case for this hypothesis. China is already the world's third leading source of greenhouse gases and is likely to become the biggest contributor before the middle of the next century. China's fast-growing, almost exclusively coal-fired power sector is responsible for roughly a third of these emissions. In the early 1990s, China opened its doors to foreign direct investment in the power sector, a development that was met with a wave of enthusiasm by multinational corporations. What impacts has this recent opening had thus far?

Data from an original RFF survey of twenty American wholly-owned or jointventure power plants in China suggest that FDI is indeed having a significant positive impact on energy

Originally published in *Resources*, No. 133, Fall 1998.

Figure 1. Sources of electric power in China

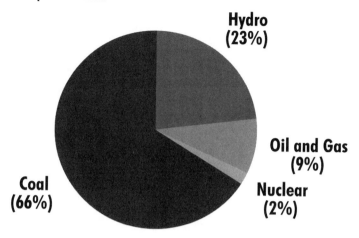

efficiency. Average rates of coal consumption per kilowatt hour of electricity generated for the plants surveyed are considerably lower rates than rates for new Chinese plants of similar scale, and are even lower than rates for new American plants. The main reason is that almost a third of the twenty plants use state-of-the-art generating technologies such as combined-cycle gas turbines (CCGTs) and circulating fluidized bed (CFB) boilers. These technologies have characteristics that make them especially attractive in China. Both CCGTs and CFBs accommodate the use of relatively cheap and plentiful low-grade fuels, a valuable feature given that the quantity and quality of fuel supply in China is uncertain.

In addition, unlike conventional steam turbines, CCGTs can be run efficiently even when started up and shut down on short notice as is often required in China. The plants in the RFF sample are even more efficient than rates of coal consumption indicate: a fifth of them use "waste" heat to generate heat or steam for industrial or residential facilities.

But not all the characteristics of FDI in China's power sector are encouraging from the standpoint of energy efficiency. To avoid the lengthy central government approval process for large plants and to minimize risk, early FDI tended to be in small-scale plants that are generally not as energy-efficient as large-scale plants. Perhaps more important, data from trade journals indicate that despite investors' early enthusiasm, the volume of FDI in China's power sector will likely fall short of government targets for the year 2000 by a substantial margin. In large part, this shortfall is the result of persistent institutional barriers to FDI. Survey data suggest that the most important barriers are the uncertainties associated with the approval process for FDI projects, the regulation of the electricity sector, and the risk of default on power purchase contracts.

Suggested Reading

Blackman, A., and X. Wu. 1999. Foreign Direct Investment in the Chinese Power Sector: Trends, Benefits, Barriers. *Energy Policy* 27(12): 695–711.

Part 11

New Horizons in Environmental Management

48 Fighting Antibiotic Resistance

Can Economic Incentives Play a Role?

Ramanan Laxminarayan

Antibiotic resistance is on the rise, imposing enormous costs on society and raising concerns about the treatment of infectious diseases. The problem can be traced to the widespread overuse of readily available drugs. Policymakers should consider introducing economic incentives to encourage both individuals and drug manufacturers to take into account the societal costs of using antibiotics.

Widespread reliance on antibiotics has spurred an alarming rise in resistant strains of bacteria, complicating the treatment of infectious diseases. Many blame the situation on doctors, patients, and livestock farmers who overuse, and sometimes even misuse antimicrobial agents. The challenge for policymakers is to promote the optimal use of antibiotics by creating economic incentives for individuals and drug manufacturers to consider the costs, as well as the benefits, of using these powerful drugs.

Resistance imposes enormous costs on society in the form of increased hospitalizations, higher mortality rates, and the diversion of resources from other medical needs into the development of new and more powerful antibiotics. Nevertheless, doctors understandably focus on the benefits to the patient, not the risks to society, when they prescribe an antibiotic. Similarly, livestock producers who use antibiotics in animal feed are motivated by the incentive of increased profits, and drug companies that encourage antibiotic use are motivated largely by the objective to profit from the antibiotic before expiration of its patent life. Such economic incentives drive the evolution of antibiotic resistance. As more antibiotics are used, bacterial resistance increases—a cycle that is exacerbated by the failure of antibiotic users to consider the full costs of their activity. Because resistance results from the selective use of drugs on sensitive strains of bacteria, it is likely to remain a pressing issue as long as we rely on antibiotics.

Although no one knows the exact costs that antibiotic resis-

Originally published in *Resources*, No. 143, Spring 2001.

tance imposes on society, the most common estimates range from $350 million to $35 billion, depending on how long resistance persists in the bacterial population, and whether or not the cost of deaths is considered. Such assessments are incomplete, however, because they fail to take into account the biological dynamics of resistance and infection. Unfortunately, limited data exist on antibiotic use and bacterial resistance, making it difficult for economists to compare costs when trying to evaluate alternatives to antibiotics.

A number of studies and reports have proposed guidelines for limiting the use of antibiotics in order to reduce resistance. But neither such guidelines nor educational efforts have been successful. Short of directly monitoring clinical practice, which would be extremely expensive, public health policymakers can do little to enforce restrictions on antibiotic use. And any attempt to admonish doctors for overusing antibiotics is likely to spark strong opposition from the medical community.

If we are to use the drugs more judiciously, it may be necessary to create a system that stresses the economic value of preserving the effectiveness of the drugs. In the language of economists, antibiotic resistance is a negative "externality" associated with antibiotic use, much as pollution is an undesirable externality associated with the generation of power at a thermal power plant. Neither the user of antibiotics nor power plants have an incentive to take into account the negative impact of their actions on the rest of society. In the case of power plants, government agencies impose emissions restraints in the form of taxes and quotas to force them to take the cost of pollution into account when determining how much power to generate. Similarly, society should devise mechanisms by which the cost of antibiotic resistance is taken into account—or, in economic terms, "internalized"—in decisions regarding the use of the drugs.

However, the externalities associated with antibiotic use are not all negative. A positive externality associated with antibiotic use is that it may cure infections, thereby reducing the likelihood of the infection being transmitted to uninfected individuals. Therefore, we need to weigh the favorable and unfavorable effects against one another to determine the optimal antibiotic use policy.

Optimal Use

Antibiotic effectiveness may be thought of as an economic or natural resource that is of value to society because it enables doctors to both prevent and treat infections. The current debate over antibiotic resistance centers on whether the current rate of depletion of this resource is greater than optimal.

From an economic perspective, the optimal use of antibiotics depends on whether the drugs are a renewable or a nonrenewable resource. This distinction relies on a biological concept (known as "fitness cost") that measures whether resistant strains of bacteria are placed at an evolutionary disadvantage when antibiotics are removed from the environment. If resistant bacteria were less likely to survive in the absence of antibiotics, one could conceive of temporarily removing an antibiotic from active use to enable it to recover its effectiveness. Antibiotic effectiveness would then be characterized as a renewable resource, much like a stock of fish that is harvested periodically and allowed to regenerate between harvest seasons. On the other hand, if the resistant strain remained prevalent, then an antibiotic would fail to regain its effectiveness even if it is temporarily removed. Effectiveness would be treated as a nonrenewable or exhaustible resource, similar to a mineral deposit. The question of renewable versus nonrenewable is difficult to answer because scientists continue to debate whether resistant bacteria endure in an environment without antibiotics.

In hospitals, where the increased use of antibiotics has contributed to a growing number of infections, officials seek to achieve optimal use by altering the menu of antibiotics available to doctors. How the antibiotics should be used to limit resistance, however, is a difficult question. If two antibiotics are available, for example, should doctors prescribe one, both, or alternate between the two?

Scientists have used mathematical modeling to show that, in the case of two antibiotics that are identical except for their modes of antibacterial activity, the optimal strategy would be to use equal fractions of both on all patients simultaneously. Recent work on the economics of resistance suggests that it may be shortsighted to use just a single antibiotic on all patients just because that antibiotic appears to be the most cost-effective option. Indeed, it may be optimal, from society's point of view, to use different drugs on different, but observationally identical, patients and include among this menu of drugs some that may not be cost-effective from the individual patient's perspective. The reason underlying this strategy is that when a greater number of patients rely on the same antibiotic, they over-congest the resource. Allocating different patients to different antibiotics, even when the costs and effectiveness of these antibiotics are different, can help minimize the likelihood that resistance will arise to any single drug.

Another issue for hospitals is the tradeoff between the costs of strengthening infection control measures and reducing antibiotic use. Infection control measures, such as sequestering nursing staff to a limited number of patients, can be effective in reducing the spread of infections in hospitals. However, recent research shows that when a number of health-care facilities (including hospitals and nursing homes) share a common pool of colonized patients, no single institution enjoys the full benefit of the infection control measures they have undertaken. Therefore, each institution has an incentive to free-ride on the efforts of others and the overall level of colonization and subsequent infection can be greater than socially optimal. Even if incentive issues are addressed, further studies are required to develop the optimal mix of strategies to reduce resistance within a single facility in an economically efficient manner.

Efforts to restrict antibiotic use in outpatient settings have been much less successful than in hospitals because no central agent (such as a hospital administrator or infection control committee) can enforce an antibiotic policy. Also, the high cost of malpractice lawsuits may induce doctors to err on the side of using stronger and broader spectrum antibiotics than may be called for. This tendency has the effect of increasing the level of resistance throughout the community, but the impact of each individual prescription is so small that the benefit perceived by the doctor of prescribing antibiotics often outweighs the small uncertain costs associated with increasing resistance. One solution would be to design guidelines that use community data to minimize the overall total cost of treatment and future resistance.

From a patient's perspective, the decision to request an antibiotic is based on two factors: the benefit of quickly recovering from an infection and the cost (minimized by insurance coverage) of taking the medication. But patients may not be aware of studies that have demonstrated conclusively that prior use of antibiotics increases a person's risk of acquiring a resistant infection.

Patients who are educated about the risks of antibiotics may be more careful about demanding such medication from the doctor. In addition, policymakers may want to consider such economic instruments as taxes, subsidies, and redesigned prescription drug insurance programs to ensure that incentives faced by both doctors and patients are aligned with the interests of society.

Livestock producers, like doctors and patients, have few incentives to consider the risks of antibiotic use. But the practice of adding antibiotics to livestock feed in order to promote growth in cattle and poultry has spurred warnings that such drugs may increase the level of bacterial resistance to antibiotics used in humans. In 1997, the World Health Organization recommended that farmers refrain from using drugs that are prescribed for humans or that can increase resistance to human medications. Policymakers need to balance the social costs of using antibiotics in animal feed against the benefits (namely more efficient livestock operations) in order to arrive at a rational policy regarding such use of antibiotics.

The Role of Patents

Firms that manufacture antibiotics face conflicting incentives with respect to resistance. On the one hand, bacterial resistance to a product can reduce the demand for that product. On the other hand, the resistance makes old drugs obsolete and can therefore encourage investment in new antibiotics.

Pharmaceutical firms are driven to maximize profits during the course of the drug's effective patent life—the period of time between obtaining regulatory approval for the antibiotic and the expiration of product and process patents to manufacture the drug. Given the paucity of tools at the policymaker's disposal, the use of patents to influence antibiotic use may be worth considering. A longer effective patent life could increase incentives for a company to minimize resistance, since the company would enjoy a longer period of monopoly benefits from its antibiotic's effectiveness.

Patent breadth is another critical consideration. When resistance is significant, other things being equal, it may be prudent to assign broad patents that cover an entire class of antibiotics rather than a single antibiotic. In such a situation, the benefits of preserving effectiveness could outweigh the cost to society of greater monopoly power associated with broader patents. Broad patents may prevent many firms from competing inefficiently for the same pool of effectiveness embodied in a class of antibiotics, while providing an incentive to develop new antibiotics.

Patent policies must take into account the global reach of antibiotic resistance. The 1999 Agreement on Trade-related Aspects of Intellectual Property Rights, sponsored by the World Trade Organization, provides for stricter enforcement of patent rights worldwide, while creating a phase-in period for developing countries that lack certain patent protections. Once antibiotic patents are enforced worldwide, pharmaceutical firms will have more incentive to research new and more effective antibiotics. Such patent rights could also have the potential to reduce the inefficient use of antibiotics by providing incentives to a single agent to conserve antibiotic effectiveness.

Future Research

The importance of scientific research in providing a reliable foundation for sound economic policy cannot be overstated. As we learn more about the relationship between antibiotic use and resistance, we can better quantify the social costs of overusing the drugs. Similarly, quantifying the relationship between antibiotic use in animal feed and resistance in humans will help us assess the economic tradeoffs involved in using the agents in livestock operations.

Further economic and scientific research could provide guidance for a number of policy issues. Such research could investigate the optimal antibiotic use in community settings, design incentives to promote the judicious use of antibiotics, and analyze the behavior of drug firms in investing in the development of new antibiotics. Finally, much research remains to be done to evaluate the costs of antibiotics in light of the biological dynamics of resistance. These efforts can help policymakers ensure that antibiotics remain a valuable resource for society.

Suggested Reading

Laxminarayan, R. (ed.). 2002. *Battling Resistance to Antibiotics and Pesticides: An Economic Approach.* Washington, DC: Resources for the Future.

Laxminarayan, R., and G.M. Brown. 2001. Economics of Antibiotic Resistance: A Theory of Optimal Use. *Journal of Environmental Economics and Management* 42: 183-206.

Fending Off Invasive Species
Can We Draw The Line Without Turning To Trade Tariffs?

Michael Margolis

Hundreds of "invasive species" have been identified, many of which have brought serious damage to ecosystems and threatened biodiversity. One response has been to propose bans or other forms of restrictions on international trade (including tariffs). But trade intervention does not promise to fend off invasive species; such trade measures are more likely to be manipulated by interest groups to their own advantage and to the detriment of the larger society. More specifically targeted control measures, responsive to the individual species, are needed to keep them out in the first place (if possible) and to prevent their proliferation where they are already present.

Next to habitat loss, the main threat to endangered species is the spread of other species. The introduction of goats onto San Clemente Island, California, for example, has led to the extinction of eight plant species and threatens at least eight more. Most of the American chestnut trees are now gone, due chiefly to a fungus known as chestnut blight. And when the great African snail became a serious garden pest in Hawaii, the rosy wolf snail was introduced to prey upon it and wound up instead eradicating several other snail species native to the islands it was supposed to protect.

Hundreds of these "invasive species" have by now been identified, and the damage they do is not limited to biodiversity. Mollusks foul industrial water-intake systems and navigation routes, numerous pests ravage crops, and the spread of European cheatgrass in the American West has contributed to the increase of fires. Other invasive species cause large changes in ecological processes such as the flow of nutrients and the amount of light reaching lake bottoms or forest floors.

This new ecological reality is part of the price of prosperity, a side effect of large-scale production, trade, and travel. Invasive species spread along the landscape disturbances created by roads, railways, and canals, and they also travel across oceans clinging to ship hulls or hiding in bilge water or packing material. While many invasive species have been introduced intentionally by people who had no idea they could cause harm, most invaders arrive in their new homes by accident. Agricultural shipments, especially those of live plants, often

Originally published in *Resources*, No. 153, Spring 2004.

include insect eggs or fungal colonies that are extremely difficult to detect. In most cases, no one thinks to look until a particular species has caused problems somewhere at least once.

In fighting the spread of invasive species, national boundaries are an appealing line of defense. Goods and people crossing those boundaries are under some scrutiny in any case and are generally restricted to a few entry points. In the United States, most trade and travel involves transportation across oceans, which means many of the life forms arriving at ports would almost never appear in the natural course of their own movements. Preventing problem species from arriving in the country has so far been the main focus of federal efforts, and environmental activists interested in the question mostly want to see more of the same. Bans are currently being sought on both the importation of logs that have not been heat-treated and on the use of solid-wood packing materials (chunks of wood used to prevent cargo from unpredictable shifting that can damage it), both of which are common pathways for forest pathogens. Others urge a crackdown on the importation of live organisms.

Guarding Our Borders

If the kind of crackdown environmental groups now want is attempted, we can expect a second generation of trade-environment conflicts, this one potentially much harder to resolve than the previous. The first time around, green groups were caught by surprise when conservation efforts ran up against trade agreements. The cases that have aroused the most passion are the 1991 ruling that the United States could not restrict tuna imports to protect dolphins and the rather belated discovery that the North America Free Trade Agreement had given investors a new venue through which to challenge regulations.

Neither of these cases is really about trade; they are about regulatory jurisdiction, and what led to the anger is that activists kept finding that the jurisdiction was not where they had thought it was. All they really needed to do was learn some new law and craft proposals with trade law in mind for the

bulk of that conflict to disappear and, by and large, that's what happened. In the invasive species case, however, the nature of the threat to the environment stands in direct opposition to the very purpose of trade agreements.

International trade is not the sole source of the problem. Trade between regions—say, the East Coast and California—can also spread invasive species, and efforts at prevention have proceeded with no special difficulty because the federal government has unquestioned authority to regulate interstate commerce and has chief responsibility for preventing the spread of invaders. On the international scale, however, that responsibility is largely in the hands of national governments, and there is no institution with equal power over trade among them.

The Trouble with Tariffs

To appreciate the difference this makes, one must consider what the World Trade Organization (WTO) and related international organizations actually do. Anyone who has casually studied international trade has probably wondered why these institutions are needed. After all, free trade is in the interest of every country; why not skip all the negotiating and just let goods flow freely?

That would, in most cases, be best for consumers, but it is politically impossible. The reason is that the benefits of a barrier to imports are enjoyed by relatively small groups of people, while the costs are spread over the whole consuming public. Few consumers can be troubled, for example, to complain to their representatives that sugar costs twice what it would under free trade, but the sugar producers happily support a full-time office dedicated to keeping it that way. The well-known virtue of free trade implies that the extra cost paid by American consumers for sugar is actually somewhat greater than the extra profit earned by the producers; however, because it amounts to only a few dollars per month for any family, the domestic political process is unlikely ever to get rid of America's sugar quota system.

International trade talks offer a way around

CHAPTER 49: FENDING OFF INVASIVE SPECIES   295

this tension. Every trade barrier harms both buyers and sellers, and if the buyers are a scattered group in one country, the sellers are probably a concentrated group in the other. The United States, in seeking to open markets for producers with influence in Washington, is also generating a benefit for foreign consumers, while American consumer interests are represented by the negotiating teams from other nations. By swapping access to each other's markets, trade negotiators have been able to move the world haltingly and partially towards the free trade ideal.

Once nations have agreed on what kind of market access to grant each other, they must agree how to guarantee that access. The simple part is to get rid of the tariffs and quotas that were explicitly designed for no other purpose than to interfere in trade. It is much harder, however, to deal with policies that discourage trade but also serve a clear social purpose—such as keeping out invasive species. Getting rid of the policies openly designed to discourage trade does nothing to get rid of the political dynamic that gave rise to those policies. So how do those political forces play out?

Interest-Group Influence

Jason F. Shogren, of the University of Wyoming, and I have developed theoretical models to answer that question, building on a model of interest-group influence that has been widely used to explain which industries get tariff protection. In these models, government officials are assumed to care both about the general welfare and campaign contributions. The cynical interpretation of this theory is that incumbents care only about getting reelected, the probability of which depends on how well-off voters feel and how much campaigns can spend on propaganda. A more charitable view is that they want to do what is right for the society, but are aware that they can lose the ability to do so by being outspent.

The damage done by invasive species that enter via imported goods alters the general-welfare component of the government's objective but has no direct impact on the private interest groups. It does,

however, indirectly alter interest-group behavior. If you represent an interest group that wants a particular import discouraged, and you know the government is going to put high tariffs on that import anyway because it carries invaders, you can save your contributions, while if you wanted free trade in that good you must contribute. The result of all these calculations is a tariff that is greater or less than the socially optimal tariff depending on the industry incentives to lobby.

What this implies is that if governments agree in trade talks to eliminate tariffs, but leave in exceptions for the goods that harbor invasive species, they almost might as well not have bothered. Unless governments can also agree on how potential invasive species damage is to be valued—which has not been contemplated in any trade agreement so far—the tariffs on those goods will rise far beyond what the damage done can really justify. And virtually every import can be a pathway for some undesirable species.

In practice, trade agreements have addressed this sort of problem not by allowing for tariffs, but by allowing regulation of the import process. For example, banning the import of logs from locations not certified as free of certain pests, or requiring that logs from such locations be heat-treated, is allowed under current agreements. Such policies have the clear advantage of focusing more narrowly on the problem than do tariffs. But it turns out that as a way to prevent protectionist abuse, this approach is not much better, and in some cases may be worse, according to our ongoing research. If there are no restrictions placed on how stringently governments set the importation standards or inspect for compliance, the import-competing lobbies will seek to have inspections increased to levels that drive import prices about as high they would have been with tariffs. Consumers may wind up even worse off than with tariffs, since the price-gauging function is now being performed by the use of real resources—to wit, the excess time spent by the inspectors.

Trade negotiators have been aware of these problems for some time. By the time the WTO was formed, there had been many cases in which importers alleged bad faith. As a result, the WTO founding documents include an Agreement on Sanitary and Phytosanitary Standards (SPS). According to the agreement, regulations intended to protect the health of animals and plants must have a scientific basis, but it turns out not to be so easy to agree on what that means. At present, the European Union is putting up with tariff retaliation rather than conceding that its ban on hormone-treated beef is unscientific, and the same may soon hold for genetically modified foods. And these issues, like almost every SPS case decided so far, have arisen in the context of agricultural trade, which is by far the least free trade on the planet. With explicit quotas and tariffs still in place, farmers have relatively little incentive to use standards as disguised protectionism. If, as seems likely, the invasive species issue begins to implicate more economic sectors where quotas and tariffs have been taken off the table, the clashes can only be louder and more frequent.

Alternatives to Trade Intervention

Once an invasive species gets started somewhere, there are a variety of strategies available to keep it in check. As already mentioned, sometimes another species is brought in to prey on the invader; often species that are innocuous in their home become invasive when transported because they escape from the predators with which they co-evolved. The snail example cited above is one of many cases in which such "biocontrol" strategies have gone awry, but the science is evolving and, in some situations, releasing predators is still deemed the best response. Other options include chemical treatments, the release of sterilized specimens of the invader itself (to distract mates from the fertile) and manual removal of invaders from the field. The last option is preferred for its minimal impact on the environment, but tends to be expensive.

For manual removal to be truly effective, it is critical to interrupt an invasive species' life cycle at just the right point. In separate work with biologists Jennifer Ruesink and Eric Buhle of the University of Washington, we have adapted an analytical strategy first developed for identifying the life cycle phase at which an endangered species most needs to be protected. The main difference is that rather than defending this weak spot in the invasive species, we wish to attack it. Typically, invasive species have short life spans and produce many offspring. Killing adults is ineffective as compared to killing the same percentage of juveniles, eggs, larvae, and so on.

There is, however, a second way in which invasive species are not just endangered species turned backwards. To protect an endangered species, we must succeed at every stage of the life cycle. To get rid of an invasive, we must only succeed at one stage. This frees us to adapt our strategy much more aggressively to the relative cost of intervention at each stage, which makes quite a difference. In many cases, killing off a given percentage of the adults is much less expensive than getting the same fraction at the other life stages, since adults tend to be larger and easier to find. This is the case for the Japanese oyster drill, a species of winkle infesting farmed and wild oysters on the West Coast. If the relative cost is ignored, analysis of the oyster drill life cycle indicates one should gather eggs; however, when cost is considered, the most effective approach is to concentrate all resources on gathering adults.

This lesson does not extend to all species—it matters greatly that in this case the adults do not move around a lot—but the analytical method does. This is but one component of a large ongoing effort by biologists and economists to design efficient strategies to combat invaders. For the foreseeable future, however, there will be no substitute for keeping them out in the first place, and the global trade system remains ill-prepared to deal with the consequences.

Suggested Reading

Buhle, Eric, Michael B. Margolis, and Jennifer L. Ruesink. April 2004. Bang for the Buck: Cost-Effective Control of Invasive Species with Different Life Histories. RFF Discussion Paper 04–06. Washington, DC: Resources for the Future. www.rff.org/Documents/RFF-DP-04–06.pdf.

Margolis, Michael B., and Jason F. Shogren. February 2004. How Trade Politics Affect Invasive Species Control. RFF Discussion Paper 04–07. Washington, DC: Resources for the Future. www.rff.org/Documents/RFF-DP-04–07.pdf.

OTA. 1993. *Harmful Non-indigenous Species in the United States.* Office of Technology Assessment: Washington, DC. http://www.wws.princeton.edu/~ota/disk1/1993/9325_n.html.

Roberts, D. 2000. Sanitary and Phytosanitary Risk Management in the Post-Uruguay Round Era: An Economic Perspective in *Incorporating Science, Economics, and Sociology in Developing Sanitary and Phytosanitary Standards in International Trade: Proceedings of a Conference.* Washington, DC: National Academy of Sciences. No. 33–50.

Shine, C., N. Williams, et al. 2000. A Guide to Designing Legal and Institutional Frameworks on Alien Invasive Species. Gland, Switzerland: World Conservation Union (IUCN).

www.issg.org. The Invasive Species Specialist Group (ISSG), part of IUCN, is a global group of 146 scientific and policy experts on invasive species from 41 countries. ISSG provides advice on threats from invasives and control or eradication methods to IUCN members, conservation practitioners, and policymakers. This website offers extensive resources.

Part 12

An Historical Perspective

50 Forty Years in an Emerging Field
Economics and Environmental Policy in Retrospect

Wallace E. Oates

Economics had little impact on the early legislation that emerged from the environmental revolution in the 1960s. Many viewed it with suspicion and hostility. But economic analysis and the use of economic incentives for environmental protection have gradually gained currency among policymakers and many environmentalists. Economists now play a regular role in the analysis of standards for environmental quality and in the design of regulatory measures. Here is one reading of how this happened.

We have seen a remarkable transformation in the role of economics in environmental policymaking over the past three decades. Coming out of the environmental revolution of the 1960s, the early federal legislation—notably the Clean Air Act Amendments of 1970 and the Clean Water Act Amendments of 1972—essentially ignored economics. In the "command-and-control" (CAC) tradition, this legislation directed environmental agencies to set air and water quality standards with little regard to their economic consequences and then to issue directives to firms for the control of their waste emissions into the environment, often specifying the technologies that were to be used.

Since those early days, however, things have changed in some quite dramatic ways. To take one example, the U.S. Congress under the 1990 CAA Amendments has adopted a wholly different regulatory strategy to tackle the troubling acid-rain problem: a market for tradable sulfur-emissions allowances. Sources throughout the nation are buying and selling entitlements to a limited quantity of sulfur discharges into the atmosphere. This approach is achieving our objective of cutting aggregate emissions in half, but it does so in a way that gives emitters discretion to determine their own levels of both emissions and abatement technology.

More generally, benefit-cost analyses of proposed environmental standards have become a routine part of the regulatory process. Although their role in the establishment of regulatory standards is, in some cases, rigidly circumscribed by existing statutes, such benefit-cost studies figure in important ways in

Originally published in *Resources*, No. 137, Fall 1999.

the debate over proposed measures (and in *ex post* reviews of policy as well).

How has environmental policymaking evolved from a process in which economics had so little relevance into one in which it plays a significant role? And what did economists have to do with this transformation? These are fascinating questions, if not easy ones to answer. But let me at least offer some reflections.

Environmental Economics Early-On

If we didn't know better, it would be natural to suppose that economics had been important in the design of environmental policy from the outset. After all, economists were, it might seem, well positioned upon the arrival of the environmental revolution. They had a coherent view of the problem of environmental degradation, one that indicated clearly the nature of the "market failure" that takes place when economic agents have free access to our scarce environmental resources. Such free access leads quite naturally to an excessive use of resources, resulting in a polluted environment. Moreover, this view of the environmental problem carries with it a direct policy prescription: government needs to introduce the correct "price" in the form of a tax on polluting waste emissions. Such a tax would represent the surrogate price that would induce polluters to cut back their emissions to the socially desired levels.

This perspective on environmental regulation, developed in the first half of the century by A.C. Pigou and others, was embedded in the academic literature by the time the amendments to the Clean Air and Water Acts were under consideration. Basic textbooks provided a standard description of the smoky factory spewing fumes over nearby residences and went on to prescribe taxes on the emissions of the offending pollutants as a corrective measure.

But this approach was completely ignored in the initial round of environmental legislation both in the United States and abroad. Why? My answer to this question comes in three parts. First, there was no constituency for whom the economist's

Update

Since the publication of this article, the TEP (or "cap-and-trade") approach has been adopted in Europe. In January 2005, the European Union initiated the largest and most ambitious emissions trading program in the world to meet its commitment under the Kyoto Protocol for the containment of greenhouse gases. Sources in all 25 nations in the EU are trading emissions allowances across national borders. The embracing of this approach by the EU is largely the result of the perceived success of the SO_2 trading program in the United States. Cap-and-trade programs are also proliferating in the United States.

view and policy proposal had much appeal. Environmentalists were decidedly hostile. The market system was the reason we had pollution in the first place, they said. The idea of putting a price on the environment was morally repugnant. Moreover, they argued, it wouldn't work: polluters would simply pay the tax and go on polluting. Environmentalists thus flatly rejected an economic approach (as I learned personally and painfully on several occasions) and called for direct controls on polluting activities.

Industry was not very sympathetic either. The idea of a new tax was, of course, not very appealing. Beyond that, some firms found that environmental controls could actually work to their advantage, because such controls were often much stricter on new industry. Many established firms welcomed the barriers to entry that command–and–control regulation was creating.

Finally, the fraternity of regulators was less than enthusiastic about discarding traditional methods of regulatory control for a largely untried system of taxes on pollution. There really was no one to champion the cause of the economic approach to environmental policy.

The second part to my answer turns to the state of environmental economics itself in the late 1960s and early 1970s. Economics had a view of the pollution problem, but it did not go much beyond a general conceptual level. It is a long way from an equation on the blackboard stating that a tax on each firm's emissions should be set equal to "marginal social damages" to the design and implementation of a workable system of pollution taxes. And few economists were working on these issues. Today there exists an active Association of Environmental and Resource Economists (AERE) with a membership approaching one thousand and with a large and energetic sister organization in Europe. But thirty years ago, only a small number of economists were seriously addressing the hard issues of policy design.

Several of them were at Resources for the Future. Allen Kneese and Blair Bower, for example, published a pathbreaking study of water quality management in 1968 that explored the scientific character of water pollution, studied the actual institutions for regulating water quality, and then turned to the design of a feasible system of fees for the control of waste emissions. But these studies were exceptions. Economists really were not in a position at that time to offer much guidance on the actual design and implementation of systems of environmental taxes.

The third part of my answer (closely related to the second) is the pervasive ignorance of the economic approach to environmental policy outside the economics profession itself. Even as late as 1981, Steven Kelman's survey of the environmental policymaking community turned up virtually no one who could even explain the basic rationale for incentive-based policy measures! Finally, it is probably a fair criticism to say that few of those who did understand the power of incentive–based approaches were willing to make the effort to educate legislators, regulators, and their staffs about this radical alternative.

Economics and the Evolution of Environmental Policy

The story of the growing role of economics in environmental policymaking is a complicated one, only imperfectly understood. Indeed, its chapters contain both serendipitous and more purposeful elements. One important facet of this story in the United States (but not in Europe) is the emergence of an alternative incentive-based policy instrument. Economists surely knew that, in principle, it is possible to attain the objective of cutting back waste emissions either by a tax or by a system of tradable emissions permits (TEP). It is straightforward to show that emissions can be reduced to the target level either by setting a sufficiently high tax on emissions or by issuing the requisite number of emissions permits and allowing trading activity to establish the market-clearing price. The outcome in the two cases is, in principle, identical.

But in the early dialogue, discussion focused primarily on the tax approach. My recollection is that most of us in our assessments of the prospects for various policy measures assumed that the so-called quantity approach involving a TEP system would encounter overwhelming opposition inasmuch as it involved literally putting the environment up for sale. Polluters would buy and sell "rights to pollute." There seemed to be little hope for such an audacious proposal.

We were wrong of course, partly, I believe, by reason of historical accident and partly because of a failure to understand the political economy of instrument choice. With the prospect of a tumultuous political confrontation in the mid-1970s over nonattainment of clean air goals in many regions of the country, the U.S. Congress introduced in 1977 a provision for "pollution offsets." Under this provision, new sources of pollution could enter nonattainment areas if existing sources cut back their emissions by more than those of the entrants. Somewhat unwittingly, I suspect, federal legislators had opened the door to what eventually became the Emissions Trading Program, under which trading of emissions allowances for air pollutants has been taking place in many areas.

Tradeable emission permit (TEP) systems turn out to have much more appeal than their tax counterpart in the policy arena. Environmentalists are much more sympathetic to them since, by restricting the number of available permits, the environmental authority can directly and unambiguously achieve its objective. Industry is also receptive. Instead of paying a tax, firms typically receive (under some kind of grandfathering provision) a valuable asset: emissions permits, which they can use either to validate their own emissions or sell for a profit. Regulators much prefer TEP systems to taxes. They can achieve their goal simply by issuing the requisite number of permits; they don't have to worry about setting and then adjusting tax rates to induce the needed reductions in pollution. It is interesting that the TEP approach has not caught on in Europe; there the use of incentive-based instruments has primarily taken the form of taxes on pollution.

The work of environmental economists has, I think, been important in this evolution. Ideas can be a powerful force in the policy arena, and economists were able to provide a compelling conceptual rationale for the new tradable-permit approach. In addition, they carried out a substantial number of careful empirical studies that documented the large cost savings available through the use of incentive-based policy instruments. Over the last thirty years, the educational void has been filled. In response to environmental concerns, courses in environmental economics have sprung up across the country. At the graduate level, the field of "Environmental and Natural Resource Economics" has emerged; Ph.D. students have written dissertations and gone on to teach, carry out research, and take positions in environmental agencies. As mentioned earlier, there now exists a large and energetic organization of environmental economists; the Association of Environmental and Resource Economics has its own journal and holds frequent conferences to help organize research efforts and disseminate the findings. At least as important has been the growing presence of economists in law schools and schools of public policy. Here, many future policymakers have received a firm grounding in the economics of environmental policy.

Resources for the Future has played an important role in this evolution. From the beginning, RFF reached the policymaking community not only through research, but through determined and patient efforts to make available and accessible to the general public not only research findings but, more generally, the basic economic principles of policy analysis and design. Indeed, this very publication, *Resources*, has a long history of doing precisely that.

Lest we go overboard with self-congratulation, however, it is important to recognize that there has been a growing receptivity in the Western world to market-based forms of regulation. The advent of Reaganomics in the United States and Thatcherism in Britain signaled the arrival of what John Kay has called a new "faith in market forces." Over this period, we have seen a basic change in the intellectual setting for social and economic policy—one that is at least as concerned with "government failure" as with "market failure." From this perspective, the evolution of environmental policy is best seen as part of a larger movement for the fundamental reform of regulatory policies, a movement that actively seeks to employ market incentives for social programs.

Much Left to Accomplish

The role of economics in environmental policy has clearly come a long way over the past thirty years. Prospective environmental programs are routinely subjected to benefit-cost assessments, and at least some attention is often given to the use of incentive-based instruments for the attainment of our prescribed standards for environmental quality. But this progress should not be exaggerated. Most of our regulatory measures, for example, are still of the command-and-control variety. Often it is not easy to design a workable and effective incentive-based mechanism. In fact, the design and implementation of such measures for different kinds of environmental problems are real challenges. An especially fascinating and difficult case is how to design a system of tradable carbon allowances on an international scale to address global climate change. This problem is the subject of widespread interest and current research. Meanwhile plenty of more mundane and localized cases of environmental management need to addressed. We have a long way to go!

While we economists can take some real satisfaction in our contributions to environmental policymaking, we must retain a certain humility. Benefit-cost analyses are a valuable component of program assessment, but we should never base decisions on environmental standards *solely* on the bottom line of a benefit-cost study. Likewise, command-and-control programs will continue to be a fundamental part of our regulatory landscape. But even here there is plenty of room for economic analysis aimed at making such CAC programs more effective in attaining their environmental targets at relatively low cost.

Suggested Reading

Oates, Wallace. 2000. From Research to Policy: The Case of Environmental Economics. *University of Illinois Law Review* 1: 1-15. Reprinted in *The Political Economy of Environmental Regulation*, edited by R. Stavins. Cheltenham, U.K.: Edward Elgar, 514-532.

Index